# Acidification of Freshwater Ecosystems

## Implications for the Future

Goal of this Dahlem Workshop:

to evaluate the causes, consequences, and reversibility of freshwater acidification—past, present, and future

Environmental Sciences Research Report ES 14

*Held and published on behalf of the*
Freie Universität Berlin

*Sponsored by:*
Funds from the Stiftungsfond Unilever in the
Stifterverband für die Deutsche Wissenschaft

# Acidification of Freshwater Ecosystems

## Implications for the Future

*Edited by*

C.E.W. STEINBERG and R.F. WRIGHT

Report of the Dahlem Workshop on
Acidification of Freshwater Ecosystems held in Berlin
September 27–October 2, 1992

*Program Advisory Committee*:
C.E.W. Steinberg and R.F. Wright, Chairpersons
P.J. Dillon, D. Landers, T. Paces, H.W. Zoettl

JOHN WILEY & SONS
Chichester • New York • Brisbane • Toronto • Singapore

Baffins Lane, Chichester,
West Sussex PO19 1UD, England

Telephone (+44) 243 779777



***Library of Congress Cataloging-in-Publication Data***

Dahlem Workshop on Acidification of Freshwater Ecosystems (1992 : Berlin, Germany)
Acidification of freshwater ecosystems : implications for the future : report of the Dahlem Workshop on Acidification of Freshwater Ecosystems held in Berlin, September 27–October 2, 1992 / edited by C.E.W. Steinberg and R.F. Wright ; program advisory committee, C.E.W. Steinberg . . . [et al.].
p. cm. — (Environmental sciences research report ; ES 14) (Dahlem workshop reports)
"Held and published on behalf of the Freie Universität Berlin; sponsored by Stiftungsfond Unilever im Stifterverband für die Deutsche Wissenschaft"—Half t.p.
Includes bibliographical references and indexes.
ISBN 0-471-94206-5
1. Acid pollution of rivers, lakes, etc.—Congresses.
I. Steinberg, Christian. II. Wright, Richard F. III. Freie Universität Berlin. IV. Stifterverband für die Deutsche Wissenschaft. Stiftungsfonds Unilever zur Wissenschafts- und Bildungsförderung. V. Title. VI. Series. VII. Series: Dahlem workshop reports.
TD427.A27D35 1992
574.5′2632—dc20 93–28725
CIP

***British Library Cataloguing in Publication Data***

A catalogue record for this book is available from the British Library

ISBN 0-471-94206-5

Dahlem Editorial Staff: J. Lupp, C. Rued-Engel
Typeset in 10/12pt Times from editor's disks by Text Processing Department, John Wiley & Sons Ltd, Chichester
Printed and bound in Great Britain by Biddles Ltd, Guildford, Surrey

# Contents

# The Dahlem Konferenzen

The purpose of the Dahlem Konferenzen is to promote international interdisciplinary exchange of scientific information and ideas and to stimulate international cooperation in research. This is achieved by arranging discussion workshops, mainly in the life sciences and environmental sciences, organized according to a model developed and tested by the Dahlem Konferenzen.

Dahlem Konferenzen was founded in 1974 by the Stifterverband für Deutsche Wissenschaft[1] in collaboration with the Deutsche Forschungsgemeinschaft[2] to promote more effective communication between scientists. It was named after the Dahlem district of Berlin, long a home of the sciences and arts. In January, 1990, it became incorporated into the Freie Universität Berlin. Financial support comes from the Senate of the Land Berlin, the Deutsche Forschungsgemeinschaft, and private foundations.

As scientific research has become increasingly interdisciplinary, a growing need for specialists in one field to understand the problems and work with the concepts of related fields has emerged. New insights can be gained when a problem is approached from the standpoint of another discipline, and because no existing form of scientific meeting provided the forum necessary for such exchanges, Dahlem Konferenzen created a new concept, which has been tested and refined over the years. Now internationally recognized as the Dahlem Workshop Model, it provides a framework for coherent interdisciplinary discussion of a topic of contemporary international interest in five working days, culminating in the draft manuscript for a book.

Dahlem Workshops provide a unique opportunity for posing questions to colleagues from different disciplines and for soliciting alternative opinions on contentious issues. The aim is not to solve problems or to reach a consensus but to identify gaps in our knowledge and possible new ways of approaching stubborn issues, and to define priorities for research. This approach is well summed up in the instruction to participants when they are asked to state what they do not know rather than what they know.

---

1 The Donors Association for the Promotion of Sciences and Humanities, a foundation created in 1921 in Berlin and supported by German trade and industry to fund basic research in the sciences

2 German Science Foundation

Workshop topics are proposed by leading scientists and are approved by an independent scientific board advised by qualified referees. Approximately one year before the workshop, an advisory committee meets to determine the scientific program and select participants, who are invited according to their scientific reputations, with the exception of a number of places reserved for junior German scientists.

The discussions at each workshop are organized around four key questions, each tackled by a group of about twelve participants with a range of expertise. There are no lectures; instead, prior to the workshop, selected participants write background papers that review particular facets of the topic and serve as the basis for the discussion. These papers are distributed to all participants before the meeting and selected participants act as referees. During the workshop, each of the four discussion groups prepares a report reflecting the ideas, opinions, and controversies that have emerged from its discussion as well as identifying problems areas and directions for future research.

The revised background papers and group reports are published as the Dahlem Workshop Reports. Each volume is edited by the chairperson(s) of the workshop and the staff of the Dahlem Konferenzen. The Reports provide multidisciplinary surveys by an international group of distinguished scientists, based on discussions of advanced concepts, techniques and models. The Dahlem Workshop Reports are published in two series: Life Sciences and Environmental Sciences (formerly Physical, Chemical, and Earth Sciences).

Jennifer Altman, Acting Director
Dahlem Konferenzen der Freien Universität Berlin
Rothenburgstr. 33, W-1000 Berlin 41, F.R. Germany

# List of Participants with Fields of Research

WOLFGANG AHLF Arbeitsbereich Umweltschutztechnik, Technische Universität Hamburg-Harburg, Eissendorferstr. 40, 21073 Hamburg, F.R. Germany

*Interactions between sediments and biota; microbial activities, sulfate reduction bioassays with sediments*

JOAN BAKER U.S. Environmental Protection Agency, 200 SW 35th St., Corvallis, Oregon 97333, U.S.A.

*Regional assessments of the effects of acidic deposition on fish communities in the United States*

SUZANNE BAYLEY B414 Biological Sciences Bldg., University of Alberta, Edmonton, Alberta T6G 2E9, Canada

*Wetland biogeochemistry, climate change in wetlands, wetland hydrology and chemical interactions, stream biogeochemistry*

JÜRGEN BÖHMER Institut für Zoologie, Universität Hohenheim, Garbenstraße 30, 70599 Stuttgart, F.R. Germany

*Biological indication of acidity and aluminium in surface waters, electron microscopic and macroscopic aluminium detection in animal tissues*

GERHARD BRAHMER Institut für Bodenkunde und Waldernährungslehre, Bertoldstr. 17, 79085 Freiburg, F.R. Germany

*Catchment manipulation experiments; water and element cycling/modeling in forested ecosystems*

DAVID BRAKKE Department of Biology, University of Wisconsin, Eau Claire, WI 54702, U.S.A.

*Limnology, chemistry of seepage lakes, long-term trends in acidic lakes in Norway, natural vs. anthropogenic contributions to acidity*

DON CHARLES Patrick Center for Environmental Research, Academy of Natural Sciences, 1900 Benjamin Franklin Parkway, Philadelphia, PA 19103, U.S.A.

*Palaeolimnological studies of lake acidification; use of diatom assemblages in sediment cores to reconstruct past water chemistry conditions; causes and mechanisms of lake acidification, particularly resulting from atmospheric deposisiton*

JACK COSBY Department of Environmental Sciences, Clark Hall, University of Virginia, Charlottesville, VA 22901, U.S.A.

*Effects of acidic deposition on terrestrial/aquatic resources; simulation modeling of ecosystems; biogeochemical cycles of headwater catchments*

NANCY DISE Norwegian Institute for Water research (NIVA), P.O. Box 69 Korsvoll, 0808 Oslo 8, Norway

*Methane emission from peatlands; nitrogen biogeochemistry*

BRIDGET EMMETT ITE, Bangor Research Unit, UCNW, Deiniol Road, Bangor, Gwynedd LL57 2UP, U.K.

*Impact of nitrogen deposition on coniferous forests and, in particular, the future role of nitrate in acidification of soil and water*

KARL-HEINZ FEGER Institut für Bodenkunde und Waldernährungslehre, Albert-Ludwigs-Universität, Bertoldstrasse 17, 79085 Freiburg, F.R. Germany

*Forest ecology, nutrient cycling, water quality in forest environment*

JESSE FORD Oregon State University, c/o U.S. Environmental Protection Agency, 200 SW 35th Street, Corvallis, OR 97333, U.S.A.

*Heavy metals, trace elements, and organochlorines in arctic ecosystems; regional assessments of resource conditions*

JAN FOTT Department of Hydrobiology, Charles University, Vinicná 7, 128 44 Prague 2, Czech Republic

*Plankton of acidified lakes, shallow lakes, and fish ponds*

JOHN GUNN Ontario Ministry of Natural Resources, Cooperative Freshwater Ecology Unit, Laurentian University, Biology Department, Sudbury, Ontario P3E 2C6, Canada

*Fisheries biology, community and population ecology, and restoration ecology in industrially stressed ecosystems*

RON HARRIMAN Freshwater Fisheries Laboratory, Faskally, Pitlochry, Perthshire PH16 5LB, U.K.

*Long-term effects of forestry practices and acidic deposition on surface water chemistry; critical loads for sulfur and nitrogen in freshwater reversibility of acidification*

ANTON HARTMANN GSF-Forschungszentrum für Umwelt und Gesundheit, Institut für Bodenökologie, Postfach 1129, 85758 Oberschleissheim, Neuherberg, F.R.Germany

*Soil microbiology, molecular microbial ecology, microbial ecotoxicology*

MICHAEL HAUHS Bayreuther Institut für Terrestrische Ökosystemforschung, BITÖK, Universität Bayreuth, Dr.-Hans-Frisch-Str. 1-3, Postfach 10 12 51, 95448 Bayreuth, F.R. Germany

*Ecosystem phenomenology; hydrology; forest growth models, links to soil processes*

MAGDA HAVAS Environment and Resource Studies, Trent University, Peterborough, Ontario, K9J 7B8, Canada

*Chemical and biological recovery of acidic, metal-contaminated lakes; aluminum toxicology in invertebrates*

HARRY HEMOND Department of Civil and Environmental Engineering 48-419, Massachusetts Institute of Technology, Cambridge, MA 02139, U.S.A.

*Biogeochemistry of catchments and peatlands, organic acidity, hydrology*

HANS HULTBERG Swedish Environmental Research Institute (IVL), P.O. Box 47086, 40258 Göteborg, Sweden

*Effects of experimental increase/decrease of mercury, sulfur and nitrogen to coniferous forested catchments (NITREX and ROOF projects) at the Lake Gärdsjön watershed in SW Sweden, liming of lakes and forested catchments*

ALAN JENKINS Institute of Hydrology, Crowmarsh Gifford, Wallingford, Oxfordshire OX10 8BB, U.K.

*Hydrochemical modeling*

INGRID JÜTTNER GSF - Forschungszentrum für Umwelt und Gesundheit, Institut für Ökologische Chemie, Postfach 1129, 85758 Oberschleissheim, Neuherberg, F.R. Germany

*Acidification chronology*

CAROL KELLY Department of Microbiology, University of Manitoba, Winnipeg, Manitoba R3T 2N2, Canada

*Microbial activities in lakes including decomposition, mercury methylation, sulfate reduction, denitrification; whole-ecosystem budgets for carbon, sulfur, mercury, nitrogen*

KARL KREUTZER Lehrstuhl für Bodenkunde, Hohenbachernstraße 22, 85354 Freising, F.R. Germany

*Forest ecosystems, interdisciplinary integrated investigations of the effects of acidification, liming, enhanced nitrogen supply by field experiments*

DIXON LANDERS US EPA, Environmental Research Laboratory, 200 SW 35th Street, Corvallis, OR 97333, U.S.A.

*Arctic contaminants, evaluating concentrations in terrestrial and aquatic ecosystems and spatial approaches to evaluating resource condition*

DIETER LEßMANN Zoologisches Institut der Universität, Berliner Strasse 28, 37073 Göttingen, F.R. Germany

*Comparison of freshwater biocoenoses under the stress of different grades of anthropogenic pollution by water acidification, industrial effluents, and communal sewage; long-term studies of acidified running waters*

ANKE LÜKEWILLE Umweltbundesamt, Bismarckplatz 1, 14193 Berlin, F.R. Germany

*Modeling of soil and water acidification processes, mapping of critical loads in the framework of the UN ECE, convention on long-range transboundary air pollution*

STEPHEN NORTON Geological Sciences, University of Maine, 123 Boardman Hall, Orono, ME 04469, U.S.A.

*Ecosystem chemical manipulations, environmental geochemistry*

STEVE ORMEROD Catchment Research Group, School of Pure and Applied Biology, University of Wales College of Cardiff, P.O. Box 915, Cardiff CF1 3XF, U.K.

*Impacts on river catchments and aquatic biology*

ALVARO OVALLE Universidade Federal Fluminense, Departamento de Geoquimica, Morro do Valonguinho, Centro, Niteroi, Rio de Janeiro 24.210, Brazil

*Fluvial hydrochemistry and its relation to land use in coastal catchments*

TOMAS PACES Czech Geological Survey, Malostranské nám. 19, 118 21 Prague, Czech Republich

*Biochemical monitoring of catchments affected by acidic deposition and agricultural practices, impact on weathering and erosion rates*

RAINER PUTZ Instituto Nacional de Pesquisas da Amazonia, Caixa Postal 478, BR-69.011 Manaus-AM, Brazil

*Benthic diatom succession, biomass and productivity on natural and artifical substrata in a river Rhine floodplain near Rastatt, FRG*

GUNNAR RADDUM Laboratory of Freshwater Ecology, Zoological Museum, University of Bergen, 5007 Bergen, Norway

*Effects of acidification on freshwater invertebrates, monitoring of acidification, effects of liming and reversibility of freshwater communities*

BJØRN OLAV ROSSELAND Norwegian Institute for Water Research (NIVA), P.O. Box 69 Korsvoll, 0808 Oslo 8, Norway

*Fish physiology & toxicology, pH/AL-toxicity to salmonids fish, population responses to acidification, strain differences of brown trout, effects of mixing zones (neutral or lime acid waters) on salmonids*

DAVID SCHINDLER Department of Zoology, CW-312 Biological Sciences, University of Alberta, Edmonton, Alberta T6G 2E9, Canada

*Ecology*

HELMUT SEGNER Environmental Research Center, Section for Environmental Chemistry and Ecotoxicology, Permoserstr. 15, 04318 Leipzig, F.R. Germany

*Assessment of toxicological effects by use of in vitro techniques*

CHRISTIAN STEINBERG GSF-Forschungszentrum für Umwelt und Gesundheit, Institut für Ökologische Chemie, Postfach 1129, Neuherberg, 85758 Oberschleissheim, F.R. Germany

*Acidification and organic contaminants, acidification chronology, ecotoxicology*

HELGA VAN MIEGROET Oak Ridge National Laboratory, Environmental Sciences Division, Building 1505, Oak Ridge, TN 37831-6038 U.S.A.

*Soil chemistry, biogeochemical cycling, nitrogen dynamics, environmental effects of forest management, watershed processes, terrestrial-aquatic linkages*

JOSEF VESELÝ Czech Geological Survey, Malostranské nám. 19, 118 21 Prague 1, Czech Republic

*Trace elements contamination, acidification, recovery of lakes, effects of nitrogen compounds on acidification, Czech surface water survey*

RICHARD F. WRIGHT Norwegian Institute for Water Research (NIVA), P.O. Box 69 Korsvoll, 0808 Oslo 8, Norway

*Water chemistry, nitrogen saturation, global change effects*

MICHAEL T. ZAHN Bayerisches Landesamt für Wasserwirtschaft, Lazarettstrasse 67, 80636 München, F.R. Germany

*Effects of deposition on quality and acidification of groundwater*

HEINZ ZOETTL Institut für Bodenkunde und Waldernährungslehre, Universität Freiburg, Bertoldstrasse 17, 79085 Freiburg, F.R. Germany

*Forest ecosystem element budgets*

# 1

# Introduction

C.E.W. STEINBERG[1] and R.F. WRIGHT[2]
[1]GSF-Forschungszentrum für Umwelt und Gesundheit, Institut für Ökologische Chemie, Postfach 1129, Neuherberg, 85758 Oberschleissheim, F.R. Germany
[2]Norwegian Institute for Water Research, P.O. Box 69 Korsvoll, 0808 Oslo 8, Norway

On a global scale, acidification is one of the major issues of freshwater pollution. In large regions of Europe and eastern North America, thousands of lakes, rivers, and streams have been damaged through acidification. Acidification of surface waters is due to a combination of natural and anthropogenic causes that occur in the catchment areas themselves and as acidifying pollutants from remote sources. Emissions of $SO_x$ and $NO_x$ into the atmosphere are carried hundreds and thousands of kilometers, with the resulting deposition affecting terrestrial and aquatic ecosystems. Measures to mitigate adverse effects include reductions in emissions and managing catchment areas and surface waters themselves (see Hauhs, this volume).

As early as 1920, freshwater acidification and problems with fisheries were reported in southern Norway, although the cause remained unexplained until the 1960s. By the 1970s, acid deposition, acidification of surface waters, and loss of fisheries had been reported in many countries in Europe as well as North America. In addition, symptoms of widespread forest decline first appeared in central Europe in the early 1980s. Since then, extensive research efforts have been devoted to the determination of the processes and mechanisms at the root of these problems. For a long time the major focus was on acidification by sulfur compounds. In several instances, however, reductions in $SO_2$ emissions and $SO_4^{2-}$ depositions did not show the anticipated recovery effects, since the lower S-load was (over-)compensated by nitrogen loadings.

Research on acidification of fresh waters and its reversibility can only be addressed in an inter- rather than multidisciplinary way. The various disciplines involved are:

1. *Aquatic geochemistry.* The ionic composition of acidified waters differs greatly from that of nonacidifed waters, e.g., the acidification in catchments releases metals from soils and sediments into the waters. Natural organic acids act both as sources of acidity and as pH buffers in low-alkalinity waters.

*Acidification of Freshwater Ecosystems: Implications for the Future*
Edited by C.E.W. Steinberg and R.F. Wright 

Processes within lakes and streams modify the chemical composition and acidification of fresh waters. These considerations have to be included in strategies for pollution control (see Van Miegroet, Hemond, Vesely, Schindler, Kelly, and Ford and Young, all this volume).

2. *Aquatic biology*. Though acidification of fresh waters is mainly the product of geochemical processes, the adverse effects are mainly on the aquatic biota. For example, acidification eliminates populations of both vertebrates and invertebrates. In some regions, habitats for sensitive organisms are being destroyed, and genetic information is lost. Clear evidence of freshwater acidification is found in paleolimnological studies, historical records of fisheries, and long-term chemical measurements (see Battarbee and Charles, Rosseland and Staurnes, and Raddum and Fjellheim, all this volume).
3. *Ecotoxicology*. Metals such as aluminum are highly toxic to fish and to several species of invertebrates and plants if present in the inorganic form. In many instances, the extinction of fish populations is attributed to these metals rather than to free acids in the water (compare arguments presented in Gunn and Belzille, and Rosseland and Staurnes, this volume). In headwater streams as well as in lakes, deposition of partly hydrolyzed and coagulated aluminum hydroxides accumulates on the ground and in the interstices of the bottom sediments, causing a high stress in the microcosm. The ability of dissolved organic acids to form soluble complexes with many metals can lower their toxicity; this must be considered in studies and mathematical models of mobilization, transport, and chemical as well as biological reactions of metal ions in surface and groundwaters.
4. *Forestry and soil science*. Acidifying compounds affect the long-term health and vitality of forests, which in turn influence freshwater quality. Forest management and natural disturbance, e.g., fire or wind throw, are other factors that influence runoff. Much of acidification and neutralization takes place in the soil through processes such as sulfate reduction, ammonium oxidation, sulfate adsorption, cation exchange, and chemical weathering. These processes, as well as the rates of acid deposition, are significantly influenced by forest management practices. Terrestrial processes and rates of change must be understood to explain freshwater acidification and to predict its reversibility (compare Feger, Harriman et al., but see also Emmett et al. and Dise et al., all this volume). What, however, do we know about acidification in the Tropics and the mechanisms involved? Is our methodology sufficiently developed for diagnosis and prognosis?
5. *Hydrology*. Adverse effects on freshwater biota are caused both by short-term acidic episodes and chronic acidification (see Ormerod and Jenkins, this volume). Changes in terrestrial ecosystems can affect hydrological pathways. Catchment hydrology is a fundamental aspect of predictive models. Other environmental factors, such as the greenhouse effect and global change, also influence catchment hydrology (see Dise et al., this volume).

6. *Political and social implications.* Acidification is an international phenomenon: emissions of $SO_2$ and $NO_x$ to the atmosphere are followed by dispersion, long-range transport across national boundaries, and deposition in areas often far from the sources. The solution to the problem is clearly to reduce emissions, but this entails high costs to society (see Paces, this volume). Thus, acidification research is politically relevant and has long been subject to unusually close scrutiny.

As we organized this workshop, our goals were (a) to integrate existing knowledge about the causes and consequences of freshwater acidification, to examine the links between terrestrial and aquatic effects, and (b) to evaluate the prospects for reversing the damage sustained so far as well as to offer a prognosis for the future. To this end, the workshop was organized into four groups, which focused on the following questions:

*Group 1.* Can we differentiate between natural and anthropogenic acidification? Here, terrestrial processes leading to water acidification were of central interest, as were the interactions between acid deposition and land use (see Emmett et al., this volume).
*Group 2.* How does acidification interact with other contaminants that affect freshwater ecosystems? The primary focus was on synergistic effects of acid, trace metals, organic compounds, and other pollutants (see Landers et al., this volume).
*Group 3.* How does acidification affect biota and what are the influences of biota on acidification? Here the focus was on changes in ecosystem structure and function as well as on feedback mechanisms due to acidification within aquatic ecosystems (see Brakke et al., this volume).
*Group 4.* Are chemical and biological changes reversible? This group focused on the future response of aquatic ecosystems, if and when acid deposition is reduced (see Dise et al., this volume).

For more than twenty years, extensive research has generated a wealth of information as to the causes and consequences of acidification of freshwater ecosystems . The very large-scale, unintentional environmental manipulation set off by the emissions of sulfur and nitrogen into the atmosphere has caused major changes in terrestrial and aquatic ecosystems. Acidification research has thrown new light on the inner mechanisms and interactions within aquatic ecosystems. The results have been central to the formulation of international environmental policy, resulting in protocols for reductions in the emission of pollutant gases (see Lükewille, this volume).

By the end of the workshop, we found that we were unable to formulate a concise prognosis for the future, since many questions remain unanswered, with new questions ever appearing on the horizon. The interactive effects of acid deposition and future global change pose new challenges for the future: the interrelationship between the thawing and oxygenation of permafrost soils, acidification, and changes in hydrological cycles. Here the role of nitrogen in acidification is of particular interest and, as more and more is understood about the historical development and present status of acidification of freshwater ecosystems, prediction of future developments becomes increasingly important. We hope that this Dahlem workshop on the *Acidification of Freshwater Ecosystems* will not only serve to highlight the current state of knowledge, but will point the way toward future research on new interdisciplinary and international levels.

We cordially thank J. Altman for planning, K. Roth for carrying out, and the staff of Dahlem Konferenzen for successfully organizing this workshop and creating a fruitful atmosphere. We also gratefully acknowledge the Stiftungsfond Uniliver in the Stifterverband für die Deutsche Wissenschaft, which provided the funds necessary for this meeting.

# 2

# Acidic Emissions and Political Systems

T. PACES
Czech Geological Survey, Malostranské nám. 19,
118 21 Prague, Czech Republic

## ABSTRACT

Statistics on the gross national product and emissions of $SO_2$, $NO_x$, and $NH_3$ in countries with different political and economic systems indicate strong relationships. Five groups of countries are tested: (a) countries with a market economy and high gross national product, (b) countries with a market economy and lower gross national product, (c) developing countries with a market economy, (d) European countries with a centrally planned economy that followed the Communistic ideology until 1989, and (e) a developing country with a centrally planned economy that follows the Communistic ideology.

Data on the acidifying emissions indicate the priorities to be solved in each country, in order for deterioration of quality in fresh water, as well as soils on a regional or continental scale, to be prevented. The first priority for the entire European community is to solve acidification due to $SO_2$ emissions in the post-Communistic countries of central and eastern Europe. Second, developed countries must lower the rates of acidification due to the emissions of $NO_x$. Third, the reduction of ammonia emissions has to be solved on a European level in countries with intensive animal husbandry. Developing countries cause severe deterioration of fresh waters due to local environmental acidification, near industrial centers.

## INTRODUCTION

Natural acidification of freshwater ecosystems is a biogeochemical process caused by dissolution of $CO_2$ and dissociation of carbonic acid, by fixation of cations (e.g., $NH_4^+$, $Ca^{2+}$, $Mg^{2+}$, $K^+$) in plants, dissociation of humic acids, oxidation of sulfide minerals, and through other chemical reactions leading to the production of hydrogen ions ($H^+$). The buffering capacity of carbonate minerals and the weathering rate of aluminosilicate minerals in rocks and soils has maintained a steady-state chemical composition of natural waters, including their pH values. Natural acidification has been intensified and accelerated by (a) industrial emissions of $SO_2$, NO, $NO_2$, and $NH_3$, (b) exhausts

*Acidification of Freshwater Ecosystems: Implications for the Future*
Edited by C.E.W. Steinberg and R.F. Wright 

of nitrogen oxides and carbon monoxide from motor vehicles, (c) incorrect applications of high doses of fertilizers, (d) ammonia emitted from animal farms, and (e) other anthropogenic activities with the production of hydrogen ions as a side effect.

Hydrogen ions generated by natural and anthropogenic biogeochemical processes dissolve aluminum from soils and leach exchangeable cations, such as the nutrients $Ca^{2+}$ and $Mg^{2+}$, out of soils. They increase chemical weathering and erosion rates, and they trigger other geochemical reactions that change the chemical composition of surface and groundwaters.

Environmental acidification is one of the fundamental processes controlling the metabolism of ecosystems. Humans have disrupted the ecological homeostasis by intensifying acidification and damaging lakes, forests, and shallow groundwater reservoirs. They have increased chemical erosion in modern landscapes.

Environmental acidification is a result of economic and agricultural development. It is related to technological achievements and depends on climatic and geological conditions. Acidification is an example of the link between science and policy (Wolters and Marseille 1992). It is an international problem (Alders 1992).

In this chapter, I examine the quantitative relationships between the rates of acidic emissions and political and economic systems of individual countries.

## METHOD

The rate of anthropogenic acidification is determined by the production rates of protons ($H^+$) through chemical reactions of acidifying compounds emitted into the environment during industrial and agricultural activities. The sources of acidification are expressed as annual acidic emission rates. These rates are related to the production of protons by the following stoichiometric equations:

$$SO_2 + H_2O + 0.5\ O_2 = SO_4^{2-} + 2\ H^+, \tag{2.1}$$

$$NO_2 + 0.5\ H_2O + 0.25\ O_2 = NO_3^- + H^+, \tag{2.2}$$

$$NH_3 + 2\ O_2 = NO_3^- + H_2O + H^+. \tag{2.3}$$

The rate of production of protons is therefore given by the sum of the rates of acidic emissions:

$$\frac{d\,[H^+]}{dt} = \frac{2\,d\,[SO_2]}{dt} + \frac{d\,[NO_2]}{dt} + \frac{d\,[NH_3]}{dt}, \tag{2.4}$$

where quantities in brackets [ ] are given in moles and time $t$ is given in years.

These are only the major reactions that cause anthropogenic acidification of environment. The rate

$$\frac{d\,[H^+]}{dt}$$

represents the potential acidification caused by the three acidifying compounds. A part of the generated protons is consumed by buffering processes, such as the weathering of rocks, which do not cause environmental acidification. Another part of protons acidifies soils and waters. The proportion between the harmless buffering and hazardous acidification of ecosystems has to be determined by an integrated monitoring of terrestrial and aquatic ecosystems (Paces 1992).

In this chapter, the sources of acidification are expressed as the annual emissions of $SO_2$, $NO_x$, and $NH_3$ by individual countries in grams. The data in the following diagrams have been compiled from the World Resources (1990), the UNEP Environmental Data Report (1991/92), Amann (1992), and French (1990).

The data on economic performance of individual countries were selected from the World Resources (1990); additional data on the gross national product (GNP) in U.S.$ were taken from data given by PlanEcon (1991). The data by World Resources were normalized with respect to GNP of the U.S.A. PlanEcon gives a value of U.S. $21,770 in 1990 while the World Resources give a value of U.S. $18,529. From these data, a recalculation coefficient of 1.175 was used to transfer the data by World Resources to the set given by PlanEcon.

The validity and precision of the emission and economic data were not tested. Hence, the general patterns in the diagrams and column graphs, rather than the exact values of data points and columns, are considered in the discussion below.

Five sets of data were compared:

1. Countries with a market economy and high GNP: United Kingdom, Sweden, Austria, West Germany, The Netherlands, Italy, Japan, and the United States of America.
2. Countries with a market economy and lower GNP: Greece, Ireland, and Portugal.
3. Developing countries with a market economy: Turkey, India, and Malaysia.
4. European countries with a centrally planned economy that followed the Communistic ideology until 1989: East Germany, Poland, Czechoslovakia, Hungary, Rumania, Bulgaria, and the European part of Soviet Union.
5. A developing country with a centrally planned economy that still follows the Communistic ideology: China.

## RESULTS

The annual rates of the major acidic emissions are given in Table 2.1. The values indicate how much each country contributes to regional acidification. However, these figures hide more subtle differences.

The relationship between annual emissions of $SO_2$ per square kilometer of individual countries and their GNP in 1987 is illustrated in Figure 2.1. The data points are distributed into three clusters according to the political and economic systems. The

**Table 2.1** Annual (1987) acidic emissions of $SO_2$, $NO_x$, and $NH_3$ in Gmol $[H^+]$ $yr^{-1}$.

| Country | Symbol in figures | $\frac{d[H^+]}{dt}$* |
|---|---|---|
| Developed market economy with higher GNP: | | |
| United Kingdom | UK | 196 |
| Sweden | S | 19 |
| Austria | A | 14 |
| West Germany | WG | 125 |
| The Netherlands | NL | 34 |
| Italy | I | 128 |
| Japan | J | (110) |
| U.S.A. | USA | (1200) |
| Developed market economy with lower GNP: | | |
| Greece | G | 21 |
| Ireland | IR | 14 |
| Portugal | P | 12 |
| Developing market economy: | | |
| Turkey | T | (30) |
| India | IN | (600) |
| Malaysia | M | (25) |
| Centrally planned economy: | | |
| East Germany | EG | 194 |
| Poland | PL | 193 |
| Czechoslovakia | CS | 120 |
| Hungary | H | 53 |
| Rumania | R | 85 |
| Bulgaria | B | 44 |
| European part of USSR | USSR | 594 |
| Developing, centrally planned economy: | | |
| China | C | (600) |

*Values in parentheses contain ammonia emissions estimated by analogy with western Europe.

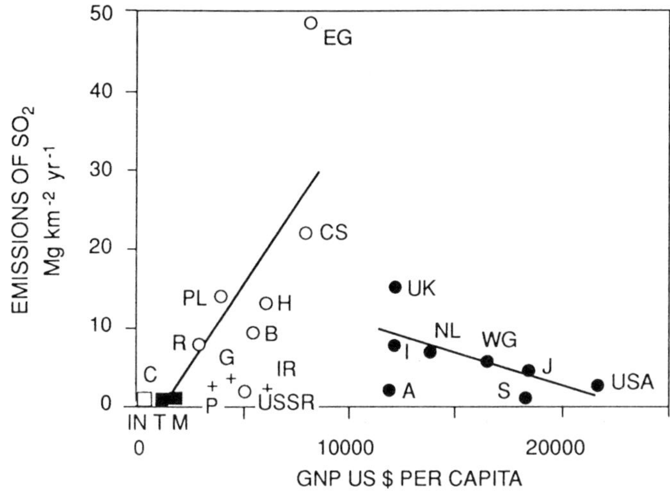


**Figure 2.1** Annual emissions of SO2 per area per year versus gross national product (GNP) per capita in 1987. Key to figures:
• = countries with a market economy and high GNP: UK (United Kingdom), S (Sweden), A (Austria), WG (West Germany), NL (The Netherlands), I (Italy), J (Japan), and USA (United States of America);
+ = countries with a market economy and lower GNP: G (Greece), IR (Ireland), and P (Portugal);
■ = developing countries with a market economy: T (Turkey), IN (India), and MA (Malaysia);
○ = European countries with a centrally planned economy that followed the Communistic ideology until 1989 (Communistic countries): EG (East Germany), PL (Poland), CS (Czechoslovakia), H (Hungary), R (Rumania), B (Bulgaria), and USSR (European part of the Soviet Union);
□ = developing country with a centrally planned economy that follows the Communistic ideology: C (China).

emissions of $SO_2$ decrease with increasing GNP in the developed countries with market economy and high GNP. In Communistic countries, the emission rate of $SO_2$ increases with their GNP. $SO_2$ emissions remain low in countries with a market economy and a low GNP, comparable to the Communistic countries. Developing countries have a low rate of acidic emissions and a low GNP. When the rate of $SO_2$ emissions is related to the population of each country, the pattern is similar. Figure 2.2 indicates that Communistic countries acidified the environment with $SO_2$ per capita at a much higher intensity than the market-oriented developed countries and the developing countries.

For the developed, market-oriented and Communistic countries, the emission rates of $SO_2$ and $NO_x$ relative to the GNP are shown in Figure 2.3. In regards to sulfur and, in some cases, even nitrogen emissions, it is clear that the Communistic countries were less efficient in reducing this kind of pollution relative to their GNPs. The column graph showing the ratio between the emissions of $SO_2$ and $NO_x$ in Figure 2.4 indicates

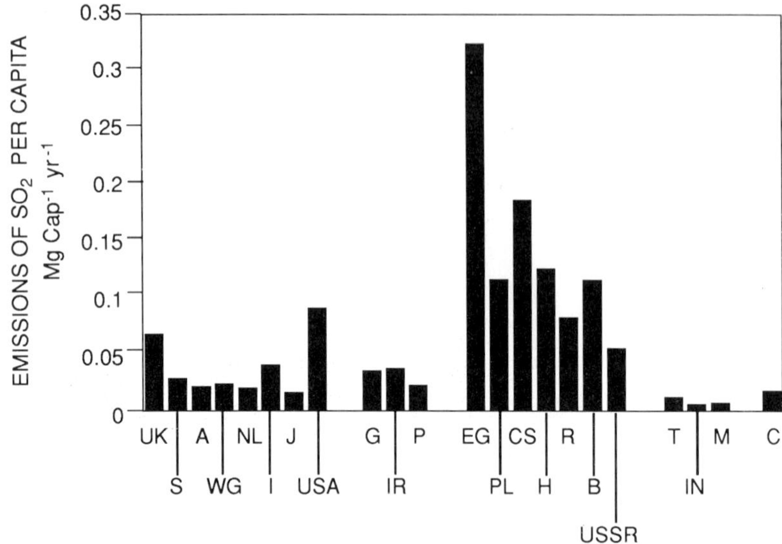


**Figure 2.2** Annual emissions of $SO_2$ in metric tonnes per capita per year in 1987.

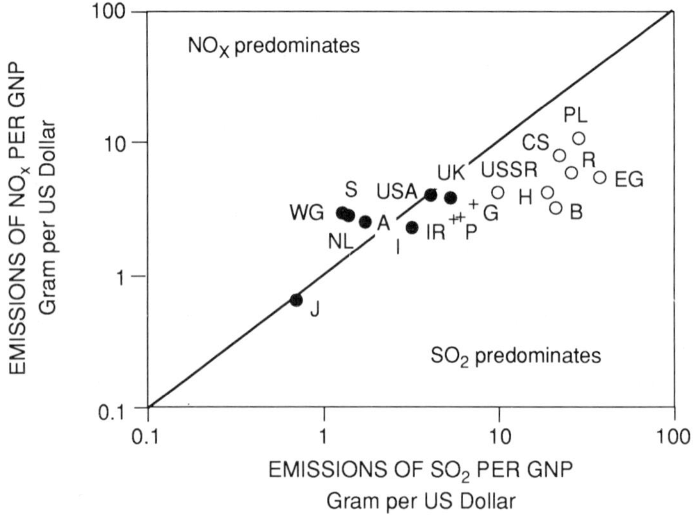


**Figure 2.3** Annual emissions of $NO_x$ per gross national product (GNP) vs. annual emissions of $SO_2$ per GNP in grams per U.S. $ in 1987; symbols are the same as in Figure 2.1.

that $SO_2$ was the major source of acidification in the Communistic countries. In the developed countries with a market economy, the ratio between the emissions of $SO_2$ and $NO_x$ varies considerably. In some countries the rates of $SO_2$ emissions still predominate; however in others, $NO_x$ emissions are almost twice as high as the

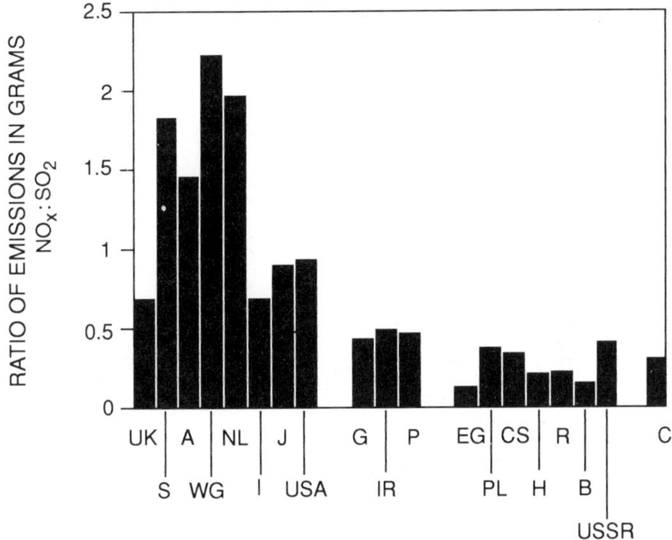


**Figure 2.4** The ratio of annual emission rates between $NO_x$ and $SO_2$ in grams per year in 1987.

emissions of $SO_2$. This is mainly due to lower rates of $SO_2$ emissions, as a result of using better quality oil and coal, and to more developed energy-saving technologies.

Economic and political systems often influence the rate of acidifying emissions independently of the population and size of individual countries. This is documented in Figure 2.5. With regard to size, the developing countries emit less acidifying compounds than their developed counterparts. Communistic countries usually, but not always, produce more acidifying compounds than the market-oriented countries of similar size. Some of the market-oriented countries have remarkably low rates of $SO_2$ emissions.

Emissions of ammonia are usually related to the intensity and type of agricultural production. The column graph in Figure 2.6 indicates large differences in the countries for which data are available, for example, between Sweden and The Netherlands.

## DISCUSSION

In spite of the fact that data on emissions and GNP may not be very accurate, their general patterns indicate that the rates of acidification are not only related to the size of countries and their populations but also to their political and economic status. Environmental acidification is buffered by rock-forming minerals in bedrock and residual minerals in soils, hence, the critical loads of acidifying compounds to aquatic and terrestrial ecosystems differ between geologically different regions (Kuylenstierna and Chadwick 1989; Sverdrup et al. 1990; Convention on Long-Range

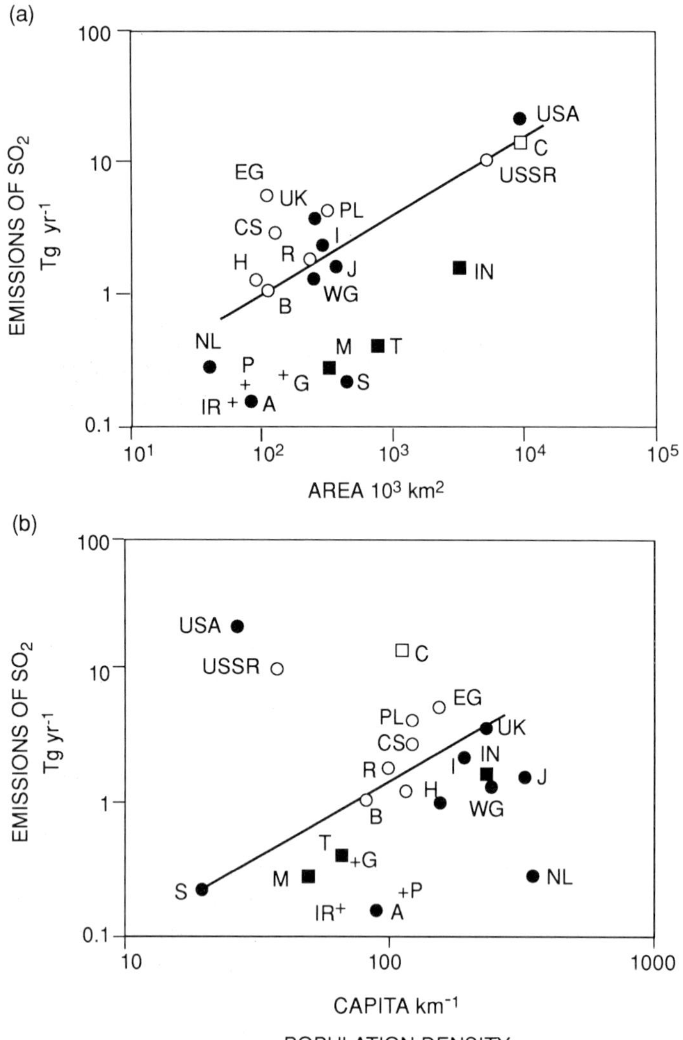


**Figure 2.5** The absolute rate of emissions of $SO_2$ versus area (a) and population density (b) of each country. Symbols are the same as in Figure 2.1; lines are drawn to separate the market-oriented from centrally planned economies.

Transboundary Air Pollution 1992). Because of the long-range transboundary transport of air pollution, large production rates of acidifying compounds in one country enhance the environmental acidification in other countries. This is the primary reason why this type of environmental deterioration has become recognized as a serious international problem.

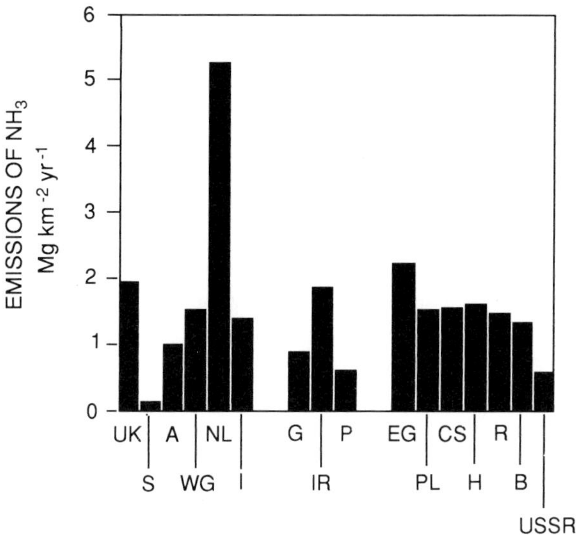


**Figure 2.6** Annual emissions of ammonia per area in 1987.

The specific rates of acidifying emissions calculated per square kilometer of the surface area or per capita of a country indicate the intensity with which individual countries contribute to the environmental problem. The data presented in Figures 2.1–2.6 are evidence that the specific rates depend on the economic and political status of each country. The reason for $SO_2$ emissions to have decreased, with respect to GNP in developed countries with a market economy (besides the obvious technological development and use of low-sulfur fossil fuels), involves the growing import of energy from other countries and, in some of the countries such as Sweden and France, the use of nuclear energy. High annual $SO_2$ emissions, with respect to GNP in the countries with centrally planned economies, were caused by the controlled low prices of energy, use of high-sulfur fossil fuels, low technological standards, and generally low concern for environmental issues under the Communistic regimes. The rates of $SO_2$ emissions per capita, shown in Figure 2.2, document the fact that Communistic countries produced more $SO_2$ per capita than the market-oriented countries did. Developing countries produce much smaller amounts of $SO_2$ per capita than the developed countries. In terms of the rates of emissions of $SO_2$ as well as $NO_x$ per GNP (Figure 2.3), the difference between the countries with a market economy and a centrally planned economy indicates that the centrally planned economy was usually less effective in preventing this type of pollution. Countries with a market economy normally emit more $NO_x$ with respect to $SO_2$ emissions compared to the countries with centrally planned economies (Figure 2.4). A larger number of cars in the countries with market economy and a better quality of fossil fuels are some of the reasons for this relatively high ratio of $NO_x$:$SO_2$.

The total rates of acidification depend on the area and population density of each country. Figures 2.5a and 2.5b indicate that the dispersion of data around the lines in the diagrams is also related to political and economic systems. Developing countries and some of the developed countries with a market economy exhibit emission rates under the lines, while most of the Communistic countries had rates above the lines.

Ammonia emissions (Figure 2.6) do not indicate any significant dependency on the political and economic system. In countries with a centrally planned economy, the emissions of ammonia were similar, while in countries with a market economy, emissions varied over a wide range. This variation probably reflects the relatively independent role that agriculture played in each of the developed countries. On the other hand, a uniform collective model of the centrally planned agriculture in Communistic countries is probably reflected by similar emissions of ammonia.

## CONCLUSIONS

Data on the acidifying emissions indicate the priorities to be addressed in each country, to prevent the further deterioration of the quality of fresh water as well as soils on regional and continental scales. It is not enough for each country to be solely responsible for its own rate of the emissions. The first priority for the whole European community is to solve acidification due to $SO_2$ emissions in the post-Communistic countries of central and eastern Europe. Second, developed countries should lower the rates of acidification due to the emissions of $NO_x$. Third, the reduction of ammonia emissions has to be solved on a European level in countries with intensive animal husbandry. Finally, acidification does not seem to be a general problem in developing countries. However, this conclusion, when confronted with the acidification damage of fresh waters in the industrial regions of China (Xia and Kuang 1991), may be valid only in a general sense. Locally, even developing countries can face severe deterioration of fresh waters due to environmental acidification.

## REFERENCES

Alders, J.G.M. 1992. Acidification: An international problem. In: Acidification Research Evaluation and Policy Applications, ed. T. Schneider, pp. 3–5. Amsterdam: Elsevier.

Amann, M. 1992. Emissions of acidifying components. In: Acidification Research. Evaluation and Policy Applications, ed. T. Schneider, pp. 65–76. Amsterdam: Elsevier.

Convention on Long-Range Transboundary Air Pollution. 1992. Evaluation of integrated monitoring in terrestrial reference areas of Europe and North America. The pilot programme 1989–1991. Environmental Data Centre. Helsinki: National Board of Waters and the Environment.

French, H.F. 1990. Green revolution: Environmental reconstruction in eastern Europe and Soviet Union. Worldwatch paper 99. Washington, D.C.: Worldwatch Institute.

Kuylenstierna, J.C.I., and M.J. Chadwick. 1989. The relative sensitivity of ecosystems in Europe to the indirect effects of acidic depositions. In: Regional Acidification Models, ed. J. Kämäri, D.F. Brakke, A. Jenkins, S.A. Norton, and R.F. Wright, pp. 3–23. Berlin: Springer.

Paces, T. 1992. Monitoring for the future: Integrated biogeochemical cycles in representative catchments. In: Acidification Research: Evaluation and Policy Applications, ed. T. Schneider, pp. 145–159. Studies in Environmental Science 50. Amsterdam: Elsevier.

PlanEcon. 1991. GNP data by PlanEcon, U.S.A., published in Czechoslovak newspapers: Lidove noviny, March 14, 1991. Prague.

Sverdrup, H., W. de Vries, and A. Henriksen. 1990. Mapping critical loads. Miljørapport 1990: 14, NORD 1990: 98. Copenhagen: Nordic Council of Ministers.

UNEP. 1991/92. Environmental Data Report. 3rd Edition. Blackwell Reference. Oxford: Basil Blackwell Ltd.

Wolters, G.J.R., and H. Marseille. 1992. Acidification is an example of the link between science and policy. In: Acidification Research, Evaluation and Policy Applications, ed. T. Schneider, pp. 7–15. Amsterdam: Elsevier.

World Resources. 1990. World Resources 1990–91. A Report by the World Resources Institute. Oxford: Oxford Univ. Press.

Xia, Y., and Q. Kuang. 1991. A preliminary study in acidification of waters in China. Proc. of Intl. Symp. on Impacts of Salinization and Acidification on Terrestrial Ecosystems and its Rehabilitation, pp. 221–216. Tokyo: Tokyo Univ. of Agriculture and Technology.

# 3

# Billion Dollar Problem, Billion Dollar Solution? Transboundary Air Pollution Calls for Transboundary Solutions

A. LÜKEWILLE[1]
Federal Environmental Agency (UBA), Bismarckplatz 1,
14193 Berlin, F.R. Germany

## ABSTRACT

Transboundary air pollution damages national resources of other countries, e.g., forests, lakes, and cultural heritage. International legal principles on state responsibility for these impacts have not yet been established. A first step to enforce emission reduction measures in the international sphere is the UN ECE Convention on Long-range Transboundary Air Pollution, which went into effect in 1983.

The protocols elaborated under the Convention on, e.g., $SO_2$ and $NO_x$, have so far been based exclusively on percentage emission reductions of these pollutants. Embedded into the UN ECE activities, the critical loads/levels concept provides a basis for definition of environmental quality standards and covers the effects of all relevant emissions on natural and cultural resources to be protected in individual countries. By incorporating this approach into integrated assessment models, it thus offers the opportunity to negotiate on the basis of these effects. The concept provides the chance to optimize not only the benefits of pollution reduction measures but also—seen from the international point of view—the cost-effectiveness of necessary expenditures.

Integrated assessment models are an important tool for evaluating the cost efficiency of different abatement strategies, including various regulatory as well as economic instruments.

[1] The views expressed are entirely those of the author and do not necessarily coincide with those of the Federal Environmental Agency (UBA).

*Acidification of Freshwater Ecosystems: Implications for the Future*
Edited by C.E.W. Steinberg and R.F. Wright © 1994 John Wiley & Sons Ltd.

## TRANSBOUNDARY AIR POLLUTION

The easiest and most widely used international emission "control" tactic for gaseous sulfur and nitrogen compounds is to dispose of them into the atmosphere. This dispersion, however, is not without consequences. Atmospheric acid deposition has received considerable attention as an international environmental problem in Europe, North America, and Asia. Polluted air masses containing sulfur and nitrogen compounds, precursors of "acid rain," travel long distances across national boundaries. They have an impact on the natural resources of other countries. Some countries, including Germany, export more pollutants than they import. For example, sources of air pollution in the United Kingdom contribute to the acid rain problem in Scandinavia and cast Britain in the role of the villain (Anon. 1977). The U.K. exports almost ten times more sulfur than it imports; Norway, however, is an example of exactly the opposite extreme (Alm 1989).

Thus, acid deposition is much more than a serious problem affecting surface waters, groundwater, and forest soils. It creates national and international regulatory problems and demands that we rethink air pollution control strategies. A high level of coordinated planning amoung countries is essential. Even so, international legal principles on state responsibility for environmental damage caused by transboundary air pollution have not yet been established.

## CONVENTION ON LONG-RANGE TRANSBOUNDARY AIR POLLUTION

In 1977, the final report of the Organization for Economic Cooperation and Development (OECD) project on emissions, transport, and deposition of sulfur pollutants in Europe was published . The results clearly showed that air pollution is a transboundary problem. Thus, emission control measures must be adopted on an international basis in order to be effective. An initial step to achieve such measures was the 1979 "Convention on Long-range Transboundary Air Pollution" (LRTAP convention), which went into effect in 1983 (Ågren 1992; Wüster 1992). Based on the 1972 UN Conference on Human Environment (held in Stockholm) and the above-mentioned OECD report, the LRTAP convention was set up within the framework of the United Nations Economic Commission for Europe (UN ECE) and was strongly supported by the Scandinavian countries. Most European and North American states are parties to this convention. The forum of the UN ECE was appropriate at the time because the countries of eastern Europe, which contribute considerably to transboundary air pollution, did not recognize the European Community.

In addition to the LRTAP convention, several protocols aimed at monitoring and evaluating the transport of long-range pollutants and at reducing air pollutant emissions and their transboundary fluxes were initiated:

**Figure 3.1** Courtesy of *Die Zeit*, October 19, 1984.

- for monitoring and evaluation of long-range transmissions (EMEP), signed by 22 states in 1984/85,
- for sulfur dioxide ($SO_2$), signed by 20 states in 1985 (see Table 3.1),
- for nitrogen oxides ($NO_x$), signed by 25 states in 1988/89 (see Table 3.1),
- for volatile organic compounds (VOCs), signed by 23 states in 1991.

One of the main objectives of the EMEP program is to model atmospheric dispersion, through the use of emission data, meteorological data, and functions describing transformation and removal processes (Wüster 1992). So-called EMEP models are used to compute deposition patterns over Europe that can then be mapped (Iversen et al. 1991). These maps are subdivided into 150 × 150 km grid cells, called "EMEP grids."

By 1993 at the latest, the $SO_2$ protocol proposes to reduce the signatory states' emissions or transboundary fluxes, based on the 1980 levels, by at least 30%. The

U.K., U.S.A., and Poland were among those countries that did not join the "30% club" (Table 3.1). Poland as well as a number of southern European countries were concerned about the possible costs of achieving the intended effects to their national economies. The U.K. and the U.S.A. cited the need for more research, greater certainty in the scientific data, alternative methods of calculation, and/or the need to take account of reductions prior to 1980 (Demidecki-Demidowicz 1984; Aniansson 1985). Several countries have gone further than required. Nine states pledged themselves to reduce $SO_2$ emissions to less than half of 1982 levels by 1995. Three states committed themselves to a two-thirds reduction: Austria, Germany, Sweden (UNEP 1992). Revision of the $SO_2$ protocol is planned for 1993, and negotiations have already started.

The protocol for nitrogen oxides, established in 1988, proposes no actual reduction. It states that as of 1994, emissions will not be allowed to exceed the 1987 levels. Twelve governments (Austria, Belgium, Denmark, Germany, Finland, France, Italy, Liechtenstein, Norway, Sweden, Switzerland, and The Netherlands) signed a declaration to implement "a reduction of national annual nitrogen emissions on the order of 30% as soon as possible and at the latest by 1998, using the level of any year between 1980 and 1986 as a basis for the calculation of the reduction" (Ågren 1988; UNEP 1988). Further, "the parties shall, as a second step, commence negotiations" on further measures to reduce national emissions "taking into account the best available scientific and technological developments, internationally accepted critical loads and other elements." The protocol so far ignores ammonia emissions from intensive animal husbandry.

The protocol for VOCs provides for a reduction of emissions and transboundary fluxes by 30% in 1999, relative to the levels of any year between 1984 and 1990. VOCs together with nitrogen oxides (and CO) take part in photochemical reactions and lead to ozone production.

## REGULATORY INSTRUMENTS

The control measures under the LRTAP convention are based exclusively on percentage emission reductions of air pollutants. Regulatory instruments widely used in air management are emission standards, often in combination with other instruments such as air quality standards (e.g., critical concentrations of certain gaseous pollutants). In many countries, limits for $SO_2$ and $NO_x$ emissions from large stationary sources, especially from new facilities, were set. One example is the regulation on $SO_2$, $NO_x$, and particulate matter emission reductions from major heating plants, which was passed by Germany and later translated into a comparable European Community directive (OECD 1989; Andersson et al. 1992). A similar approach is the setting of emission limits for mobile sources. These standards are often based on "best available technology" (BAT) approaches and take into account new process technologies and their cost-effectiveness.

**Table 3.1** Status of the sulfur and $NO_x$ protocol.

| Country | Sulfur Protocol Signature | Sulfur Protocol Ratification | $NO_x$ Protocol Signature | $NO_x$ Protocol Ratification |
|---|---|---|---|---|
| Austria | 9.7.85 | 4.6.87 (R) | 1.11.88 | 15.1.90 (R) |
| Belarus | 9.7.85 | 10.9.86 (At) | 1.11.88 | 8.6.89 (At) |
| Belgium | 9.7.85 | 9.6.89 (R) | 1.11.88 | |
| Bulgaria | 9.7.85 | 26.9.86 (Ap) | 1.11.88 | 30.3.89 (R) |
| Canada | 9.7.85 | 4.12.86 (R) | 1.11.88 | 25.1.91 (R) |
| Czech and Slovak Federal Republic | 9.7.85 | 26.11.86 (Ap) | 1.11.88 | 17.8.90 (Ap) |
| Denmark | 9.7.85 | 29.4.86 (R) | 1.11.88 | |
| Finland | 9.7.85 | 24.6.86 (R) | 1.11.88 | 1.2.90 (R) |
| France | 9.7.85 | 13.3.86 (Ap) | 1.11.88 | 20.7.90 (R) |
| Germany (5) | 9.7.85 | 3.3.87 (R, 2) | 1.11.88 | 16.11.90 (R) |
| Greece | | | 1.11.88 | |
| Holy See | | | | |
| Hungary | 9.7.85 | 11.9.86 (R) | 3.5.89 | |
| Iceland | | | | |
| Ireland | | | 1.5.89 | |
| Italy | 9.7.85 | 5.2.90 (R) | 1.11.88 | |
| Liechtenstein | 9.7.85 | 13.2.86 (R) | 1.11.88 | |
| Luxembourg | 9.7.85 | 24.8.87 (R) | 1.11.88 | 4.10.90 (R) |
| The Netherlands | 9.7.85 | 30.4.86 (At, 3) | 1.11.88 | 11.10.89 (At, 3) |
| Norway | 9.7.85 | 4.11.86 (R) | 1.11.88 | 11.10.89 (R) |
| Poland | | | 1.11.88 | |
| Portugal | | | | |
| Rumania | | | | |
| San Marino | | | | |
| Spain | | | 1.11.88 | 4.12.90 (R) |
| Sweden | 9.7.85 | 31.3.86 (R) | 1.11.88 | 27.7.90 (R) |
| Switzerland | 9.7.85 | 21.9.87 (R) | 1.11.88 | 18.9.90 (R) |
| Turkey | | | | |
| Ukraine | 9.7.85 | 2.10.86 (At) | 1.11.88 | 24.7.89 (At) |
| USSR | 9.7.85 | 10.9.86 (At) | 1.11.88 | 21.6.89 (At) |
| United Kingdom | | | 1.11.88 | 15.10.90 (R, 4) |
| United States | | | 1.11.88 | 13.7.89 (At) |
| Yugoslavia | | | | |
| European Community | | | | |
| TOTAL | 20 | 20 | 26 | 18 |

Notes: *R = ratification, Ac = accession, Ap = approval, At = acceptance
(1) With declaration upon signature
(2) With declaration upon ratification
(3) For the Kingdom in Europe
(4) Including the Bailiwick of Guemsey, the Isle of Man, Gibraltar, the United Kingdom Sovereign Base Areas of Akrotiri and Dhekhelia on the Island of Cyprus
(5) The former GDR signed the Sulfur Protocol on 9.7.85 and the $NO_x$ Protocol on 1.11.88

## THE CRITICAL LOADS CONCEPT

Environmental quality standards cover the effects of all emission sources on the receptor to be protected (OECD 1989). Within the framework of the LRTAP convention, the critical loads concept provides the basis for definition of such environmental standards (Bull 1991; Hettelingh et al. 1992).

A critical load is defined as *"a quantitative estimate of an exposure to one or more pollutants below which significantly harmful effects on specified sensitive elements of the environment do not occur according to present knowledge."* This definition was adopted by the UN ECE workshop on Critical Loads for Sulfur and Nitrogen, held in Sweden in 1988 (Nilsson and Grennfelt 1988). Sensitive elements include terrestrial ecosystems, forest soils, groundwater, and surface waters. Thus, a critical load is the maximum "no significant effect" level of a pollutant (see Figure 3.2).

The setting of target loads (Figure 3.2) includes technical, social, and economic considerations, and thus is a political process. Theoretically, target loads can be set below the critical loads ("safety factor"). Target loads above the critical loads allow some damage to sensitive receptors.

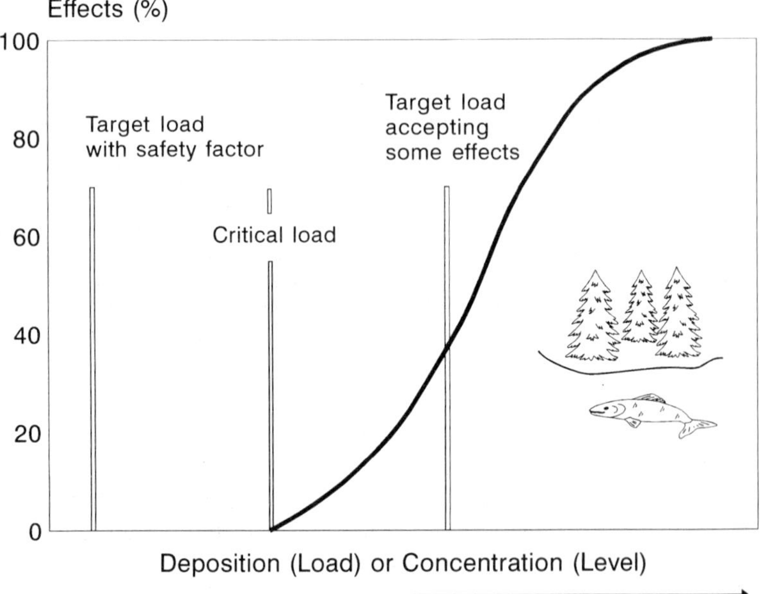


**Figure 3.2** Theoretical curve showing alternative "target loads" compared with "critical loads." After Bull (1991).

Critical and target loads can be aggregated to single values for EMEP grid cells (150 × 150 km) and compared with the present deposition loads, as well as with results of different emission reduction strategies (for details, see Hettelingh et al. 1992).

To promote such mapping activities under the LTRAP convention, a Task Force on Mapping Critical Levels and Loads (TFM) was established under the UN ECE Working Group on Effects in 1988. The TFM is led by Germany (e.g., UN ECE 1990). Figure 3.3 shows the most important bodies of the LRTAP convention connected with the critical loads concept.

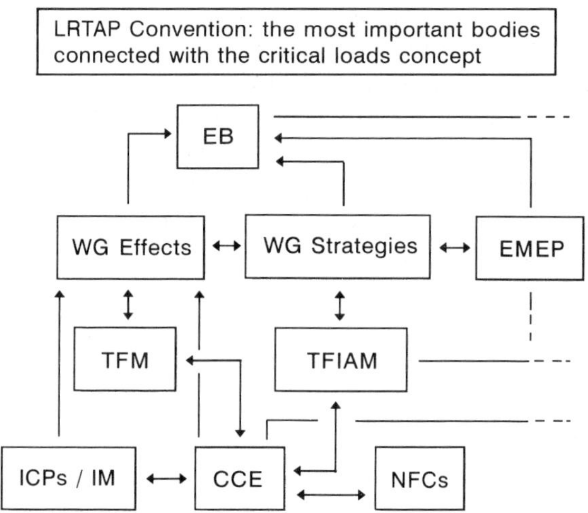


**Figure 3.3** Interrelation between important bodies of the LRTAP convention: EB = Executive Body; WG = Working Group; TFM = Task Force on Mapping; TFIAM = Task Force on Integrated Assessment Modeling; EMEP = The Cooperative Program for Monitoring and Evaluation of Long-range Transmissions of Air Pollutants in Europe; ICPs = International Cooperative Programs on assessment and monitoring air pollution effects on (1) rivers and lakes (leading country: Norway), (2) forests (leading country: Germany), (3) materials, including historic and cultural monuments (leading country: Sweden), and (4) agricultural crops (leading country: United Kingdom); IM = Pilot Program on Integrated Monitoring of Air Pollution Effects and Ecosystems (leading country: Sweden); CCE = Coordination Center for Effects (The Netherlands); NFCs = National Focal Centers.

In collaboration with National Focal Centers (NFCs), the Coordination Center for Effects (RIVM-CCE, The Netherlands) has so far produced European maps showing the critical loads for acidity, sulfur, and the corresponding excesses (Hettelingh et al. 1991). Figure 3.4 shows these excesses of the critical loads for acidity and sulfur due to present emissions of acidic precursors. Such maps serve as a useful, scientifically derived basis for the current negotiations on a revised LRTAP sulfur protocol. Areas with the greatest excesses (> 2000 eq $ha^{-1}$ $yr^{-1}$) are to be found primarily in central Europe.

(a)

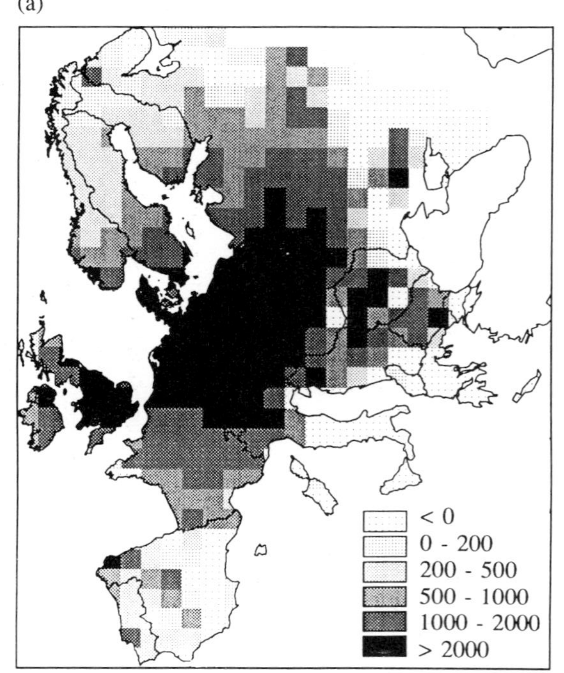


(b)

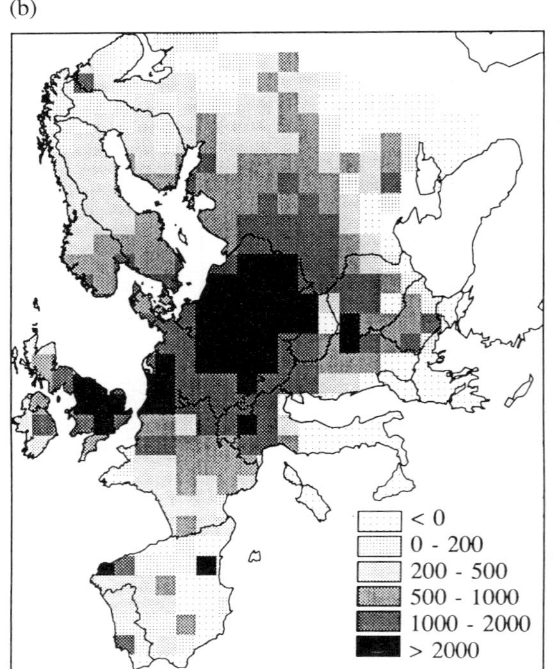


**Figure 3.4** Excesses of the critical loads for (a) acidity and (b) sulfur due to present emissions of acidic precursors or sulfur (1 percentile); units = eq ha$^{-1}$ yr$^{-1}$ (Hettelingh et al. 1991, 1992).

A UN ECE workshop on Critical Loads for Nitrogen (held in Sweden in 1992) began to lay the foundations for assessing critical loads for nitrogen (N) and thus for the production of European critical loads maps for N by 1993 (Grennfelt and Thörnelöf 1992). As mentioned above, the parties to the convention will, as a second step within the framework of the N protocol, commence negotiations on further measures to reduce national N emissions by 1995, at the latest. Specific reference is made to internationally accepted critical loads.

A critical level is *"the concentration of pollutants in the atmosphere above which direct adverse effects on receptors, such as plants, ecosystems or materials may occur according to present knowledge,"* a definition adopted by the UN ECE workshop on Critical Levels in Germany in 1988 (UN ECE 1988).

The mapping of critical levels of $SO_2$, $NO_x$, $NH_3$, $O_3$ (ozone) and eventually VOCs (volatile organic compounds) for materials, plants, plant communities, and/or ecosystems is planned, using the results from the UN ECE workshop on Critical Levels in the U.K. (1992) as well as from forthcoming workshops in the U.K. (target levels for buildings and materials) and Switzerland (ozone) in 1993. Another future activity may be the assessment and mapping of critical loads for heavy metals and persistent organic compounds.

## INTEGRATED ASSESSMENT OF COSTS AND BENEFITS

The costs of abatement may be outweighed by the benefits of reducing emissions (Wüster 1992). Under the LRTAP convention, the assessment of costs and benefits is covered by the Task Force on Integrated Assessment Modeling (TFIAM), led by The Netherlands (see Figure 3.3). The **R**egional **A**cidification **IN**formation and **S**imulation (RAINS) model has played an important role in evaluating the cost efficiency of different abatement strategies in the TFIAM. RAINS and other models (e.g., CASM [Swedish Environmental Institute at York] and ASAM [Imperial College, U.K.]) are being used by the TFIAM in international negotiations on air pollution reductions in Europe. The RAINS model has been developed by the International Institute for Applied Systems Analysis (IIASA, Austria) and is "integrated in the sense that it links public policy alternatives with their consequences in nature, and also because it brings information about costs of control, emissions, atmospheric transport, and ecological impact together in one place in a consistent manner" (Alcamo et al. 1990). Figure 3.5 shows schematically the structure of RAINS. The model incorporates $SO_2$, $NO_x$, and $NH_x$ transfer matrices based on the EMEP models.

Concerning the calculations of certain sulfur emission reductions and costs, RAINS contains options for fuel substitution, use of low-sulfur fuels, and desulfurization during and after fuel combustion. Energy conservation strategies have been recently included. Cost estimates take into consideration the country- and sector-specific factors, such as labor costs (Alcamo et al. 1990).

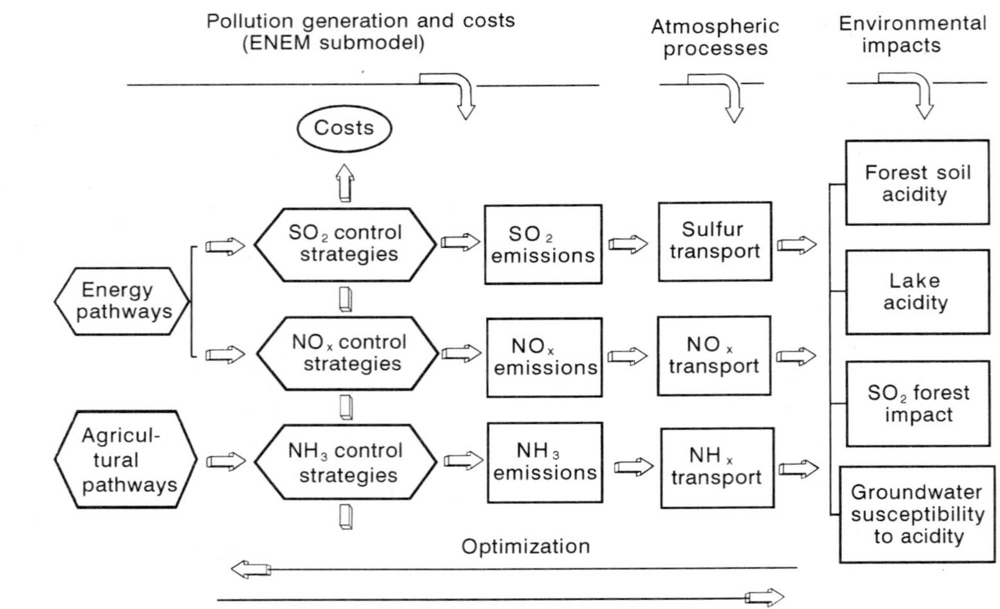


**Figure 3.5** Schematic diagram of the RAINS model; rectangles represent calculations and hexagons the inputs (Amann 1992, pers. comm.); ENEM = *en*ergy/*em*ission/cost.

Using the options included in RAINS, scenarios of different abatement strategies can be analyzed, with the current loads of acidity and sulfur as reference scenarios. The resulting deposition patterns can be compared to the corresponding critical loads for acidity and sulfur, and the excesses can be mapped on a European scale. One example is the scenario of "maximum feasible reductions of emissions of acid precursors" analyzed by Hettelingh et al. (1992). It is "assumed that removal efficiencies of 90 to 98 percent are achieved by applying flue gas desulfurization to large boilers in refineries, power plants and industry," and that "small boilers are supplied with low-sulphur fuels, and best available techniques are applied to reduce $NO_x$ emissions." Figure 3.6 shows the expected pattern of excess for the critical loads of acidity. The area recieving deposition higher than critical loads is reduced significantly. However, the map of excesses shows that even using the best available abatement technology everywhere, it is presently impossible to reach critical loads of acidity for each EMEP grid: target loads must be used to optimize emission reductions to obtain maximum benefits. Within the UN ECE, an increasing importance is being attached to economic instruments as a part of emission control strategies.

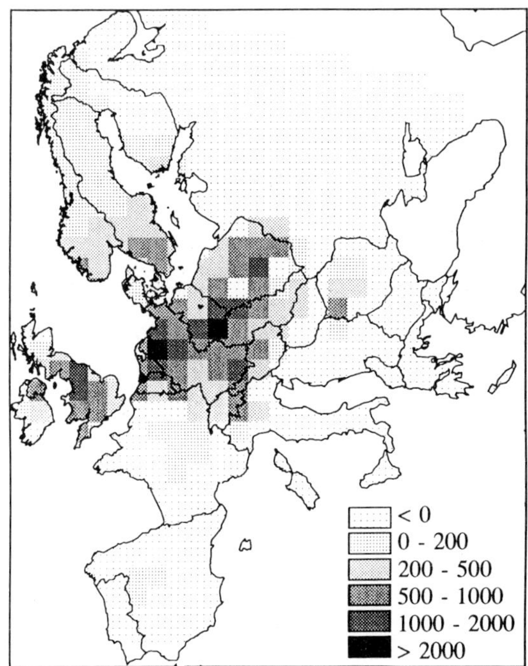


**Figure 3.6** Excesses of the critical load for acidity after maximum feasible reductions of $SO_2$ and $NO_x$ emissions in Europe (1 percentile); units = eq $ha^{-1}$ $yr^{-1}$ (Hettelingh et al. 1992).

## ECONOMIC INSTRUMENTS

"We must make better use of the market to give us a cleaner environment, more quickly, and at less cost" (Brundtland 1991). This message seems to be the environmental challenge of the 1990s. Less costs mean more environmental benefit per unit expenditure since the costs of abatement (emission reduction), on one hand, or of treatment (emission amelioration), on the other, will continue to be high. To cite an example of the former, German investments have been made to reduce emissions of sulfur and nitrogen, amounting to DM 14,000 million and DM 12,000 million, respectively, over a period of six years (Schärer and Haug 1990).

The Swedish government has spent SEK 780 million on its operational liming program over the period 1977–1990 (Lövgren et al. 1992). Of course, pollution effects themselves are costly: loss of potential timber production due to the effects of air pollution may be equivalent to 3,000 million U.S. $ annually (Wiseman 1991). It is clear that acidification represents a more than billion dollar problem that will require a more than billion dollar solution.

Two major sorts of instruments can be distinguished: command-and-control regulations and economic incentives. Regulations may include, for example, standards for sulfur emissions for power plants or bans on motorcars without catalytic converters.

Economic instruments for changing the relative prices on various goods are subsidies (to pollution abatement), charges (on pollution activities), fees (e.g., for waste removal), and marketable emission permits (Lindeneg 1990; Cropper and Oates 1992). Some examples for establishing such instruments on the national scale include:

1. In Sweden, charges are used as an economic supplement to administrative (regulatory) instruments (Duer et al. 1990). Fuel and emission standards are planned to be combined with economic incentives and disincentives. A sulfur tax is imposed on coal, peat, and oil. If sulfur emissions are reduced, e.g., through flue gas desulfurization, a portion of the sulfur tax is refunded (Lövgren et al. 1992).
2. In the U.S.A., a measure for the trading of sulfur emissions to control acid rain was introduced with the new 1990 Amendments to the Clean Air Act (Cropper and Oates 1992). Similar to emission standards, the total quantities are set administratively; prices and individual quantities are set on the market of emission permits (Lindeneg 1990).

Indeed, there *is* already a whole "tool box" containing various types of instruments that can be used in environmental politics, depending on the goals and the specific environmental problem.

The *polluter pays principle* says that those who use society's scarce environmental resources should compensate the public for their use (e.g., OECD 1991). However, due to the economic situation of countries in central and eastern Europe it may be more effective, in the international sphere, to consider "burden sharing," i.e., to provide economic and technical assistance to these countries in order to reduce their share of air pollution.

## CONCLUSIONS

Transboundary air pollution and its impact on natural resources of other countries demands the rethinking of air pollution control strategies. The critical loads concept, based on standards related to effects of pollutants at specified sensitive elements of the environment, offers the chance to negotiate on the basis of these effects. Additionally, the critical loads approach thus provides opportunities to optimize not only the benefits of pollution control measures but also the cost-effectiveness of necessary expenditures.

## ACKNOWLEDGEMENTS

I wish to thank Bernd Schärer (Berlin) and Ron Harriman (Perthshire) for their critical and constructive comments, James Burn (Bilthoven) and Julia Lupp (Berlin) for editorial assistance. Thanks are also due to Jean-Paul Hettelingh, Bob Downing, and Peter de Smet (Bilthoven) for permission to use their "scenario maps."

**Figure 3.7** Sketch by Chris Rose, reprinted from *Acid News* **3-4**:22-23 (Dec. 1986).

## REFERENCES


Anon. 1977. Million-dollar problem — billion-dollar solution. *Nature* **268**:89.

Ågren, C. 1988. Nitrogen oxides freeze. *Acid Magazine* **7**:18–19.

Ågren, C. 1992. The UN ECE Convention on Long-range Transboundary Air Pollution. In: Third International NGO Strategy Seminar on Air Pollution, ed. C. Ågren and P. Elvingson, pp. 6-12. Göteborg: Swedish NGO Secretariat on Acid Rain.

Alcamo, J., R. Shaw, and L. Hordijk. 1990. The Rains Model of Acidification, Science and Strategies in Europe. Dordrecht: Kluwer.

Alm, H. 1989. Emissions are falling, but is it enough? *Acid Magazine* **8**:5–6.

Anderson, M., G. Bennekou, and H. Schroll 1992. Environmental problems and environmental regulations in western Europe, 1980–1989. *Environ. Manag.* **16(2)**:187–194.

Aniansson, B. 1985. A firm commitment. *Acid Magazine* **3**:29–30.

Brundtland, G.H. 1991. Environmental challenges of the 1990s: Our responsibility towards future generations. *Nord. J. Econ* **2**:4–9.

Bull, K.B. 1991. The critical loads/levels approach to gaseous pollutant emission control. *Environ. Pollut.* **69**:105–123.

Cropper, M.L., and E.E. Oates 1992. Environmental economics: A survey. *J. Econ. Lit.* **30**:675–740.

Demidecki-Demidowicz, M.R. 1984. Acid rain report and the goverment's reply. *Environ. Pol. Law* **13/3/4**:107–108.

Duer, H., K. Lindeneg, and T. Parkkinen 1990. Economic instruments in environmental policy in the Nordic countries. *Nord. J. Econ.* **1**:7–8

Grennfelt, P., and E. Thörnelöf, eds. 1992. Critical loads for nitrogen. A workshop report. *Nord 1992*:**41**. Copenhagen: Nordic Council of Ministers.

Hettelingh, J.-P., R.J. Downing, and P.A.M. de Smet, eds. 1991. Mapping critical loads for Europe, CCE Technical Report No. 1. Bilthoven: National Inst. of Public Health and Environmental Protection (RIVM).

Hettelingh, J.-P., R.J. Downing, and P.A.M. de Smet. 1992. The critical loads concept for the control of acidification. In: Acidification Research, Evaluation and Policy Applications, ed. T. Schneider, pp. 161–174. Amsterdam: Elsevier.

Iversen,T., N.E. Halvorsen, S. Mylona, and S. Sandnes. 1991. Calculated budgets for airborne acidifying compounds in Europe, 1985, 1988, 1989, 1990, MEMEP. *MSC-W Report* **1/91**. Oslo: Norwegian Meteorological Institute.

Lindeneg, K. 1990. Instruments in environmental policy. *Nord. J. Environ. Econ.* **1**:4–6.

Lövgren, K., G. Persson, and E. Thörnelöf. 1992. Acidification policy: Sweden. In: Acidification Research, Evaluation and Policy Applications, ed. T. Schneider, pp. 241–246. Amsterdam: Elsevier.

Nilsson, J., and P. Grennfelt. 1988. Critical loads for sulphur and nitrogen: Report from a workshop held at Skokloster, Sweden 19–24 March 1988. *Nord 1988*:**97**. Copenhagen: Nordic Council of Ministers.

OECD (Organization for Economic Cooperation and Development). 1989. Energy and the Environment: Policy Overview. Paris: OECD/International Energy Agency.

OECD. 1991. Recommendation of the Council on the Use of Economic Instruments in Environmental Policy. Paris: OECD.

Schärer, B., and N. Haug. 1990. Bilanz der Großfeuerungsanlagenverordnung. *Staub–Reinhaltung der Luft* **50**:139–144.

UN ECE. (United Nations Economic Commission for Europe). 1988. ECE critical levels workshop, Bad Harzburg 14–18 March 1988, final draft. Geneva: UN ECE.

UN ECE. 1990. Draft manual on methodologies and criteria for mapping critical levels/loads. Geneva: UN ECE.

UNEP. 1988. Protocol concerning the emissions of nitrogen oxides or their transboundary fluxes. *Environ. Pol. Law* **18/6**:228–234

UNEP. 1992. The state of the global environment. *Our Planet* **4**:4–5.

Wiseman, R. 1991. European forest production cut by air pollution. *Prev. Pollut.* **18**:33–35.

Wüster, H. 1992. The Convention on Long-range Transboundary Air Pollution: Its achievements and its potentials. In: Acidification Research, Evaluation and Policy Applications, ed. T. Schneider, pp. 221–239. Amsterdam: Elsevier.

# 4

# The Relative Importance of Sulfur and Nitrogen Compounds in the Acidification of Fresh Water

H. VAN MIEGROET*
Environmental Sciences Division, Oak Ridge National Laboratory, Oak Ridge, TN 37831–6038, U.S.A.

## ABSTRACT

The acidity of streams and lakes is influenced by the amount and form of nitrogen (N) and sulfur (S) deposition from the atmosphere and by biogeochemical S and N cycling in the catchment and within lakes and streams. Hydrologic flowpaths and water residence times play a key role in determining the relative importance of atmospheric vs. internal processes in the pH and acid-neutralizing capacity (ANC) of fresh waters. Sulfate movement through the catchment is primarily regulated by adsorption-desorption reactions in the soil, whereas microbial reduction reactions within the sediment are the main means by which the $SO_4^{2-}$ concentration is decreased and the ANC increased in the water column. Nitrogen transformation and transport in terrestrial and aquatic environments are, for the most part, biologically controlled. Nitrate export from the watershed is an important component of episodic freshwater acidification associated with snowmelt and stormflow, whereas $SO_4^{2-}$ is generally a more important component in chronic water acidification. Historical trends in surface water chemistry suggest an increasing contribution of $NO_3^-$ to water acidification. Current surface water acidification models are primarily based on the $SO_4^{2-}$ adsorption characteristics of the soils in the drainage basin; less effort has been directed toward the assessment and quantification of terrestrial N dynamics or toward the relative role of in-stream and in-lake processes. Models also remain limited in their ability to predict and quantify $SO_4^{2-}$ desorption. Accurate prediction of surface water acidification under various N and S deposition scenarios depends on the extent to which we are able to quantify S and N release and retention processes in terrestrial and aquatic systems under current conditions and in the future, as a result of changes in acid deposition and global environmental conditions.

* Present address: Utah State University, Dept. Forest Resources, Logan UT 84322-5215, U.S.A.

*Acidification of Freshwater Ecosystems: Implications for the Future*
Edited by C.E.W. Steinberg and R.F. Wright 

## INTRODUCTION

The acid-base chemistry of streams and lakes is regulated by the amount and composition of atmospheric deposition and by biogeochemical processes in the catchment and within streams and lakes. In this chapter, I focus on the influence of nitrogen (N) and sulfur (S) compounds because (a) they are major constituents of atmospheric deposition and have recently become the focus of critical loads assessments in Europe and North America, and (b) they are essential nutrients for most biota and cycle naturally through terrestrial and aquatic ecosystems. To evaluate the relative impact of atmospheric inputs on freshwater acidity, it is necessary to define the term acidification clearly and to understand the mechanisms of change caused or mediated by natural and anthropogenic processes.

## ACIDITY AND ACIDIFICATION

The acidity of water is typically expressed by the pH or proton activity, an intensity factor that reflects the relative balance between $H^+$ sources and sinks at any given point in time and space. From this perspective, the effect of anthropogenic emissions or of different biological and geochemical processes on freshwater acidity will depend on the extent to which $H^+$ ions are produced or consumed (Van Breemen et al. 1983).

Acidity can also be expressed in terms of alkalinity or acid-neutralizing capacity (ANC), a capacity factor that is related to pH. In surface waters, ANC is considered a more suitable index of acid-base chemistry compared to pH because it is less subject to transient variations induced by changes in $CO_2$ partial pressure (Hemond 1990). Acidification is an increase in acidity measured by a decrease in pH or ANC. As discussed by Hemond (1990), several conceptual definitions of ANC are used with regards to natural waters, and this causes considerable confusion when comparing and evaluating data. The classical definition of ANC or alkalinity of natural waters, for example, only considers the carbonate system (Stumm and Morgan 1981):

$$\mathrm{ANC} = [\mathrm{HCO_3^-}] + 2[\mathrm{CO_3^{2-}}] + [\mathrm{OH^-}] - [\mathrm{H^+}]. \tag{4.1}$$

However, complications in ANC measurement and definition arise in surface waters that contain organic acids or ionic Al species. Depending on the pK of these constituents relative to the water pH, they protonate or release $H^+$ ions, contributing to or reducing the ANC (Hemond 1990; Reuss and Johnson 1986; Turner et al. 1990).

An alternative, and perhaps more useful definition of ANC within the context of this chapter, is:

$$\begin{aligned}\mathrm{ANC} = {} & 2[\mathrm{Ca^{2+}}] + 2[\mathrm{Mg^{2+}}] + [\mathrm{K^+}] + [\mathrm{Na^+}] + [\mathrm{NH_4^+}] \\ & -2[\mathrm{SO_4^{2-}}] - [\mathrm{NO_3^-}] - [\mathrm{Cl^-}] = C_B - C_A ,\end{aligned} \tag{4.2}$$

where $C_B$ is the sum base cations and $C_A$ the sum of strong acid anions.

Al or organics will be discussed only to the extent that thcy are relevant to N and S fluxes. On the basis of Equation 4.2, any process that increases $SO_4^{2-}$ or $NO_3^-$ concentrations in the water without an equivalent increase in base cations causes a decrease in ANC; any process that removes these anions without similar reduction in base cations generates alkalinity (Van Breemen et al. 1983; Reuss and Johnson 1986). Because of the typically high $CO_2$ partial pressures in the soil, alkalinity depressions through strong acid anion input have only a minor effect on soil solution pH, but the pH will decrease more after soil water enters surface waters and equilibrates with atmospheric $CO_2$ (Reuss and Johnson 1985). Below, the importance of atmospheric deposition and terrestrial and aquatic processes involving N and S compounds to freshwater acidification will be discussed both in terms of $H^+$ budgets and the retention or release of the strong acid anions $NO_3^-$ and $SO_4^{2-}$.

## THE ROLE OF HYDROLOGY IN FRESHWATER ACIDIFICATION

The relative role of atmospheric deposition and internal biological and geochemical processes on freshwater acidification is influenced not only by the rate and composition of atmospheric deposition, watershed characteristics (such as soil type and depth, geology, and vegetation), and the size of the watershed relative to the surface water area, but also by watershed hydrology (sources, flowpaths, residence times). Shorter, quicker surficial flowpaths within the watershed generally allow less interaction between the solution and the weatherable base-rich materials deeper in the soil or with the biological components of the system and result in outflow that more closely reflects the composition of the incoming water. Such flows, for example, may play an important role in the episodic freshwater acidification often observed during snowmelt and heavy rainstorms. Likewise, rapid flow rates in headwater streams and shorter lakewater residence times provide less time for in-stream and in-lake processes or interactions with the sediments.

Pathways through organic-rich and reducing environments, such as peatlands and riparian zones, can substantially alter the chemistry of the drainage water (e.g., Hemond 1980; Wieder 1985; Bayley et al. 1986; Bayley et al. 1987). In such systems, hydrology is important to N and S biogeochemistry because (a) changes in the redox potential associated with fluctuations in the extent and location of water-saturated zones affect retention/reduction and release/oxidation processes (e.g., Bayley et al. 1993) and (b) longer residence times also appear to increase anion retention (Urban et al. 1987).

The direct role of atmospheric inputs in freshwater acidification is expected to be higher in systems and at periods characterized by short water-residence times, whereas biogeochemical processes in the catchment and the water column become more important regulators of water chemistry as flowpaths lengthen and water residence times increase. Hydrologic flowpaths may also influence the relative role of N vs. S inputs in solution acidification because terrestrial $SO_4^{2-}$ retention depends on adsorption reactions in the deeper soil, whereas N dynamics and $NO_3^-$ leaching are mostly regulated by biological processes in the upper soil.

## BIOGEOCHEMICAL SULFUR PROCESSES

Atmospheric S deposited to forested watersheds as particulates, $SO_2$ gas, or dissolved in cloudwater and precipitation undergoes several reactions that ultimately result in the formation of the $SO_4^{2-}$ anion (Calvert et al. 1985). The influence of nonmarine S deposition on freshwater acidity is regulated by the degree to which this anthropogenic $SO_4^{2-}$ is retained in the watershed or removed from the water column.

Many of those processes and the extent to which they are influenced by acidic deposition have been the subject of several assessments in the U.S. (National Acid Precipitation Assessment Program [NAPAP]) and Europe (Scientific Committee on Problems of the Environment [SCOPE], United Nations Economic Commission for Europe [UN–ECE] Critical Loads Program). Although broad regional S deposition regimes have been established for Europe and the U.S.A., considerable uncertainty still exists regarding dry and wet deposition fluxes at the scale of individual watersheds (see discussion in Hicks et al. 1990).

### Catchment Processes

*Biological S Retention*

Biological processes play a limited role in long-term $SO_4^{2-}$ retention in watersheds (except perhaps in peatlands) because S requirements and annual uptake rates by the vegetation are generally modest compared to atmospheric S inputs, especially in high deposition areas (Johnson and Lindberg 1992). Some microbial immobilization of exogenous S may occur in the organic-rich upper soil; however, this represents only temporary storage and buffering of acidity because the release of similar amounts of $SO_4^{2-}$ during decomposition of organic matter counteracts alkalinity produced with previous strong anion uptake (Figure 4.1; see also discussion in Turner et al. 1990). Because microbial S requirements are closely tied to C and N dynamics, microbial immobilization potential is greater in forests with high organic C pools (see discussion in Turner et al. 1990). Even though biological processes may not be directly responsible for net retention of external $SO_4^{2-}$, they may nevertheless affect $SO_4^{2-}$ adsorption by causing temporal and spatial variations in soil-water $SO_4^{2-}$ concentrations.

Organic-rich peatlands can serve as an important sink of atmospheric $SO_4^{2-}$, primarily through biological processes. The fraction of the input S retained varies among bogs, with the time of the year, and with the total atmospheric S load, and is generally lower than for atmospheric N (Bayley et al. 1986; Urban et al. 1987). Plant uptake in the surface peat accounts for part of the removal of $SO_4^{2-}$, whereas microbial reduction in the anaerobic peat beneath the water table is thought to be the major mechanism of $SO_4^{2-}$ retention (Hemond 1980; Bayley et al. 1986; Wieder and Lang 1988). During this process, $S^{2-}$ is formed, which can either volatilize as $H_2S$ or react with organic matter to form carbon-bonded S. The alkalizing effect of this $SO_4^{2-}$ removal/reduction on the acidity of the drainage water is counteracted when S is

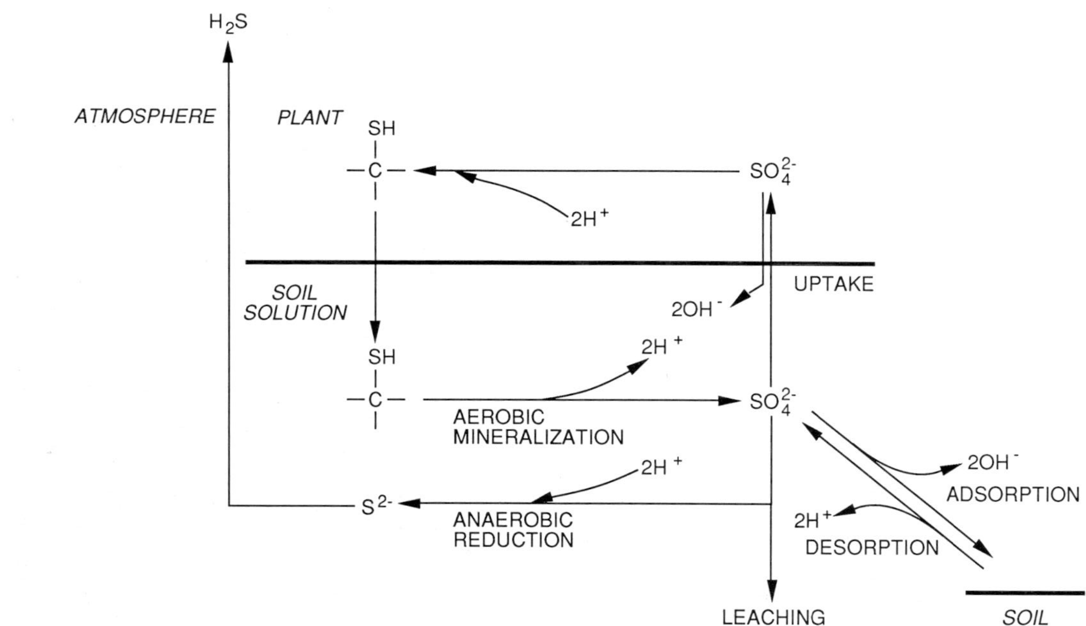


**Figure 4.1** Simplified sulfur cycle in the soil. Adapted from Reuss (1977).

reoxidized and $SO_4^{2-}$ is released into solution. This occurs, for example, with a drop of the water table level due to drought (e.g., Bayley et al. 1993).

*Adsorption and Precipitation*

Generally, the main process of soil $SO_4^{2-}$ retention in soil is adsorption, a term used generically to indicate any physicochemical $SO_4^{2-}$ retention mechanism, including adsorption to Al and Fe oxide surfaces in the soil (Harrison et al. 1989) and precipitation reactions between $SO_4^{2-}$ and Al (e.g., Khanna et al. 1987). No practical way has been found to distinguish clearly between these processes of $SO_4^{2-}$ retention in the field, and both processes appear to follow similar dynamics. During the adsorption process, $SO_4^{2-}$ exchanges with one or more $OH^-$ ions on the Al oxide surfaces, causing a rise in alkalinity and soil solution pH (Mitchell et al. 1992). Precipitation of $AlOHSO_4$ (e.g., jurbanite, basaluminite) has essentially the same effect (e.g., removal of strong acid anions from solution and $H^+$ consumption), although the pathway and kinetics are different (Khanna et al. 1987; Mitchell et al. 1992).

Sulfate adsorption capacity varies widely between soils, depending on pH, clay content, organic mater content, and quantity of Fe and Al oxides (often related to weathering status). Based on these soil characteristics, broad regional assessments of $SO_4^{2-}$ retention capacity of watersheds using a variety of models have been made for the eastern U.S. (e.g., Church et al. 1989; Baker et al. 1990). Waters draining watersheds that predominantly contain nonadsorbing soils are obviously more susceptible to direct acidification by atmospheric S input. In those instances, in-stream or in-lake S transformations become critical in curtailing deposition-induced changes in surface water acidity.

Sulfate adsorption increases with increasing $SO_4^{2-}$ concentration in solution. Consequently, watersheds with $SO_4^{2-}$-adsorbing soils (even those in dynamic equilibrium with current atmospheric S inputs) should be able to retain more S since the $SO_4^{2-}$ concentration in the solutions entering from the atmosphere increases (i.e., with increased S loads or $SO_4^{2-}$ concentration through evaporation). Also, a decrease in soil pH caused by natural or anthropogenic processes tends to enhance $SO_4^{2-}$ adsorption by increasing the positive charge of the Al and Fe oxide surfaces through protonation. This phenomenon is thought to be responsible for the depression in $SO_4^{2-}$ solution concentrations associated with $NO_3^-$ peaks (e.g., Nodvin et al. 1988; Johnson and Lindberg 1992). When soils become even more acidic, $SO_4^{2-}$ retention may decline again because of dissolution of the Al-oxide coatings, which reduces the number of adsorption sites (Mitchell et al. 1992), or because of a breakdown of previously formed $AlOHSO_4$ minerals (Khanna et al. 1987). Organic matter appears to have a negative effect on $SO_4^{2-}$ adsorption, presumably by blocking the adsorption sites with organic ligands; hence the generally lower adsorption capacity of the upper soil horizons (Johnson and Todd 1983).

*Desorption and Reversibility*

Because $SO_4^{2-}$ adsorption is concentration-dependent, reductions in atmospheric S inputs may actually cause a release or desorption of previously adsorbed $SO_4^{2-}$ until a new equilibrium is reached between adsorbed and dissolved $SO_4^{2-}$ concentrations. This degree of reversibility and the relative rate of desorption directly influences the longevity and magnitude of atmospheric $SO_4^{2-}$ effects on solution ANC and the rate of recovery of fresh waters after a decreases in atmospheric $SO_4^{2-}$ deposition (Figure 4.2; Reuss and Johnson 1986; Harrison et al. 1989; Turner et al. 1990). If $SO_4^{2-}$ does not desorb with lower $SO_4^{2-}$ input concentrations (i.e., irreversible adsorption), an immediate decrease to the new lower equilibrium $SO_4^{2-}$ concentrations should result, and alkalinity generation associated with prior adsorption is considered permanent. If adsorption is reversible, however, a net $SO_4^{2-}$ release from the soil will follow until a new equilibrium is attained between the adsorbed and solution $SO_4^{2-}$ levels. Such desorption of $SO_4^{2-}$ reduces solution ANC and effectively extends the period during which high atmospheric S inputs influence freshwater acidity and may cause a considerable delay in the potentially mitigating effect of S pollution abatement policies.

To date, little quantitative or even qualitative information is available on the desorption characteristics of forest soils or the factors controlling reversibility. A review of the available data on adsorption-desorption indicates (a) that most soils exhibit some degree of "permanent" or "irreversible" $SO_4^{2-}$ retention (Harrison et al.

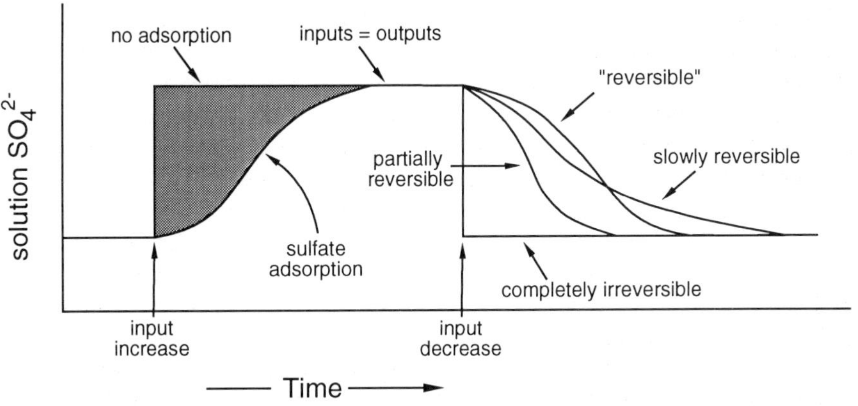


**Figure 4.2** Paths of $SO_4^{2-}$ concentration in the soil water with and without adsorption following S deposition increases and with different levels of desorption following S deposition decreases (Harrison et al. 1989).

1989), and (b) that soil properties leading to higher initial $SO_4^{2-}$ adsorption capacity decrease reversibility (Mitchell et al. 1992). Adsorption and desorption tend to increase as the equilibration time increases between the solution and the exchanger (i.e., as water residence time increases). This indicates, once again, the crucial role of watershed hydrology on solution chemistry, particularly since solution concentrations measured in the field during high flow seldom reflect the equilibrium conditions predicted by adsorption isotherms.

*Other Sources*

Bacterial oxidation of sulfides to $H_2SO_4$ constitutes a source of acidity within the catchment that is thought to have only a local influence on water chemistry (Baker et al. 1990). This process may result in an increase in $SO_4^{2-}$ export from soils that contain S-bearing minerals. Observations in the southeastern U.S. and Europe indicate that oxidation of pyrite contributes only a small fraction of the $SO_4^{2-}$ flux and acidity in upland streams compared to atmospheric deposition (Paces 1985; Cook et al. 1993).

Perhaps a more important (albeit periodic) source of $SO_4^{2-}$ originates from organic-rich substrates that have accumulated large amounts of reduced sulfur under water-logged (anoxic) conditions (e.g., wetlands, peats, alluvial soils) or are located in areas that are intermittently waterlogged and aerated (e.g., riparian zones). Water desaturation or a dropping water table during periods of drought may allow for the influx of sufficient $O_2$ to cause reoxidation of stored S to $SO_4^{2-}$. Subsequent storms may then flush out this $SO_4^{2-}$ into the streams, causing an increase in $SO_4^{2-}$ and potentially a decline in freshwater ANC (e.g., Bayley et al. 1986; Bayley et al. 1993).

**In-stream and In-lake Processes**

In watersheds that are at a steady state with the atmospheric $SO_4^{2-}$ inputs (e.g., based on the adsorption characteristics of the catchment soils), regional surface water surveys have indicated that biogeochemical processes within the lake water column may represent a major source of acid neutralization, removing as much as 50% of the total $SO_4^{2-}$ input to the watershed (Cook et al. 1986; Baker et al. 1990; Cook and Kelly 1992).

*Sulfur Processes in the Water Column*

Assimilatory $SO_4^{2-}$ reduction by the algae and bacteria in the water column and subsequent sedimentation of the organic matter is not considered an important process of sulfate removal in lakes where high $SO_4^{2-}$ inputs easily exceed biological S demands, with the latter largely governed by the trophic status (N and P availability) of the lake (Cook et al. 1986; Turner et al. 1990; Cook and Kelly 1992). In low $SO_4^{2-}$ lakes, however, where S availability to aquatic biota may be low, $SO_4^{2-}$ uptake by algae and the concurrent $H^+$ consumption appears to be a relatively more significant mechanism of ANC generation.

*Sediment Interactions*

Most of the $SO_4^{2-}$ removal that takes place in fresh waters, especially those with high $SO_4^{2-}$ concentrations, occurs in the anoxic zones within the sediments via dissimilatory reduction by bacteria (Figure 4.3). This process, which is more important in summer than in winter, uses $SO_4^{2-}$ as an oxidant during anaerobic decomposition of organic matter and converts $SO_4^{2-}$ into $S^{2-}$, producing one equivalent of ANC for every equivalent $SO_4^{2-}$ consumed. The effect of $SO_4^{2-}$ removal on freshwater acidity is permanent only to the extent that $H_2S$ gas is lost from the aquatic system into the atmosphere or that the reduced S is stored permanently in the sediments as carbon-bonded S compounds or FeS. Reoxidation of $H_2S$ in the water column before permanent gas exchange or sediment storage counters the ANC produced during earlier reduction reaction.

Bacterial $SO_4^{2-}$ reduction is concentration-dependent (following first-order kinetics) and increases with increasing $SO_4^{2-}$ concentration. The significance of this process in the overall S and ANC budget depends on two factors: (a) the rate of material flux across the water-sediment interface, and (b) the amount of time the water remains in contact with the sediments. Sulfate retention is not always complete. It is linked to lake hydrology and generally increases with increasing water residence time (Cook et al. 1986; Kelly et al. 1987; Cook and Kelly 1992). Sedimentation of reduced S via the

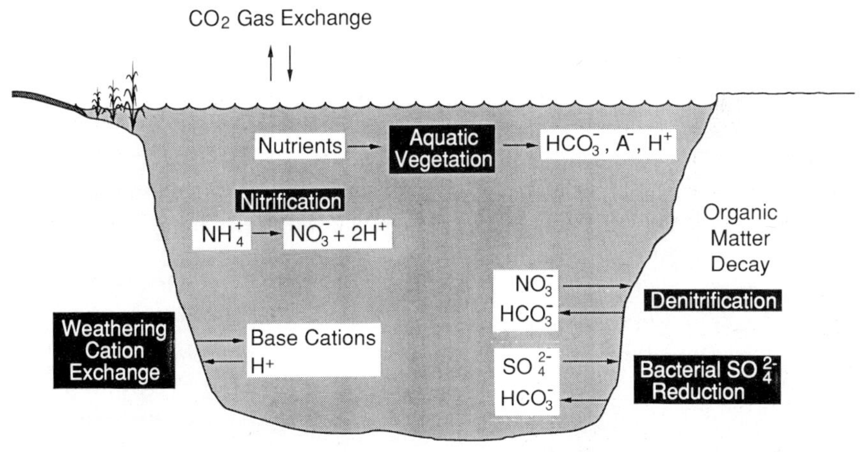


**Figure 4.3** Schematic representation of in-lake and in-stream processes affecting freshwater acidity. Courtesy of R. Turner, Oak Ridge National Laboratory, Oak Ridge, TN, U.S.A.

above pathway also requires a steady flux of organic carbon (a) to maintain anoxic conditions, (b) as a source of energy to the bacteria, and (c) as a substrate of organic S formation. Precipitation of reduced S as FeS in the sediment may be limited by Fe availability (see Turner et al. 1990).

## BIOGEOCHEMICAL NITROGEN PROCESSES

There are a wide variety of N compounds in the atmosphere in gaseous, particulate, or dissolved form, and considerable uncertainty still exists regarding their deposition rates (see Hicks et al. 1990). The forms that are thought to contribute the greatest amounts of N deposition to watersheds are $HNO_3$ vapor (which in turn forms $NO_3^-$ upon dissolution in water), particulate $NO_3^-$ and $NH_4^+$, and dissolved $NO_3^-$ and $NH_4^+$ in precipitation and cloudwater (see Johnson and Lindberg 1992). Other N oxides are also present in the atmosphere but their concentrations are not well quantified (Singh 1987). Among those, NO and $NO_2$ are known to oxidize into $HNO_3$, ultimately forming $NO_3^-$ upon dissolution in water (Singh 1987).

### Interactions with the Soil

As illustrated in Figure 4.4, belowground N dynamics are rather complex. They are essentially biologically controlled, and $NO_3^-$ fluxes within the watershed are governed by the relative size of N sources and sinks, and particularly by the temporal and spatial variability in N input and retention mechanisms.

#### *Nitrogen Retention*

Nitrogen retention in soils is primarily the result of plant uptake and microbial immobilization, while geochemical processes, such as $NH_4^+$ adsorption, generally play a less important role. In many terrestrial ecosystems where productivity is N-limited, atmospheric N is largely retained with little or no $NO_3^-$ leaching out of the soil, thus minimizing the potential ANC depression by anionic $NO_3^-$ inputs from the atmosphere. However, inorganic N essentially enters the soil in two ionic forms, namely $NH_4^+$ and $NO_3^-$, and their retention has opposite effects on solution acidity: $NH_4^+$ uptake generates acidity by releasing $H^+$, while the immobilization of $NO_3^-$ decreases the $H^+$ concentration in solution (Figure 4.4). In terrestrial ecosystems that are clearly N-deficient, the acidifying effect of atmospheric N inputs is thus directly linked to the ratio of $NO_3^-$ to $NH_4^+$ in the deposition: when $NO_3^-/NH_4^+ > 1$ (i.e., $NO_3^-$ input > $NH_4^+$ input), net alkalinization is expected; when $NO_3^-/NH_4^+ < 1$ (i.e., $NO_3^-$ input < $NH_4^+$ input), net acidification is expected.

#### *Nitrogen Transformations and Release*

Because many forest ecosystems in the world are N-deficient, it is often assumed that N retention in the catchment is nearly complete and that terrestrial N dynamics play

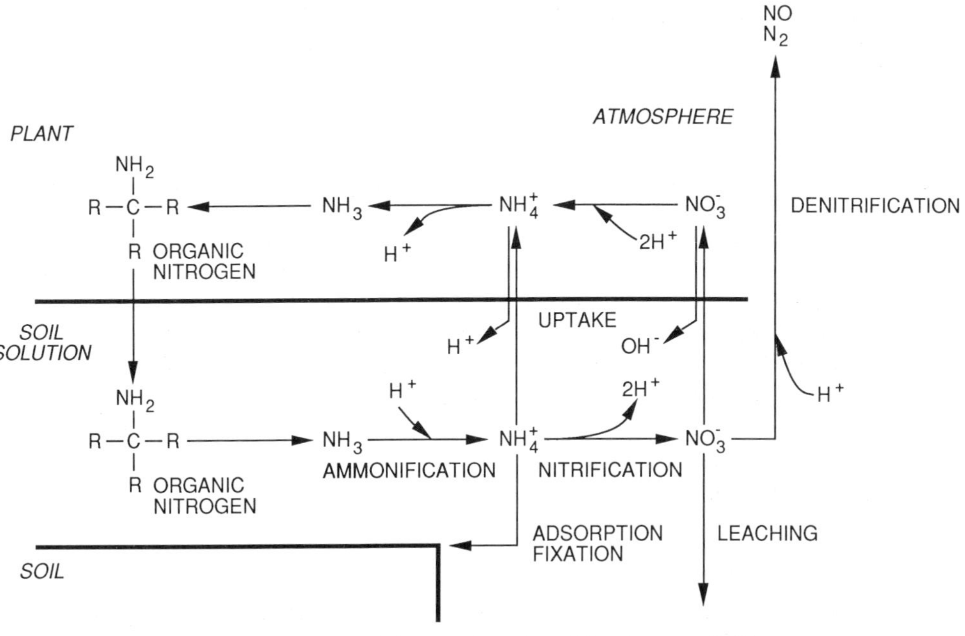


**Figure 4.4** Simplified nitrogen cycle in the soil. Adapted from Reuss (1977).

a minor role in freshwater acidification. However, there is increasing evidence that forested watersheds in eastern North America and Europe have reached or are approaching N saturation, i.e., show significant $NO_3^-$ leaching losses (Henriksen and Brakke 1988; Stoddard 1991; Johnson and Lindberg 1992; Murdoch and Stoddard 1992; Stoddard 1993). This condition is caused by an interplay between increasing atmospheric N inputs, vegetation composition and dynamics, disturbance or land-use history, and factors controlling the mineralization of accumulated N pools.

In systems that are N-enriched, there is less competition for available N between microorganisms and plants, and the total N input from atmospheric deposition and internal mineralization may exceed biological N requirements, allowing nitrification to occur. This process acidifies the solution by simultaneously releasing $NO_3^-$ and 1 or 2 $H^+$ ions per $NO_3^-$ anion, depending on whether the $NO_3^-$ originates from the decomposition of organic N or the oxidation of external $NH_4^+$, respectively (Figure 4.4). Once formed, mobile $NO_3^-$ moves quickly in the soil and its acidifying influence is neutralized by the rate at which base cations are released from the exchange complex and through weathering or by the extent to which $NO_3^-$ is removed from the soil by denitrification in anoxic zones of the watershed and through biological uptake. The latter is highly unlikely in N-rich systems.

In N-rich soils, the role of atmospheric deposition in water acidification is related to (a) the rate of $NO_3^-$ input, (b) the rate of nitrification of atmospheric $NH_4^+$, and indirectly to (c) the extent to which N inputs stimulate mineralization and nitrification of native organic N pools. Internal processes are also important to the extent that they (a) determine the relative rate of biological N uptake vs. nitrification of atmospherically deposited N, (b) regulate mineralization and nitrification rates of internally stored N, and (c) induce seasonality in the biological source-sink relationships, which contribute to episodic acidification pulses. A final factor that must also be considered is that the relative role of atmospheric inputs and internal N transformation processes in $NO_3^-$ export from terrestrial to the aquatic systems can change significantly with management practices or disturbances in the watershed (Swank 1986).

## Wetland Processes

In many northern landscapes, wetlands and organic-rich peats form the hydrologic and biogeochemical link between upland soils in the catchment and drainage waters. Because N is often a limiting nutrient in bog ecosystems, N inputs are largely retained in such systems, and $NO_3^-$ levels in the drainage water are generally low (e.g., Hemond 1983; Bayley et al. 1987). The removal of $NO_3^-$ from the solution occurs primarily through plant uptake and denitrification (Hemond 1983), generating one equivalent of ANC per molecule $NO_3^-$ removed. A small fraction of net $NO_3^-$ removal has also been ascribed to the reduction of $NO_3^-$ into $NH_4^+$. Reoxidation of stored reduced N during drought periods and flushing of mobile $NO_3^-$ in subsequent storms could potentially have adverse effects on freshwater ANC. However, in instances where periodic $NO_3^-$ oxidation pulses are observed in peatlands impacted by low atmospheric N inputs, they are generally small and short-lived, exerting little influence on drainage water ANC (Bayley et al. 1987; Bayley et al. 1993). It has been suggested that the N-retention capacity of peats may decline as high atmospheric pollution negatively impacts the health and growth of the peat vegetation and N inputs become supra-optimal (Lee et al. 1987).

## Processes in Lakes and Streams

### *Biotic Assimilation and Transformation of N in the Water Column*

Many of the processes that take place in the water column are similar to those discussed for the terrestrial component of the watershed, with effects on freshwater acidity strongly influenced by the form and rate of N input. Microbial oxidation of $NH_4^+$ into $NO_3^-$ or microbial assimilation of $NH_4^+$ are net sources of acidity; assimilatory reduction of $NO_3^-$ by the biota in the water column generates ANC. The net ANC balance is determined by the relative abundance of $NO_3^-$ vs. $NH_4^+$ in the water column and by biological N requirements (e.g., Schindler et al. 1985).

Any significant N input from the watershed (chronic or episodic) generally occurs predominantly as $NO_3^-$, while atmospheric N can be deposited as $NH_4^+$ and $NO_3^-$. Acid deposition thus indirectly influences the amount and form of N in surface water by influencing the degree of N saturation and the amount of $NO_3^-$ leaching from the drainage basin; it does so more directly by the ratio of $NH_4^+$ to $NO_3^-$ in the atmospheric inputs.

In streams and lakes where the primary production of algae and bacteria is limited by nutrients other than N (such as P) and N inputs exceed biological N requirements, N transformations in the water column are not important in the overall ANC balance. Under those circumstances, the acidifying effect of atmospheric deposition and terrestrial processes will be proportional to the $NO_3^-$ (strong acid anion) input via each pathway (e.g., Henriksen and Brakke 1988). In lakes where biological activity is limited by N availability, uptake of $NH_4^+$ and $NO_3^-$ should be more complete and the resulting gain or loss in ANC is determined by the relative ratio of $NH_4^+$ to $NO_3^-$ uptake. The relative significance of $NO_3^-$ removal through biological uptake and sedimentation is expected to decline as fresh waters become increasingly loaded with $NO_3^-$ (Kelly et al. 1990; Turner et al. 1990).

### *Sediment Interactions*

Biochemical N reactions in the sediments of lakes and streams are linked to the decomposition of organic matter in the anoxic sediment layers. During dissimilatory $NO_3^-$ reductions, $NO_3^-$ is converted into $N_2$ (and possibly also $NH_4^+$) by microorganisms while $H^+$ is simultaneously consumed. Denitrification is a permanent source of ANC to the extent that the $N_2$ gas is lost to the atmosphere. The process is driven by $NO_3^-$ levels in solution, organic matter availability, and redox potential. Because denitrification is a less efficient removal process than $NO_3^-$ assimilation by algae, $NO_3^-$ is more likely to accumulate and affect the acid-base chemistry in surface waters receiving large N inputs (Turner et al. 1990).

The rate of ANC production via denitrification is a function of water residence time: it is greatest in lakes during stagnation periods and lowest during fall and spring overturn. Spring overturn in some areas coincides with potential high water flows and peak $NO_3^-$ inputs from the watershed associated with snowmelt, increasing the probability of $NO_3^-$-induced ANC depression in surface waters during that period. Some denitrification may also occur in the water column, but it is generally limited to periods when the hypolimnion is anoxic (e.g., summer).

## EVALUATION OF THE RELATIVE ROLE OF SULFUR AND NITROGEN

Conceptually, the mechanisms underlying surface water acidification are reasonably well understood and have formed the basis for several watershed acidification models (e.g., Birkenes, MAGIC, ILWAS, RAINS, Trickle-down; Church et al. 1989; Thornton et al. 1990). However, our ability to draw general conclusions regarding the relative role of S vs. N dynamics and of watershed vs. in-lake processes in surface water acidification and buffering has remained limited due to the following reasons:

1. There is still considerable uncertainty concerning atmospheric N and S fluxes to individual watersheds.
2. Complete ANC budgets have been determined for relatively few freshwater systems.
3. Not all key processes have been adequately quantified, especially under a changing atmospheric deposition regime.
4. The ANC balance and the relative role of N and S processes are closely linked to hydrologic pathways and water residence times.
5. The net outcome of such evaluation strongly depends on the time scale (chronic vs. episodic; seasonal vs. annual vs. long-term).

Atmospheric impact research and modeling have typically focused on the effects of S deposition and the rate of $SO_4^{2-}$ adsorption in the soil on water acidification, while less effort has been directed toward the assessment and the quantification of terrestrial N dynamics or toward the role of in-stream and in-lake processes. The decision to exclude N from these models in the past was partly based on the assumption that most terrestrial ecosystems (including peatlands) are N-limited and strongly retain N, and that the effect of atmospheric N inputs on watershed acidification should therefore remain small (e.g., Henriksen and Brakke 1988; Baker et al. 1990). This assumption is probably correct in low N-pollution areas, where the uptake of small amounts of external N by terrestrial, wetland, or aquatic biota should only have a small effect on the overall $H^+$ budget and total anion flux. In high N-pollution areas or areas where plant growth and vitality are low or vegetation has been damaged, N inputs may exceed biological N demands, resulting in nitrification and excessive $NO_3^-$ leaching. Even complete retention of this incoming N could influence solution acidity, depending on the relative ratio of $NH_4^+$ to $NO_3^-$ in the input and the differential $H^+$ flux associated with the immobilization of both N forms.

Furthermore, lake and stream survey data from the eastern U.S. and Europe indeed indicate significant $NO_3^-$ losses from some watersheds, especially in areas where N deposition is high (Henriksen and Brakke 1988; Stoddard 1991; Murdoch and Stoddard 1992; Stoddard 1993) or where forests have been damaged by pollution (Paces 1985). In some cases, $NO_3^-$ has been found to account for more than 50% of the total anion flux exiting the soil (e.g., Johnson and Lindberg 1992), while $NO_3^-$ concentrations in some U.S. streams approach 20–30% of the $SO_4^{2-}$ concentrations (Kaufmann et al. 1991; Stoddard 1993). Since the 1970s, historical water quality data from several areas in the eastern U.S. and Europe indicate a considerable increase in the $NO_3^-:(NO_3^- + SO_4^{2-})$ ratio (e.g., Nilsson and Grennfelt 1988; Stoddard 1991; Stoddard 1993). Such a ratio is at best a conservative estimate of the relative role of $NO_3^-$ in water acidification because the calculations are often based on annual or multi-year averages, i.e., they express anion contributions to *chronic* acidification. They do not account for the *episodic* acidification pulses associated with snowmelt or storm events, during which $NO_3^-$ is known to play a generally more important role than $SO_4^{2-}$ (Malanchuk and Nilsson 1989; Baker et al. 1990; Wigington et al. 1990).

Although $SO_4^{2-}$ clearly remains a major contributor to chronic water acidification in many areas of the U.S. and Europe (Nilsson and Grennfelt 1988; Baker et al. 1990), the relative role of $NO_3^-$ is expected to increase as more forested watersheds become N-saturated and abatement policies for S pollution take effect. Even if S inputs remained unaltered, there is evidence suggesting that nitrification within N-saturated watersheds may also reduce $SO_4^{2-}$ solution concentrations and $SO_4^{2-}$ exports by increasing the soil $SO_4^{2-}$ adsorption capacity through acidification and protonation of the exchange sites (e.g., Nodvin et al. 1988; Johnson and Lindberg 1992).

In summary, current acidification models are still largely based on a broad understanding of $SO_4^{2-}$ adsorption capacity of soils, while S desorption, soil N dynamics, the factors controlling N saturation, and the relationship between hydrology and relative S and N retention are still poorly quantified. Little emphasis has been placed on the role of wetland processes or in-lake and in-stream processes in the overall ANC balance, the influence of water residence time on $SO_4^{2-}$ and $NO_3^-$ removal from the water column, or the difference in acidifying potential between nitric and sulfuric acid. Unless these knowledge gaps are bridged, it will remain difficult to assess the relative role of N and S compounds in the acidification of freshwater systems accurately.

## ACKNOWLEDGEMENTS

This work was supported by Oak Ridge National Laboratory, which is managed by Martin Marietta Energy Systems, Inc., under contract DE–AC05–84OR21400 with the U.S. Department of Energy. Publication No. 3917, Environmental Sciences Division, Oak Ridge National Laboratory.

## REFERENCES

Baker, L.A., J.M. Eilers, R.B. Cook, P.R. Kaufmann, and A.T. Herlihy. 1990. Interregional comparisons of surface water chemistry and biogeochemical processes. In: Acidic Deposition and Aquatic Ecosystems—Regional Cases Studies, ed. D.F. Charles, pp. 567–613. New York: Springer.

Bayley, S.E., R.S. Behr, and C.A. Kelly. 1986. Retention and release of S from freshwater wetland. *Water, Air, Soil Pollut.* **31**:101–114.

Bayley, S.E., D.W. Schindler, B.R. Parker, M.P. Stainton, and K.G. Beaton. Effects of forest fire and drought on acidity of base-poor boreal forest stream: Similarities between climatic warming and acid precipitation. *Biogeochem.* **17**:191–204..

Bayley, S.E., D.H. Vitt, R.W. Newbury, K.G. Beaty, R. Behr, and C. Miller. 1987. Experimental acidification of Sphagnum-dominated peatland: First-year results. *Can. J. Fish. Aquat. Sci.* **44(1)**:194–205.

Calvert, J.G., A. Lazrus, G.L. Kok, B.B. Heikes, J.G. Walega, L. Lind, and C.A. Cantrell. 1985. Chemical mechanisms of acid generation in the troposphere. *Nature* **317**:27–35.

Church, M.R. et al. 1989. Future effects of long-term sulfur deposition on surface water chemistry in the Northeast and Southern Blue Ridge Province. Report of the Direct Delayed Response Program. EPA/600/3–89/061. Washington, D.C.: U.S. Environmental Protection Agency.

Cook, R.B., J.W. Elwood, R.R. Turner, M.A. Bogle, P.J. Mulholland, and A.V. Palumbo. 1993. Acid-base chemistry of high-elevation streams in the Great Smoky Mountains. *Water, Air, Soil Pollut.*, in press.

Cook, R.B., and C.A. Kelly. 1992. Sulphur cycling and fluxes in temperate dimictic lakes. In: Sulfur Cycling on the Continents, ed. R.W. Howarth, J.W.B. Stewart, and M.V. Ivanov, pp. 145–188. New York: Wiley.

Cook, R.B., C.A. Kelly, D.W. Schindler, and M.A. Turner. 1986. Mechanisms of hydrogen ion neutralization in an experimentally acidified lake. *Limn. Ocean.* **31**:134–148.

Harrison, R.B., D.W. Johnson, and D.E. Todd. 1989. Sulfate adsorption and desorption reversibility in a variety of forest soils. *J. Environ. Qual.* **18**:419–426.

Hemond, H.F. 1980. Biogeochemistry of Thoreau's Bog, Concord, Massachusetts. *Ecol. Monogr.* **50**:507–526.

Hemond, H.F. 1983. The nitrogen budget of Thoreau's bog. *Ecology* **64**:99–109.

Hemond, H.F. 1990. Acid neutralizing capacity, alkalinity, and acid-base status of natural waters containing organic acid. *Environ. Sci. Technol.* **24**:1486–1489.

Henriksen, A., and D.F. Brakke. 1988. Increasing contributions of nitrogen to the acidity of surface waters in Norway. *Water, Air Soil Pollut.* **42**:183–201.

Hicks, B.B., R.R. Draxler, D.L. Albritton, F.C. Fehsenfeld, J.M. Hales, T.P. Meyers, R.L. Vong, M. Dodge, S.E. Schwartz, R.L. Tanner, C.I. Davidson, S.E. Lindberg, and M.L. Wesely. 1990. Atmospheric processes research and process model development. National Acid Precipitation Assessment Program, Acidic Deposition: State of the Science and Technology, Report 2. Washington D.C.: NAPAP.

Johnson, D.W., and S.E. Lindberg. 1992. Atmospheric Deposition and Forest Nutrient Cycling. New York: Springer.

Johnson, D.W., and D.E. Todd. 1983. Relationship among iron, aluminum, carbon and sulfate in a variety of forest soils. *Soil Sci. Soc. Am. J.* **47**:792–800.

Kaufmann, P.R., A.T. Herlihy, M.E. Mitch, J.J. Messer, and W.S. Overton. 1991. Stream chemistry in the eastern United States: 1. Synoptic survey design, acid-base status and regional patterns. *Water Resour. Res.* **27**:611–627.

Kelly, C.A., J.M.W. Rudd, R.H. Hesslein, D.W. Schindler, P.J. Dillon, C.T. Driscoll, S.A. Gherini, and R.E. Hecky. 1987. Prediction of biological acid neutralization in acid-sensitive lakes. *Biogeochem.* **3**:129–140.

Kelly, C.A., J.W.M. Rudd, and D.W. Schindler. 1990. Acidification by nitric acid: Future considerations. *Water, Air, Soil Pollut.* **50**:49–61.

Khanna, P.K., J. Prenzel, K.J. Meiwes, and E. Matzner. 1987. Dynamics of sulfate retention by acid forest soils in an acidic deposition environment. *Soil Science Soc. Am. J.* **51**:446–452.

Lee, J.A., M.C. Press, S. Woodin, and P. Ferguson. 1987. Responses to acidic deposition in ombrotrophic mires in the U.K. In: Effects of Atmospheric Pollutants on Forests, Wetlands, and Agricultural Ecosystems, ed. T.C. Hutchinson and K.M. Meema, pp. 549–560. Berlin: Springer.

Malanchuk, J.L., and J. Nilsson. 1989. The role of nitrogen in the acidification of soils and surface waters. *Nord* **1989**:10

Mitchell, M.J., M.B. David, and R.B. Harrison. 1992. Sulfur dynamics of forest ecosystems. In: Sulfur Cycling on the Continents, ed. R.W. Howarth, J.W.B. Stewart, and M.V. Ivanov, pp. 215–254. New York: Wiley.

Murdoch, P.S., and J.L. Stoddard. 1990. The role of nitrate in the acidification of streams in the Catskill Mountains of New York. *Water Resour. Res.* **28:**2707.

Nilsson, J., and P. Grennfelt. 1988. Critical loads for sulphur and nitrogen. *Nord* **1988**:15.

Nodvin, S.C., C.T. Driscoll, and G.E. Likens. 1988. Soil processes and sulfate loss at the Hubbard Brook Experimental Forest. *Biogeochem.* **5**:185–199.

Paces, T. 1985. Sources of acidification in central Europe estimated from elemental budgets in small basins. *Nature* **315**:31–36.

Reuss, J.O. 1977. Chemical and biological relationships relevant to the effect of acid rainfall on the soil-plant system. *Water, Air, Soil Pollut.* **7**:461–478.

Reuss, J.O., and D.W. Johnson. 1985. Effect of soil processes on the acidification of water by acid deposition. *J. Environ. Qual.* **14**:26–31.

Reuss, J.O., and D.W. Johnson. 1986. Acid Deposition and the Acidification of Soils and Waters. New York: Springer.

Schindler, D.W., M.A. Turner, and R.H. Hesslein. 1985. Acidification and alkalinization of lakes by experimental addition of nitrogen compounds. *Biogeochem.* **1**:117–133.

Singh, H. 1987. Reactive nitrogen in the troposphere. *Environ. Sci. Technol.* **21**:320–327.

Stoddard, J.L. 1991. Trends in the Catskill stream water quality: Evidence from historical data. *Water Resour. Res.* **27**:2855–2864.

Stoddard, J.L. 1993. Long-term changes in watershed retention of nitrogen: Its causes and aquatic consequences. In: Environmental Chemistry of Lakes and Reservoirs, Adv. Chem. Ser. No. 237, Washington, D.C.: Am. Chemical Soc., in press.

Stumm, W., and J.J. Morgan. 1981. Aquatic Chemistry, 2nd ed. New York: Wiley.

Swank, W.T. 1986. Biological control of solute losses from forest ecosystems. In: Solute Processes, ed. S.T. Trudgill, pp. 85–136. Chichester: Wiley.

Thornton, K.W., D. Marmorek, and P.F. Ryan. 1990. Methods for projecting future changes in surface water acid-base chemistry. National Acid Precipitation Assessment Program, Acidic Deposition: State of the Science and Technology, Report 14. Washington, D.C.: NAPAP.

Turner, R.S., R.B. Cook, H. Van Miegroet, D.W. Johnson, J.W. Elwood, O.P. Bricker, S.E. Lindberg, and G.M. Hornberger. 1990. Watershed and Lake Processes Affecting Surface Water Acid-Base Chemistry. National Acid Precipitation Assessment Program, Acidic Deposition: State of the Science and Technology, Report 10. Washington, D.C.: NAPAP.

Urban, N.R., S.J. Eisenreich, and E. Gorham. 1987. Proton cycling in bogs: Geographic variation in northeastern America. In: Effects of Atmospheric Pollutants on Forests, Wetlands, and Agricultural Ecosystems, ed. T.C. Hutchinson and K.M. Meema, pp. 577–598. Berlin: Springer.

Van Breemen, N., J. Mulder, and C.T. Driscoll. 1983. Acidification and alkalinization of soils. *Plant and Soil* **75**:283–308.

Wieder, R.K. 1985. Peat and water chemistry at Big Run Bog, a peatland in the Appalachian mountains of West Virginia, U.S.A. *Biogeochem.* **1**:277–302.

Wieder, R.K., and G.E. Lang. 1988. Cycling of inorganic and organic sulfur in peat from Big Run Bog, West Virginia. *Biogeochem.* **5**:221–242.

Wigington, P.J., T.D. Davies, M. Tranter, and K. Eshleman. 1990. Episodic acidification of surface waters due to acidic deposition. National Acid Precipitation Assessment Program, Acidic Deposition: State of the Science and Technology, Report 12. Washington, D.C.: NAPAP.

# 5

# Lake Acidification and the Role of Paleolimnology

R.W. BATTARBEE[1] and D.F. CHARLES[2]
[1]Environmental Change Research Centre, University College London,
26 Bedford Way, London WC1H 0AP, U.K.
[2]Patrick Center for Environmental Research, Academy of Natural Sciences, 1900 Benjamin Franklin Parkway, Philadelphia PA 19103, U.S.A.

## ABSTRACT

Paleolimnological studies of recent and long-term acidification have proven to be some of the most effective, and sometimes only, ways to evaluate the relative importance of causal mechanisms: natural processes, anthropogenic land-use change, and atmospheric deposition of strong acids. Analyses of a wide variety of characteristics from dated sediment cores are used to infer biological change, water chemistry changes (from direct sediment chemistry measurements and calculated from diatom and chrysophyte assemblages), and change in atmospheric deposition and catchment processes.

Paleolimnological studies of low-alkalinity lakes receiving acidic deposition in both Europe and North America demonstrate clearly, on a regional basis, that (a) long-term natural processes can cause slow acidification of lakes (decrease of one pH unit in several hundred to thousands of years), but not rapid acidification, (b) land-use change by itself (with the exception of afforestation) cannot account for the observed recent acidification of lakes, and (c) acidic deposition is the only logical explanation for recent rapid acidification. Paleolimnological approaches are being used effectively to demonstrate where reversibility has and has not occurred, to evaluate computer models of catchment acidification processes, and to develop critical load maps for acid deposition.

## INTRODUCTION

Slow environmental changes, such as surface water acidification, are often undetected in their early stages, becoming apparent only when key ecological thresholds are crossed. This lack of recognition is due to (a) the past (and present) inadequacy of appropriate long-term records, with occasional notable exceptions (e.g., reviewed in National Research Council 1986 and by Sullivan 1990), and (b) the difficulty in

*Acidification of Freshwater Ecosystems: Implications for the Future*
Edited by C.E.W. Steinberg and R.F. Wright 

distinguishing longer-term trends that occur over decadal time-scales from shorter, natural seasonal and interannual fluctuations.

Such trends, however, can be identified using paleoecological techniques. In studies of acid deposition and its effects, the lake sediment record has been of great importance, and it has been crucial in evaluating the causes of lake acidification.

## THE LAKE SEDIMENT RECORD

Organic sediments accumulate gradually in lakes, incorporating a physical, chemical, and biological record of material derived from the lake, its catchment, and from the atmosphere. The record is usually continuous, it is potentially available at all lacustrine sites, and at most sites it accumulates at a sufficient rate for it to be used, as desired, to focus on annual, decadal, or longer time-scales (Anderson and Battarbee 1993).

In the "acid rain" debate, this approach has been used successfully by both European and American paleolimnologists (Charles et al. 1989; Battarbee et al. 1990; Sullivan 1990; Whitehead et al. 1990; Cumming et al. 1992) to answer the questions:

1. Has recent lake acidification occurred?
2. When did acidification begin? At what rate did it occur?
3. What was the chemical and biological status of the lake prior to acidification?
4. Why has acidification occurred?
5. What is the extent of acidification within specific regions?
6. Have acidification trends reversed in regions where deposition of strong acids has decreased?

In this chapter, we present the main conclusions of the paleolimnological research on lake acidification carried out so far and highlight uncertainties and new challenges.

## METHODS

### Dating

Recent lake sediments are dated using the $^{210}$Pb (half-life 22.26 years) method. They are usually supplemented by the stratigraphic record of radioisotopes produced from the testing of nuclear weapons in the atmosphere between 1954 and 1963.

Older sediments can be dated using the $^{14}$C method, although in many European lakes, the erosion of older carbon from catchment soils over the last 1000 years or so causes dates to have older-than-expected ages. Consequently, except where lake sediments are annually laminated, sediments below the oldest $^{210}$Pb date and above the youngest reliable $^{14}$C date (usually about 500 years) can only be dated approximately through extrapolation.

**Diatoms and pH Reconstruction**

Diatoms are algae and occur in great abundance and diversity in aquatic environments. The mixture of species growing in a lake is strongly influenced by water chemistry, especially pH. Since they possess a resistant siliceous cell wall, they accumulate in lake sediments after death and can be used to reconstruct past lakewater pH, and, to a lesser extent, dissolved organic carbon (DOC) and Al (Birks, Line et al. 1990; Kingston et al. 1992). Chrysophytes—algae covered with siliceous scales—are also valuable for quantitatively inferring acidification trends (e.g., Dixit et al. 1989).

To quantify the pH changes that have occurred, a transfer function is used. The pH preferences of species or groups of species are measured or calculated from a modern ecological dataset and then used to infer the pH of the lake represented by the successive diatom samples in the sediment core (Charles and Smol 1993). The most recent and robust statistical approach for pH reconstruction is based on calculations of the pH optima of individual species using a weighted averaging technique (Birks, Juggins et al. 1990). Reconstructions for DOC, Al, and acid-neutralizing capacity (ANC) are treated in a similar way. Weighted averaging consistently performs better than any other pH reconstruction procedure, in terms of the lowest standard error of prediction (typically plus or minus 0.3 pH units).

**Trends in the Biology of Acidified Lakes**

Lake sediments contain a range of microfossils in addition to diatoms. An analysis of the complete fossil record, including chrysophytes, Cladocera, and chironomids in association with diatom-based reconstructions of pH, DOC and Al, can provide both direct and indirect evidence for the biological response of lakes to acidification (cf. Charles et al. 1990; Battarbee et al. 1990). At some sites, fish history may be inferred from *Chaoborus* (distinctly different taxa occur in lakes with and without fish; e.g., Uutala 1990) and compared with reconstructed monomeric Al (Kingston et al. 1992).

To date most paleolimnological studies of lake acidification have used biota as a means to reconstruct water chemistry change. The opportunities to use this approach to study overall community response to acidification has been underutilized but offers great potential.

**Causal Hypothesis Testing**

Dating and diatom-based pH reconstruction are the key techniques used in answering the questions whether and when acidification has taken place. To use the sediment record to test hypotheses about the causes of lake acidification, additional paleoecological techniques and good research design are needed. Additional techniques include the use of pollen, carbonaceous particle, trace metal, and other analyses to evaluate changes in catchment vegetation and atmospheric contamination at acidified sites (e.g., Battarbee et al. 1985; Charles et al. 1989). Appropriate research design involves the careful selection of sites within and between regions (spatial comparisons and controls) and, unique to paleolimnology, the use of the historical sediment record to make comparisons within and between sites. The combination of space and time comparisons afforded by this approach is especially powerful (Figure 5.1) and is illustrated by several studies (e.g., Kreiser et al. 1990; Renberg and Battarbee 1990; Birks, Berge et al. 1990; Charles et al. 1990; Sullivan et al. 1990; Cumming et al. 1992).

## CAUSES OF LAKE ACIDIFICATION

The main alternative explanations for surface water acidification are natural processes, land-use change, and acid deposition (Figure 5.1).

### Natural Acidification

Most lakes that have current acidification problems occur in the glaciated regions of Europe and North America and as such are young lakes, i.e., somewhat over 10,000 years old. Diatom analysis of the full postglacial sediment record has been carried out at a considerable number of low alkalinity sites throughout these regions; however, quantitative pH reconstructions and assessments of the record in the context of the "acid rain debate" are available for only a few (e.g., Whitehead et al. 1989; Jones et al. 1989; Renberg 1990; Ford 1990). The rate of acidification over this time period varies considerably from site to site. At some sites, the earliest phase was characterized by circumneutral to alkaline water. The length of this phase varies, but often it is confined to the very early Holocene. It is followed by a period of relatively rapid acidification as base cations are leached from the newly created unweathered soils and as acid organic soil horizons develop and thicken (e.g., Whitehead et al. 1989; Ford 1990; Steinberg et al. 1991). At some sites, gradual acidification continues throughout the Holocene period; at others there is no further change until the 19th to 20th century. At the Round Loch of Glenhead, which has very little mineral soil in its catchment, the early alkaline phase is absent and the lake is characterized by very low (pH 5.5) values throughout. Lille Öresjön (Renberg 1990) shows a complex history, with the gradual pH decline being interrupted about 2300 B.P. by a sudden increase in pH as agriculture, indicated by pollen analysis, expanded in the catchment area.

Despite these variations in long-term history, it is apparent that long-term acidification does occur, but not at all sites. Where it does occur, it is an extremely slow process when compared with the rate of change in recent decades. It tends to be associated with increasing organic acidity and dystrophication, and, for most sites, pH is never below 5.0 until the post-1800 period.

These data, and others in the literature, show that natural acidification processes are not the cause of present-day problems; however, they do indicate that the sensitivity of lakes to the impact of acid deposition can be increased over the time of the Holocene through a long-term loss in alkalinity.

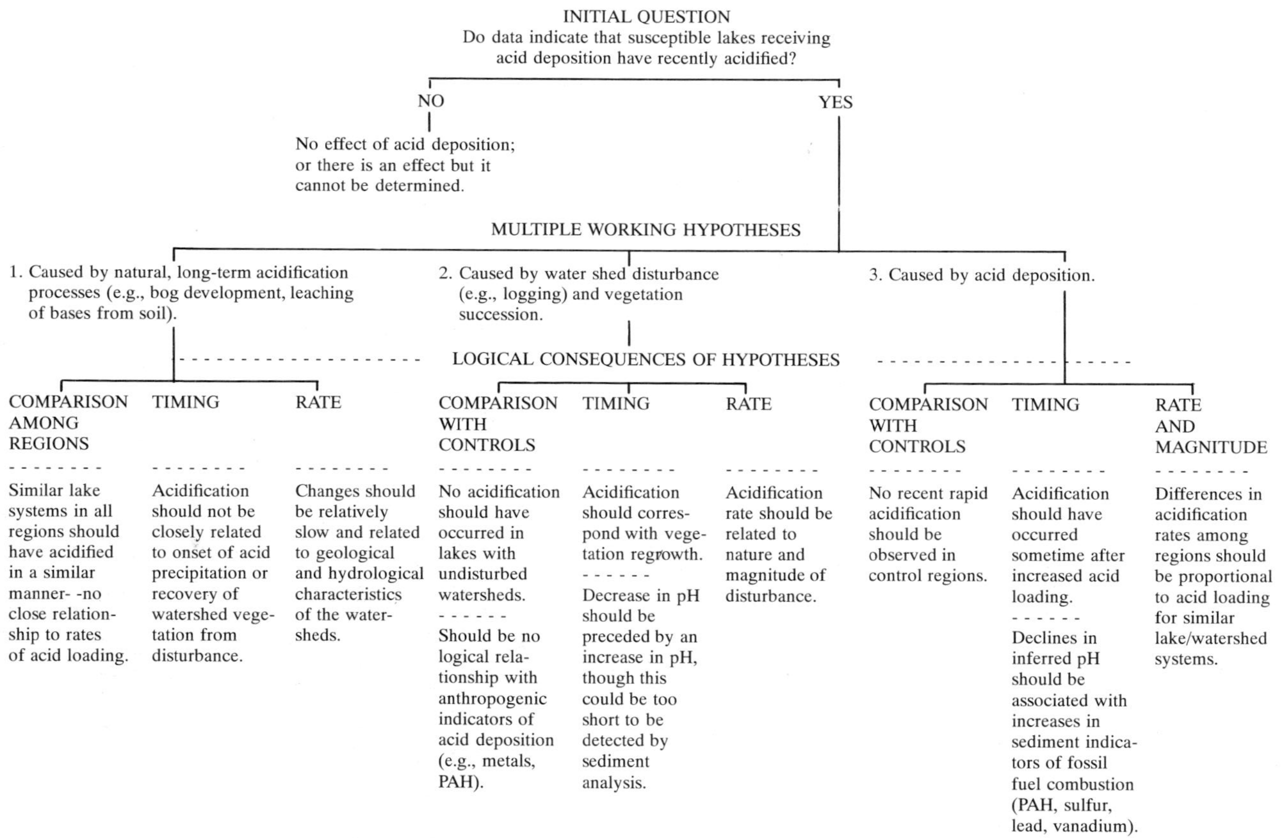

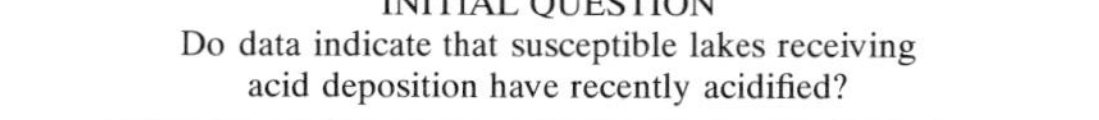


**Figure 5.1** Questions, multiple (alternative) working hypotheses, and logical consequences of the hypotheses evaluated in paleoecological studies of lake acidification (from Charles and Whitehead 1986).

**Land Use**

Although diatom analysis of acidified lakes invariably shows that recent acidification has only occurred in the last century or so, during the time of rapidly expanding fossil fuel combustion, this association in time is insufficient, by itself, to ascribe the acidification to acid deposition. Indeed, Rosenqvist (1978) has argued that a decline in pastoral agriculture in Norway occurred over the same period and that resulting vegetation changes were the primary cause of recent acidification. This hypothesis cannot be regarded as an alternative explanation since many sites that have not undergone such land-use change have been severely acidified (e.g., Battarbee et al. 1985; Charles et al. 1990; Birks, Berge et al. 1990). Nonetheless, it could be regarded as an additional factor at some sites acting separately or in combination with acid deposition where:

- historical agricultural and forestry practices (e.g., grazing, tillage, liming, deforestation) have raised pH and the abandonment of such practices has caused a return to earlier naturally lower pH levels (e.g., Renberg 1990; Davis et al. 1988; Renberg et al. 1993);
- forests have replaced grass and heathlands in areas of high acid deposition and trap higher quantities of dry and occult deposition (e.g., Harriman and Morrison 1982; Stoner et al. 1984; Flower 1987; Kreiser et al. 1990).

Yet to test whether shifts in land use per se can cause lake acidification, it is necessary to find situations where the appropriate soil and vegetation changes have occurred independently of acid deposition (Battarbee 1990). This can be done either by studying the effect of recent land-use changes on lakes in regions with low acid deposition or by using analogs in the postglacial record (Jones et al. 1986; Anderson and Korsman 1990; Renberg et al. 1990).

In both Norway and Sweden, one main land-use change that has taken place in the region of acidified lakes has been the expansion of spruce forest planted on or colonizing abandoned grazing land. As a past analog, Renberg et al. (1990) took the natural spruce immigration into northern and central Sweden, where it occurred about 3000 years ago. They selected lakes that were currently either acidified or sensitive to acidification, identified the spruce immigration in the sediment core through pollen analysis, and used diatom analysis to assess the possible effects of the vegetation change on water quality. None of the eight sites studied showed any indication of acidification.

So far there has been no clear demonstration in the literature of the acidifying effect of land-use changes on surface waters, lasting more than a few years, independent of other factors. This assessment is supported by analysis of detailed sediment core stratigraphies of well over 100 lakes in North America and Europe (e.g., Charles et al. 1990). Although more studies from a wider range of sites and vegetation types would be desirable, it is apparent that the only major importance of land-use change is the role of forests in scavenging acidity from the atmosphere. New plantations and existing spruce forests in sensitive but clean areas of Scotland and Norway have no apparent impact on lake water acidity.

### Acid Deposition

In contrast to the absence of evidence for the impact of land-use change on surface water acidification, the evidence implicating the role of acid deposition is abundant. Recent acidification occurs only in areas of high sulfur deposition, within which temporal responses are related to the capacity of the lake and its catchment to neutralize acid deposition (Charles et al. 1990; Figure 5.2). Especially convincing is the temporal correlation between the pH changes in acidified areas of North America and Europe and the sediment record of pollutants from fossil-fuel combustion, in particular the fly-ash particle, trace metal, and polycyclic aromatic hydrocarbon (PAH) records (Charles et al. 1990; Battarbee et al. 1989). For the hundreds of lakes studied, recent acidification occurs after the onset of acidic deposition, as indicated by fossil-fuel combustion products; the magnitude of acidification is generally greatest for systems receiving the greater acid loading and having soil, geological, and hydrological characteristics with the least ability to neutralize acids. Also convincing is the strong relationship between the magnitude of acidification that has occurred in a region and both the ANC of the lakes and the magnitude of atmospheric loading the region receives (Battarbee 1990; Sullivan et al. 1990; Whitehead et al. 1990; Cumming et al. 1992). This relationship would not exist if land-use change were the most important causal factor.

Figure 5.3 illustrates diatom changes and the carbonaceous particle (coal plus oil) profile for the Round Loch of Glenhead in Galloway, southwest Scotland. Substantial recent contamination during the twentieth century, and especially since about 1940 (the period of rapid lake acidification), is apparent.

## CRITICAL LOADS

The paleoecological evidence of the importance of acid deposition as the cause of surface water acidification is so strong that the data can be used to derive an empirical dose-response model for the acidification status of any water body (Figure 5.4). For U.K. sites, this model indicates that acidification is probable when the ratio of calcium in lake water (in $\mu eq\ L^{-1}$) to acidic deposition (in $Keq\ ha^{-1}\ yr^{-1}$) is less than 94:1. This ratio can then be used to define the critical load of sulfur for a site and to produce critical load maps (Battarbee et al. 1993).

## REVERSIBILITY

The ability to understand and predict reversibility of, and recovery from, acidification is very important for managing sensitive aquatic resources and setting policy for

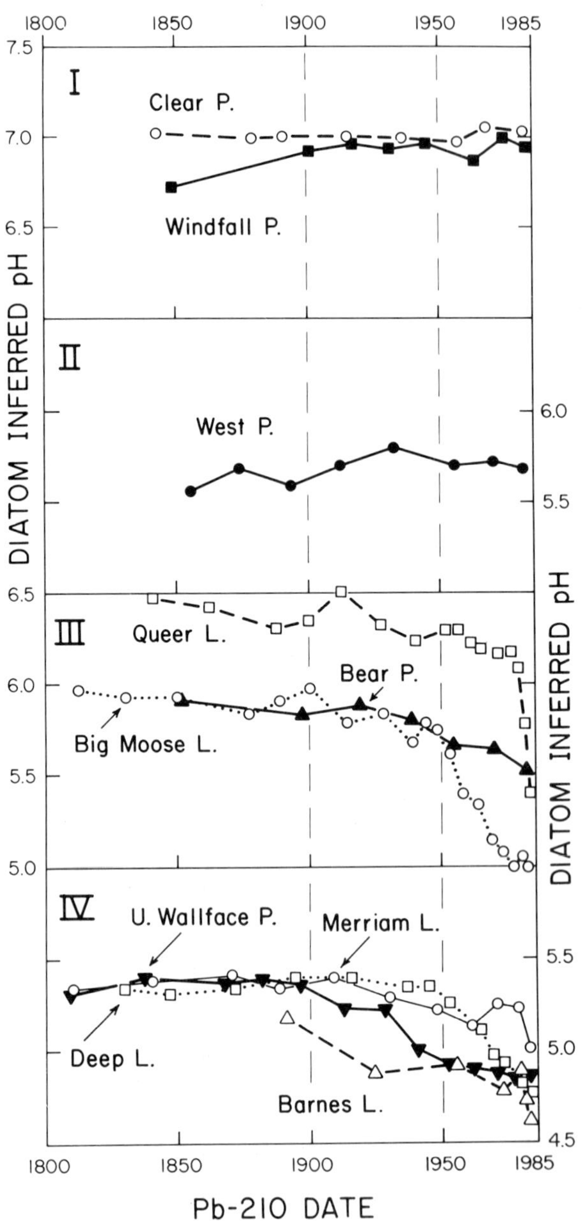


**Figure 5.2** Diatom inferred pH vs. $^{210}$Pb date for 12 lakes in the Adirondack Mountains, New York. Lakes are grouped according to pattern of response to increased atmospheric deposition. Category I: no or small acidification trend, lakes with pH > 6.5, lake and catchment well-buffered. Category II: no or small acidification trend, lakes with pH < 5.0–5.5, organic acid and Al buffering dominant. Category III: definite acidification trend, switch from $HCO_3^-$ buffering (pH > 5.5) to organic and Al buffering. Category IV: definite acidification trend, lakes with current and pre-1850 pH < 5.5, organic and Al buffering (reprinted by permission of Kluwer Academic Publishers; from Charles et al. 1990).

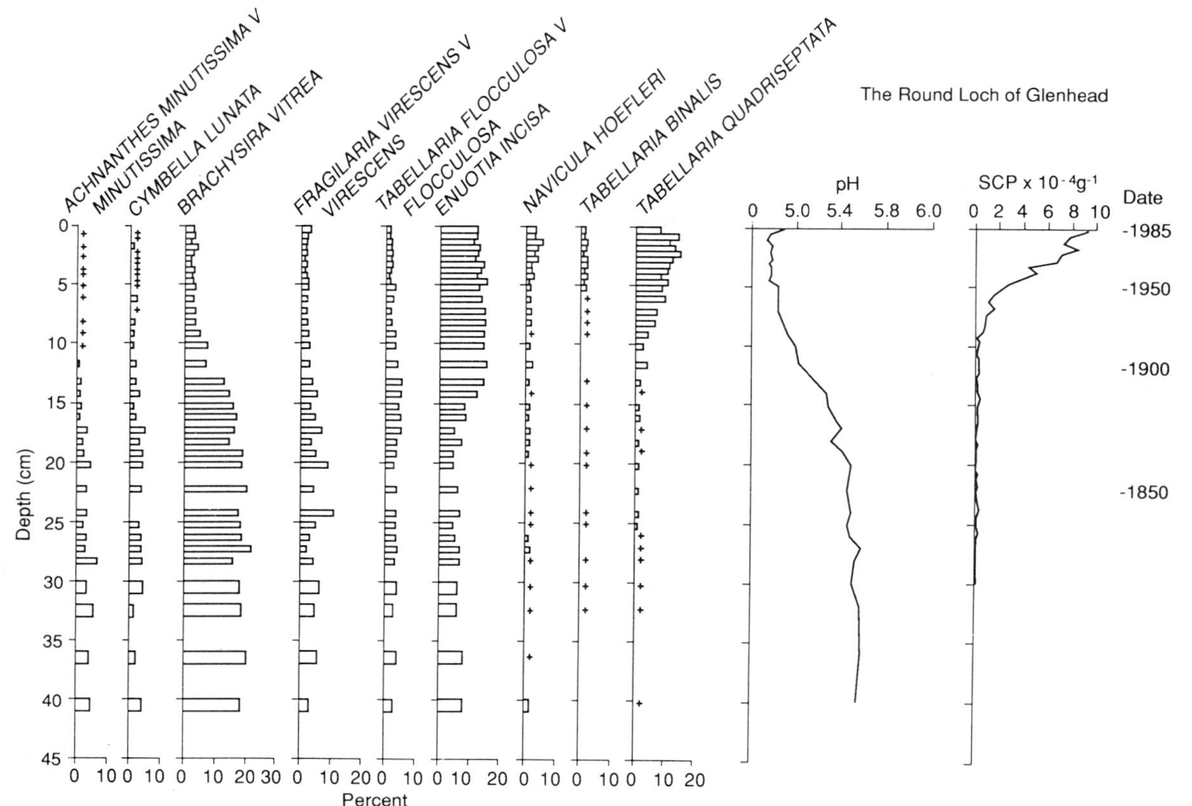


**Figure 5.3** Summary diatom diagram, pH reconstruction, concentration of spheroidal carbonaceous particles (SCP), and $^{210}Pb$ dates vs. core depth for the Round Loch of Glenhead, Galloway, southwest Scotland (from Jones et al. 1989; Battarbee et al. 1989; Birks et al. 1990b).

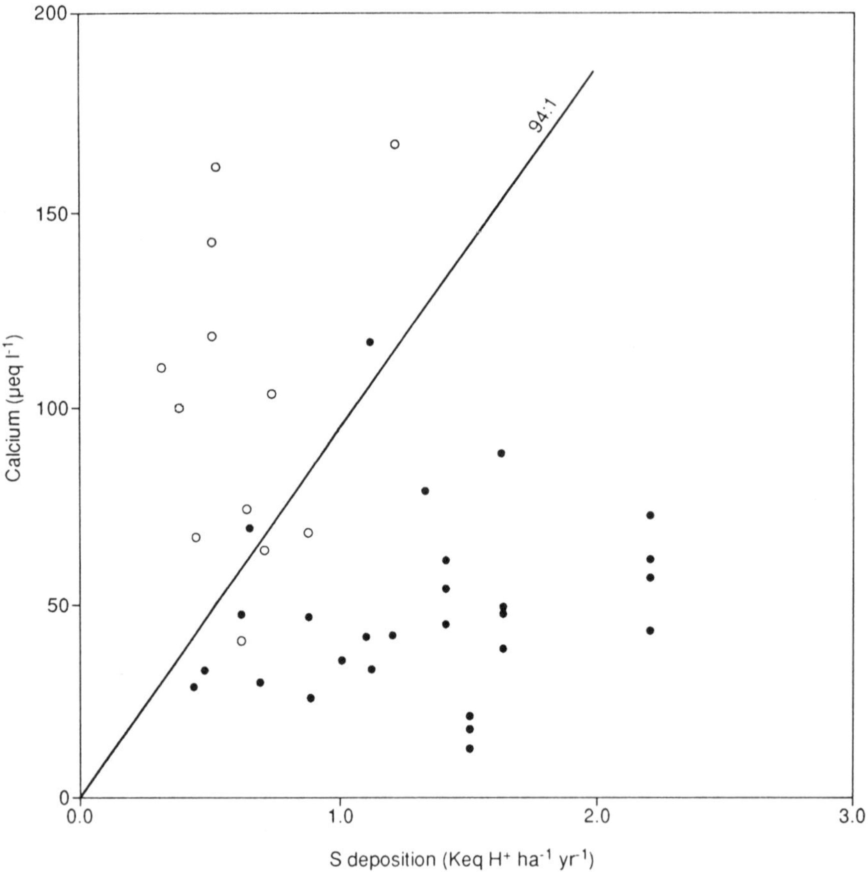


**Figure 5.4** Lake water calcium vs. acidic deposition for sites with diatom-inferred pH profiles in the United Kingdom (• acidified lakes; o nonacidified lakes).

controlling emissions (see Dise et al., this volume). Although process-driven dynamic models are the primary tool for predicting change in surface water chemistry in response to emission reductions, there are significant uncertainties in the models, and decadal-length records of water chemistry that can be used to validate models are rare. Paleolimnological studies are of critical importance because they can provide information on both chemical conditions prior to the onset of acidification and the pattern of response of lakes that have a variety of chemical and catchment characteristics. These data can be used to assess the reversibility hypotheses and to evaluate the predictive ability of models (e.g., Sullivan et al. 1992).

Already, paleolimnological studies have provided significant findings and insights concerning reversibility. The most striking results come from Ontario (Dixit et al. 1989; 1992a,b), for example, Swan Lake (Figure 5.5.). These data show significant

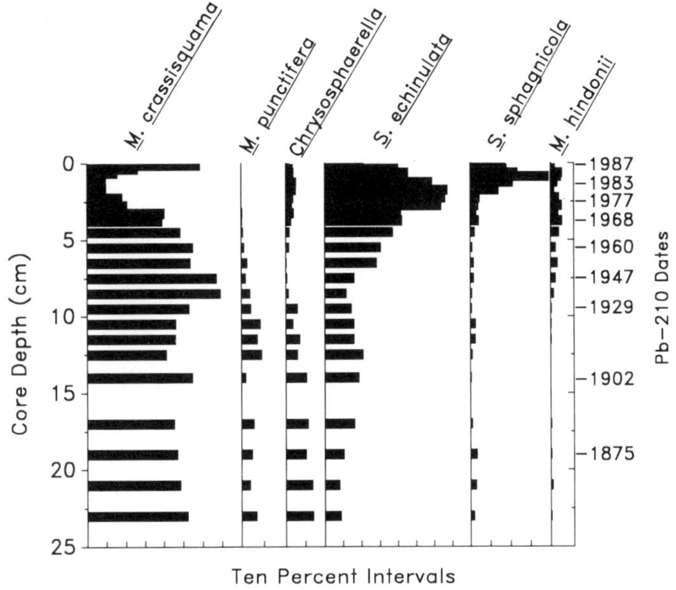


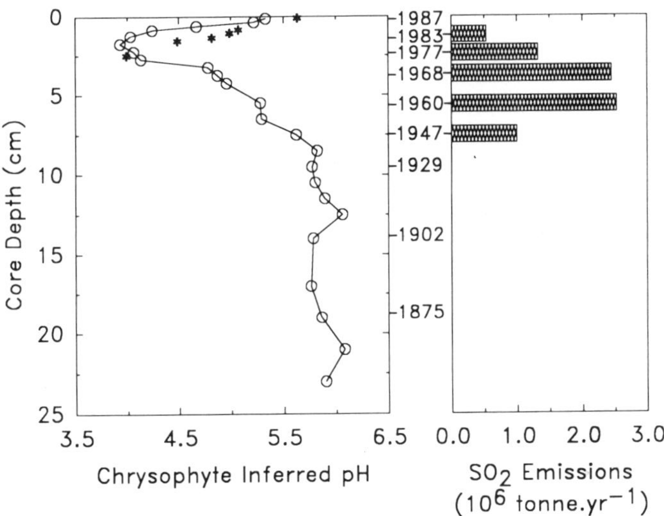


**Figure 5.5** Percent abundance of scales of common chrysophyte taxa, $^{210}$Pb dates, and inferred pH for a sediment core from Swan Lake compared with recorded sulfur dioxide emissions from the Nickel smelter installations at Sudbury, Ontario. * = measured pH (reprinted by permission of the National Research Council of Canada; from Dixit et al. 1989).

increase in lakewater pH following reduction of emissions from Sudburyt smelters. They are not, however, typical of the response of lakes in other regions because of the dominant influence of the Sudbury smelter as a point source. However, evidence from Galloway, southwest Scotland, shows more typical sites beginning to recover following a reduction in S deposition of about 40% (Allott et al. 1992). This confirms the view that initial response to S reduction can be quite rapid. The response of low ANC lakes in the Adirondack Mountains, New York, based on paleolimnological studies is mixed. Following S deposition decreases of about 25% since 1970, some lakes have continued to acidify, some have increased in inferred pH, and some show no trend (Sullivan 1990). Diatoms are also being used to monitor reversibility of acidification in the Black Forest, Germany (C. Steinberg, pers. comm.).

## FUGURE ISSUES AND RESEARCH NEEDS

Reversal of acidification may be complicated by changes in catchment and atmospheric factors, such as increases in forest cover and age, increases in nitrate leaching, and changes in climate. Any one of these factors may be sufficient to offset expected improvement based on sulfur reduction alone, and all need to be taken into account in calculation of critical load exceedance. At present, such factors cannot be satisfactorily built into predictive models. It is essential, therefore, that key sites in regions sensitive to acidification are carefully monitored to create long-term datasets that can be used not only to monitor changes but also to test paleoecologically based reconstruction of the past and model-based predictions of the future.

Usefulness of paleolimnological approaches for assessing lake acidification issues can be improved in several areas; the most important for increasing accuracy of water chemistry inference is to expand ecological knowledge of the organisms that leave remains in the sediment. The most effective steps would be to (a) combine the many separate existing data sets; (b) collect additional data for selected regions and habitat types, additional ecological characteristics, and more groups of biota; and (c) perform laboratory culture studies to determine dose-response relationships for individual taxa.

Paleolimnological approaches offer great opportunities for investigating several ongoing issues, including the role and importance of N and organic acids as factors influencing acidification, reversibility and recovery, response of community composition and structure to acidity levels, the rates of loss of base cations from soils, effects of land-use change on water chemistry, and evaluation of acidification models. Paleolimnological investigations will be most effective if they are integrated with studies involving models and field experiments designed to test specific hypotheses.

## CONCLUSIONS

Time-space comparisons of acidification trends derived from recent lake sediment records clearly show the importance of acid deposition as the main cause of surface water acidification, and analyses of earlier sediments help to separate the roles of natural and anthropogenic factors in the acidification process. Paleolimnological evidence demonstrates that land-use change (other than afforestation) can cause increases in lake pH, but not regional-scale decreases.

Current challenges for paleolimnologists, as for others involved in acidification research, include assessing site responses to sulfur reduction and evaluating the influences of land-use change, increased nitrogen deposition, climate change, and the relationships between them.

## ACKNOWLEDGEMENTS

R.W. Battarbee recognizes ECRC, SWAP colleagues, and funding bodies (CEGB, NERC, Royal Society and DoE) for their contributions to the development of this paper. We thank John Birks, Ingemar Renberg, John Smol, and Christian Steinberg for reviewing the manuscript and making many helpful comments.

## REFERENCES

Allott, T.E.H., R. Harriman, and R.W. Battarbee. 1992. Reversibility of lake acidification at the Round Loch of Glenhead, Galloway, Scotland. *Environ. Pollut.* **77**:219–225.

Anderson, N.J., and R.W. Battarbee. 1993. Aquatic community persistence and variability: A palaeolimnological perspective, in press.

Anderson, N.J., and T. Korsman. 1990. Land use change and lake acidification: Iron-age desettlement in northern Sweden as a pre-industrial analogue. *Phil. Trans. Roy. Soc. Lond. B* **327**:373–376.

Battarbee, R.W. 1990. The causes of lake acidification with special reference to the role of acid deposition. *Phil. Trans. Roy. Soc. Lond. B* **327**:339–347.

Battarbee, R.W., T.E.H. Allott, A.M. Kreiser, and S. Juggins. 1993. Setting critical loads for UK surface waters: The diatom model. In: Critical Loads: Concepts and Applications, ed. M. Hornung and R.A. Skeffington, pp. 99–103. London: HMSO.

Battarbee, R.W., R.J. Flower, A.C. Stevenson, and B. Rippey. 1985. Lake acidification in Galloway: A palaeoecological test of competing hypotheses. *Nature* **314**:350–352.

Battarbee, R.W., J. Mason, I. Renberg, and J.F. Talling. 1990. Palaeolimnology and Lake Acidification. London: The Royal Society, 219 pp.

Battarbee, R.W., A.C. Stevenson, B. Rippey, C. Fletcher, J. Natkanski, M. Wik, and R.J. Flower. 1989. Causes of lake acidification in Galloway, south-west Scotland: A paleoecological evaluation of the relative rates of atmospheric contamination and catchment change for two acidified lakes with non-afforested catchments. *J. Ecol.* **77**:651–672.

Birks, H.J.B., F. Berge, J.F. Boyle, and B.F. Cumming. 1990. A palaeolimnological test of the land use hypothesis for recent lake acidification in south-west Norway using hill-top lakes. *J. Paleolimnol.* **4**:69–65.

Birks, H.J.B., S. Juggins, and J.M. Line. 1990. Lake surface-water chemistry reconstructions from palaeolimnological data. In: The Surface Waters Acidification Programme, ed. B.J. Mason, pp. 301–313. Cambridge: Cambridge Univ. Press.

Birks, H.J.B., J.M. Line, S. Juggins, A.C. Stevenson, and C.J.F. ter Braak. 1990. Diatoms and pH reconstruction. *Phil. Trans. Roy. Soc. Lond. B* **327**:263–278.

Charles, D.F., R.W. Battarbee, I. Renberg, H. van Dam, and J.P. Smol. 1989. Paleoecological analysis of lake acidification trends in North America and Europe using diatoms and chrysophytes. In: Acid Precipitation, ed. S.A. Norton, S.E. Lindberg, and A.L. Page, pp. 207–276. New York: Springer.

Charles, D.F., M.W. Binford, E.T. Furlong, R.A. Hites, M.J. Mitchell, S.A. Norton, F. Oldfield, M.J. Paterson, J.P. Smol, A.J. Uutala, J.R. White, D.R. Whitehead, and R.J. Wise. 1990. Paleoecological investigation of recent lake acidification in the Adirondack Mountains, N.Y. *J. Paleolimnol.* **3**:195–241.

Charles, D.F., and J.P. Smol. 1993. Long-term chemical changes in lakes: Quantitative inferences from biotic remains in the sediment record. In: Environmental Chemistry of Lakes and Reservoirs, ed. L. Baker. Washington, D.C.: Am. Chemical Soc., in press.

Charles, D.F., and D.R. Whitehead. 1986. The PIRLA project: Paleoecological investigations of recent lake acidification. *Hydrobiologia* **143**:13–20.

Cumming, B.F., J.P. Smol, J.C. Kingston, D.F. Charles, H.J.B. Birks, K.E. Camburn, S.S. Dixit, A.J. Uutala, and A.R. Selle. 1992. How much acidification has occurred in Adirondack region lakes (New York, USA) since preindustrial times? *Can. J. Fish. Aquat. Sci.* **49**:128–141.

Davis, R.B., D.S. Anderson, D.F. Charles, and J.N. Galloway. 1988. Two-hundred-year pH history of Woods, Sagamore, and Panther Lakes in the Adirondack Mountains, New York State. In: Aquatic Toxicology and Hazard Assessment, ed. W.J. Adams, G.A. Chapman, and W.G. Landis, vol. 10, ASTM STP 971. Philadelphia: Am. Soc. for Testing and Materials.

Dixit, S.S., A.S. Dixit, and J.P. Smol. 1989. Lake acidification recovery can be monitored using chrysophycean microfossils. *Can. J. Fish. Aquat. Sci.* 46:1309–1312.

Dixit, A.S., S.S. Dixit, and J.P. Smol. 1992a. Changes in lakewater pH and metal concentrations in Killarney Provincial Park lakes, near Sudbury, Ontario. *Can. J. Fish. Aquat. Sci.* **49**, in press.

Dixit, S.S., J.P. Smol, J.C. Kingston, and D.F. Charles. 1992b. Diatoms: Powerful indicators of environmental change. *Environ. Sci. Tech.* **26**:23–32.

Flower, R.J., R.W. Battarbee, and P.G. Appleby. 1987. The recent palaeolimnology of acid lakes in Galloway, south-west Scotland: Diatom analysis, pH trends and the role of afforestation. *J. Ecol.* **75**:797–824.

Ford, M.S. 1990. A 10,000 year history of natural ecosystem acidification. *Ecol. Monogr.* **60**:57–89.

Harriman, R., and B.R.S. Morrison. 1982. Ecology of streams draining forested and non-forested catchments in an area of central Scotland subject to acid precipitation. *Hydrobiologia* **88**:251–263.

Jones, V.J., A.C. Stevenson, and R.W. Battarbee. 1986. Lake acidification and the land use hypothesis: A mid-postglacial analogue. *Nature* **322**:157–158.

Jones, V.J., A.C. Stevenson, and R.W. Battarbee. 1989. The acidification of lakes in Galloway, south-west Scotland: A diatom and pollen study of the postglacial history of the Round Loch of Glenhead. *J. Ecol.* **77**:1–23.

Kingston, J.C., H.J.B. Birks, A.J. Uutala, B.F. Cumming, and J.P. Smol. 1992. Assessing trends in fishery resources and lakewater aluminium from paleolimnological analyses of siliceous algae. *Can. J. Fish. Aquat. Sci.* **49**:116–127.

Kreiser, A.M., P.G. Appleby, J. Natkanski, B. Rippey, and R.W. Battarbee. 1990. Afforestation and lake acidification: A comparison of four sites in Scotland. *Phil. Trans. Roy. Soc. Lond. B* **327**:377–383.

National Research Council. 1986. Acid Deposition: Long-term Trends. Washington, D.C.: Natl. Academy Press.

Renberg, I. 1990. A 12 000 year perspective of the acidification of Lilla Öresjön, south-west Sweden. *Phil. Trans. Roy. Soc. Lond. B* **327**:357–361.

Renberg, I., and R.W. Battarbee. 1990. The SWAP Palaeolimnology Programme: A synthesis. In: The Surface Waters Acidification Programme, ed. B.J. Mason, pp. 281–300. Cambridge: Cambridge Univ. Press.

Renberg, I., T. Korsman, and N.J. Anderson. 1990. Spruce and surface water acidification: An extended summary. *Phil. Trans. Roy. Soc. Lond. B* **327**:371–372.

Renberg, I., T. Korsman, and H.J.B. Birks. 1993. Prehistoric increases in the pH of acid-sensitive Swedish lakes caused by land use changes. *Nature* **362:**824–826.

Rosenqvist, I.T. 1978. Alternative sources for acidification of river water in Norway. *Sci. Total Environ.* **10**:39–49.

Steinberg, C.E.W., S. Saumweber, and J. Kern. 1991. Paleolimnological trends in total organic carbon indicate natural and anthropogenic sources of acidity in Grosser Arbersee, Germany. *Sci. Total Environ.* **107**:83–90.

Stoner, J.H., A.S. Gee, and K.R. Wade. 1984. The effects of acidification on the ecology of streams in the upper Tywi catchment in west Wales. *Environ. Pollut.* **35**:125–157.

Sullivan, T.J. 1990. Historical changes in surface water acid-base chemistry in response to acidic deposition. In: Acidic Deposition: State of Science and Technology Report 11, pp. 11–3–11–CP–18. Washington, D.C.: Natl. Acid Precipitation Assessment Program.

Sullivan, T.J., D.F. Charles, J.P. Smol, B.F. Cumming, A.R. Selle, D. Thomas, J.A. Bernert, and S.S. Dixit. 1990. Quantification of changes in lakewater chemistry in response to acidic deposition. *Nature* **345**:54–58.

Sullivan, T.J., R.S. Turner, D.F. Charles, B.F. Cumming, J.P. Smol, C.L. Schofield, C.T. Driscoll, H.J.B. Birks, A.J. Uutala, J.C. Kingston, S.S. Dixit, J.A. Bernert, and P.F. Ryan. 1992. Use of historical assessment for evaluation of process-based model projections of future environmental change: Lake acidification in the Adirondack Mountains, New York, U.S.A. *Environ. Pollut.* **77**:253–262.

Uutala, A.J. 1990. *Chaoborus* (Diptera: Chaoboridae) mandibles: Paleolimnological indicators of the historical status of fish populations in acid-sensitive lakes. *J. Paleolimnol.* **4**:139–151.

Whitehead, D.R., D.F. Charles, and R.A. Goldstein. 1990. The PIRLA project (Paleoecological Investigation of Recent Lake Acidification): An introduction to the synthesis of the project. *J. Paleolimnol.* **3**:187–194.

Whitehead, D.R., D.F. Charles, S.T. Jackson, J.P. Smol, and D.R. Engstrom. 1989. The developmental history of Adirondack (NY) lakes. *J. Paleolimnol.* **2**:185–206.

# 6

# Influence of Soil Development and Management Practices on Freshwater Acidification in Central European Forest Ecosystems

K.H. FEGER
Institute of Soil Science and Forest Nutrition, Albert-Ludwig-University, 79085 Freiburg, F.R. Germany

## ABSTRACT

Acidic soils and mobile anions in soil solution are preconditions for the acidification of fresh waters. Mobile anions, of which $SO_4^{2-}$ and $NO_3^-$ are the most important, originate both from atmospheric deposition and internal sources. This chapter reviews aspects of natural and anthropogenic acidification of forest soils in central Europe. Due to their development from formerly weathered periglacial sediments and a long-lasting intensive export of biomass, many soils have been deeply acidified and have extremely low base saturation. The extent of historical and recent soil acidification is discussed based on the proton budget of selected forest ecosystems. A considerable potential for freshwater acidification evolves from a temporal and spatial discoupling of nutrient uptake and mineralization, resulting in excess nitrification and subsequent nitrate leaching. The rate of this process depends on the silvicultural system (harvesting system, e.g., clear-cut, thinning regime and choice of tree species). In addition, management practices influence soil-water pathways, which markedly affect proton buffering and anion mobility. Furthermore, the interactions between forestry and atmospheric deposition are stressed. Since deposition to central European forests varies markedly on a regional scale and anion mobility is highly controlled by internal processes in the terrestrial system, the separation between various causes of acidification is difficult, particularly at sites where atmospheric deposition is low or moderate.

*Acidification of Freshwater Ecosystems: Implications for the Future*
Edited by C.E.W. Steinberg and R.F. Wright 

## CONCEPTS OF ACIDIFICATION OF SOILS AND WATERS

Regional surveys show that acidic surface waters are generally restricted to watersheds with acidic soils underlain by a base-poor bedrock. In central Europe, most of the acidic waters are located in forested areas (Wieting 1986), and thus the focus is on acidification processes in forest ecosystems. Acidic soils (low pH and base saturation <15%) and elevated concentrations of mobile anions in soil solution are preconditions for the acidification of fresh waters. Before discussing the relative importance of atmospheric vs. ecosystem internal causes of freshwater acidification, I will briefly discuss the definitions of terms and processes that will be used in this chapter.

Soil acidification is defined as a decrease in the acid-neutralizing capacity of the solid soil ($ANC_{(s)}$) and is accomplished by removal of cationic components from the mineral phase. $ANC_{(s)}$ can be expressed as the sum of basic components minus the strongly acidic components (Van Breemen et al. 1983):

$$ANC_{(s)} = 6(Al_2O_3) + 2(CaO) + 2(MgO) + 2(K_2O) + 2(Na_2O) + 4(MnO_2) + 2(MnO) + 6(Fe_2O_3) + 2(FeO) - 2(SO_3) - 2(P_2O_5) - (HCl) \quad (6.1)$$

Soil acidification is a process driven by proton fluxes and occurs naturally during soil formation. However, it can be strongly accelerated by land management and acid deposition. The regulation of proton fluxes in forests encompasses numerous interconnected processes distributed throughout a variety of ecosystem compartments. Since nearly all biogeochemical reactions are coupled with $H^+$ transfers, the total element cycling has to be considered (Van Breemen et al. 1983; Bredemeier 1987; Feger 1993). A conceptual framework of the $H^+$ budget of an ecosystem is given in Figure 6.1. Management or, in a broader context, all measures of forest utilization influence the $H^+$ budget of a forest ecosystem by altering (a) internal $H^+$ production, (b) water pathways, and (c) the rate and chemical composition of atmospheric deposition.

In contrast to $ANC_{(s)}$, the acid-neutralizing capacity in aqueous systems ($ANC_{(aq)}$) is defined as the aqueous base equivalence minus the strong acid equivalence:

$$ANC_{(aq)} = [HCO_3^-] + 2[CO_3^{2-}] + [A^-] + [OH^-] - [H^+], \quad (6.2)$$

where $[A^-]$ is the molar concentration of a potential proton-accepting anion (e.g., a natural organic anion). Water acidification is a loss of $ANC_{(aq)}$. Soil processes, such as base cation weathering and/or desorption, affect the acidification of aqueous system by releasing cations and $ANC_{(aq)}$.

Reuss and Johnson (1985) have described the mechanisms whereby elevated concentrations of strong mineral acid anions can cause acidification of surface waters (depression in $ANC_{(aq)}$ and pH) via equilibrium processes in the soil solution. Hence,

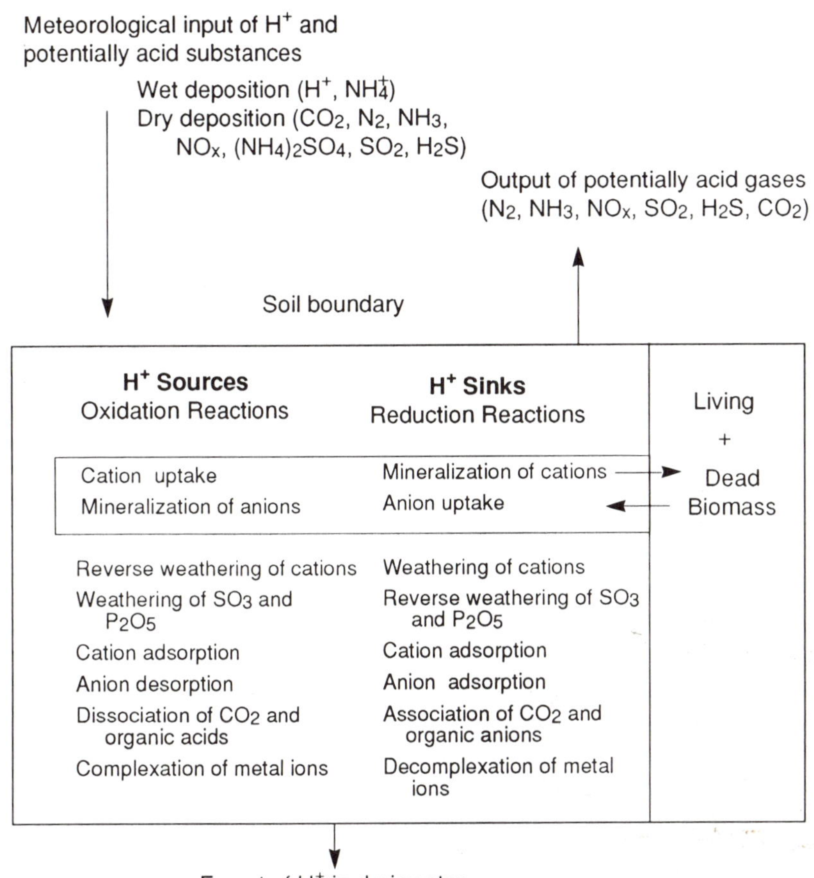


**Figure 6.1** Conceptual model of the $H^+$ budget of an ecosystem (modified from van Breemen et al. 1983).

the mobility of anions is the controlling factor in the transport of protons and inorganic Al species from the soil to the surface water system. This mechanism is schematically shown in Figure 6.2. Mobile anions, of which $SO_4^{2-}$ and $NO_3^-$ are the most important, originate from both atmospheric deposition and internal sources (cf. Van Miegroet, this volume). In the context of surface water acidification, the transformation and/or retention of atmospherically derived anions as well as the internal production of anions has to be considered.

Because the aggrading biomass takes up base cations, tree growth is an $H^+$-producing process (Figure 6.1). If all biomass returns to the soil and undergoes

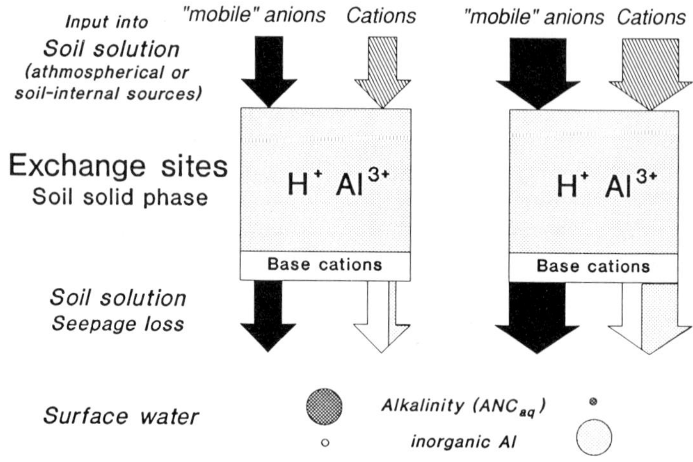


**Figure 6.2** Schematic diagram of the mechanism of cation transport in acidic soils with (a) low concentrations and (b) high concentrations of "mobile" anions of strong mineral acids.

decomposition, no net soil acidification would result. Managed forests, however, are characterized by an export of biomass representing a loss of $ANC_{(s)}$. The temporal discoupling between the $H^+$ fluxes associated with biomass accumulation and decomposition leads to phases of acidification in the soil where more base cations are taken up than released from decomposition. Excess nitrification is another important process. It occurs when the rate of uptake or microbial immobilization of N is smaller than the supply from mineralization. Such nitrification pulses with subsequent $NO_3^-$ leaching can strongly decrease $ANC_{(s)}$ (Van Miegroet and Cole 1984; Ulrich and Matzner 1986; Kreutzer 1988). With regard to surface water acidification, excess nitrification is relevant as it produces mobile anions, thereby enhancing the transport of soil acidity to the aqueous system.

## CONDITIONS OF SOIL DEVELOPMENT

### Geological and Mineralogical Aspects

In areas sensitive to surface water acidification, soils are characterized by low levels of base cations on the exchange sites. Such soils are highly saturated with $H^+$ and $Al^{3+}$. In large areas of central Europe, periglacial solifluction has been a key factor in soil development (Stahr 1979; Rehfuess 1990). This contrasts with northern and alpine regions, where pre-glacial soils and part of the bedrock were removed by the glaciers. Accordingly, soils in central Europe have developed not directly from the underlying bedrock but from periglacial solifluction layers several meters thick (Figure 6.3). Since these detrital layers include a varying amount of formerly weathered soil material, sometimes even from the Tertiary (Hofmann and Fiedler 1986), the exchangeable acidity in lower parts of the soil and regolith can be considerable. In addition, the fractures in the bedrock are often mantled with highly weathered material. With respect to freshwater chemistry, a deeply acidified substrate means a strongly reduced capacity to neutralize the acidity ($H^+$ and $Al^{3+}$) in the water percolating from the rooting zone. Acidic subsoils are naturally abundant in central Europe and occur more or less independent from present deposition rates. The extent and vertical distribution of solid phase acidity depends primarily upon the mineralogical properties of the different solifluction layers. Accordingly, a clear relationship between the soil-forming substrate and chemical composition of surface waters was found in hydrochemical surveys in the Black Forest (Zoettl et al. 1985; Feger and Brahmer 1987).

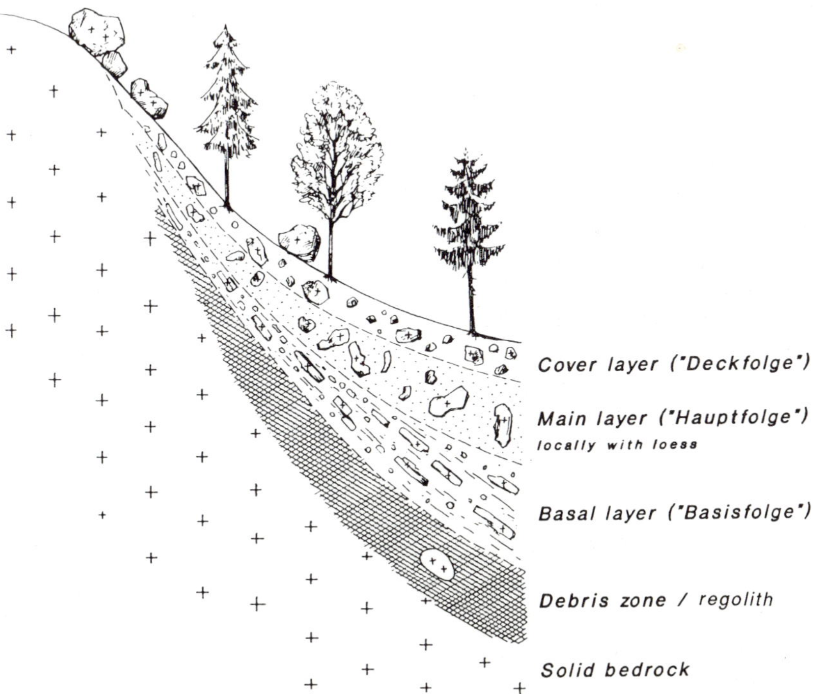


**Figure 6.3** Schematic diagram of the buildup of periglacial solifluction layers from which soils in mountainous areas in central Europe have commonly developed (modified from Rehfuess 1990).

Ulrich and Malessa (1987) claimed that the acidity of the deeper solum is primarily due to the recent deposition of mineral acids. However, there has been clear evidence of subsoil pH values of < 4 as early as the 1920s (Fiedler and Hofmann 1985). Numerous studies have shown a general pH decrease in the upper soil of central European forest sites during recent decades (cf. review by Rehfuess 1990). One of the most likely explanations is acid deposition. While there is clear evidence for a decrease in top soil pH, recent changes in base saturation have yet to be confirmed (cf. Rehfuess 1990). This is due to the fact that pH and exchangeable cations are only weakly correlated. The current depth distribution of cation exchange properties in a profile may not directly be explained in terms of a recently formed "acidification front" (Ulrich and Malessa 1987) moving towards the groundwater and subsequently to the surface water.

## Hydraulic Pathways and Soil-Water Regime

Soil stratification is another important property of soils derived from solifluction material; it strongly influences hydraulic pathways. In many hillslope profiles, a compact basal layer ("Basisfolge") has been found. It is typically cemented and shows a parallel-oriented skeleton (Figure 6.3). At sites where such an indurated basal layer is on top, close to the surface, the drainage pattern is highly superficial (Förster 1988; Feger 1989). As a consequence of shallow drainage, surface waters are mainly fed by acidic soil water. Subsoil horizons and the bedrock, which are crucial in anion retention and acid buffering, are then of minor importance. Under such conditions, surface water chemistry is characterized, for most of the year, by low pH values, elevated concentrations of DOC, Al, Mn, and Fe (Feger and Brahmer 1987). Only a shallow lateral transport in the upper soil can explain the export of these components, which under vertical drainage conditions are immobilized in the B horizon. Under the condition of lateral flow in the organic soil horizons, dissociated organic acids (mainly fulvic acids from the percolation of raw humus type O horizons) can markedly affect the runoff acidity. In addition, lateral drainage can explain the quick response in surface water chemistry due to changes in the ion exchange properties in the upper, but not the lower, soil.

Episodic acidification of stream water is closely linked to the soil-water regime and associated chemical and microbial processes. In soils that are almost permanently waterlogged (e.g., stagnogleys) the re-wetting after an extended drought often leads to sulfide oxidation and/or mineralization of S and N compounds. Since these soils are characterized by shallow lateral drainage, the washout of $SO_4^{2-}$ and $NO_3^-$ leads to an acid pulse in the runoff water. The interrelation between water pathways, soil chemical and microbial processes, and surface water chemistry has been described for the Black Forest (Zoettl et al. 1985; Feger and Brahmer 1987), the Bavarian Forest (Förster 1988), and the Erzgebirge (Fiedler and Katzschner 1989).

## FORMER FOREST UTILIZATION PRACTICES

Before systematic forestry commenced in the 19th century, many central European forests were subjected to intensive land use (forest pasturing, cutting for fuelwood, litter raking, and resin boxing). The duration and cumulative effect of these activities (in particular, the devastating clear-cuttings that took place during the 17th and 18th centuries) caused a gradual but drastic change in tree species composition and a progressive degradation (podzolization) of the soils. The originally dominant species beech (*Fagus sylvatica* L.) and silver fir (*Abies alba* Mill.) were replaced by spruce and pine (Zeitvogel and Feger 1990). These developments resulted in raw humus formation and a strongly enhanced base cation depletion in the upper soil (eluvial) horizons. According to Ulrich and Matzner (1986), the intensity of this historical human-made acidification of soils shows a clear relationship with elevation. Thus, acidification rates were maximal in mountainous regions, with high cation leaching due to a cool and humid climate, which favored podzolization.

The acidification rates that result from intensified removal of biomass can far exceed the acidification rates related to modern long-rotation forestry and acid deposition (Figure 6.4). In central Europe, litter raking and forest grazing were the dominant practices by which nutrients were transferred from forests to farmland. Even though the detrimental effects of litter removal had been recognized by the end of the last century (Ebermayer 1876), this practice continued in many areas until the middle of this century. Glatzel (1991) reported a significantly lower base saturation on base-poor sites, where litter had formerly been removed, as compared to those without litter raking. The same trends were found for forests where grazing took place. The highest $ANC_{(s)}$ depletion rates result from short-term rotation forestry mainly in combination with fuelwood coppicing (Figure 6.4).

Table 6.1 gives an estimate of the proton budget for a heavily utilized forest in the Black Forest for the time period since A.D. 1400. The present total atmospheric $H^+$ deposition is 40 mmol $m^{-2}$ $yr^{-1}$. This is low compared to other, more polluted sites in central Europe. "Natural" $H^+$ deposition is assumed to be 15 mmol $m^{-2}$ $yr^{-1}$, based on a natural pH in rainfall of 5.0 in continental areas (Delmas and Gravenhorst 1983). For the 600-year period, the total $H^+$ load is roughly 4,800 mmol $m^{-2}$, of which only a minor percentage is due to acid rain. This total $H^+$ load equals a mean annual $ANC_{(s)}$ loss of 80 mmol $m^{-2}$. Given a silicate weathering rate of roughly 40 mmol $m^{-2}$ $yr^{-1}$ at such sandstone sites (Ulrich and Matzner 1986; Feger 1993), a reduction in exchangeable base cations of maximal 2,400 mmol $m^{-2}$ can be estimated. The storage of exchangeable base cations in the soil profile 0–80 cm is today only 2,500 mmol $m^{-2}$, representing a base saturation of 5%. These crude numbers and $ANC_{(s)}$ losses up to 500 mmol $m^{-2}$ $yr^{-1}$, as exemplified in Figure 6.4, reveal the relative importance of intensive biomass utilization for the low base saturation of many forest soils in central Europe.

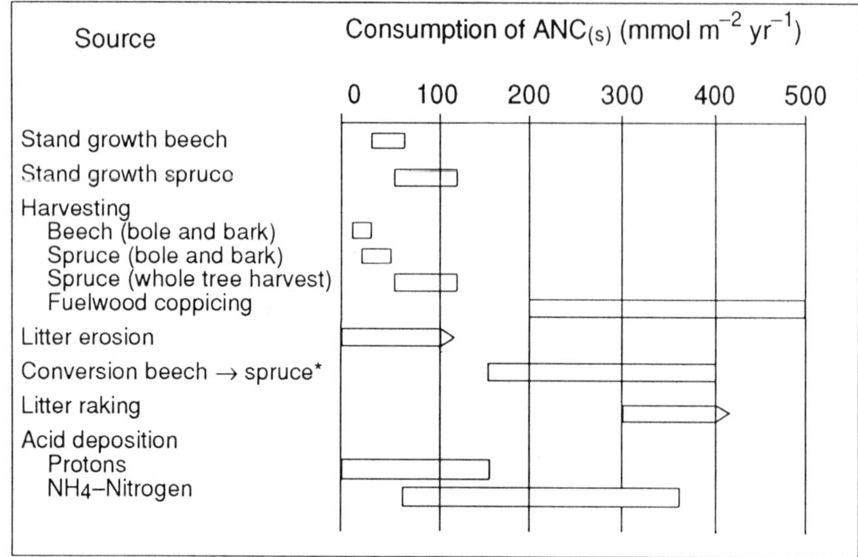


**Figure 6.4** Contribution of various processes to the depletion of acid buffering capacity of forest soils in Austria. Figures for the conversion of beech to spruce and for litter raking are peak rates that decline as the system reaches equilibrium or declines productivity (modified from Glatzel 1991).

## MODERN FORESTRY

Management practices of the modern long-rotation forestry can also add to the acidification resulting from natural processes, former land use, or from acid deposition. For freshwater acidification, both the potential increase in soil acidity (decrease in base saturation) and the internal production of mobile anions are relevant. In Figure 6.5 the internal and external $H^+$ load of four typical forest ecosystems are compared. Internal $H^+$ formation results from tree growth (uptake of a surplus amount of cations) and internal production of anions, mainly $NO_3^-$. External $H^+$ sources consist of total $H^+$ deposition and the $H^+$ production in the soil resulting from the amount of $NH_4^+$–N in excess of $NO_3^-$–N in deposition (cf. Bredemeier 1987; Feger 1993). Soil acidification results either from $NH_4^+$ uptake or nitrification. Sites in Figure 6.5 are characterized by marked differences in their present atmospheric deposition as well as in their management history.

**Table 6.1** Estimated proton budget of a forest site in the Black Forest (SW Germany) for the time since A.D.1400 (municipal forest of Villingen, soil type: dystric cambisol).

| | $H^+$ | (mmol $m^{-2}$) |
|---|---|---|
| Deposition of $H^+$ in rainfall | | |
| natural | 5,900 | (12.2%) |
| anthropogenic since ca. 1900 ("acid rain") | 2,700 | (5.6%) |
| Internal production of $H^+$ | | |
| former utilization of woody biomass (1400–1800) (rotation 80 yr, whole-tree utlilization | 30,000 | (63.2%) |
| litter removal (3 rotations, repetition 6 yr) | 4,300 | (9.0%) |
| forest grazing | ? | |
| utilization of precursor stand (1800–1900) (stemwood without bark) | 1,800 | (3.7%) |
| cation accumulation present stand | 3,000 | (6.3%) |
| **Total load of $H^+$** (during the 600-yr period) | 47,700 | (100%) |
| = *mean annual $H^+$ load* | 80 | mmol $m^{-2}$ $yr^{-1}$ |
| *potential rate of silicate weathering* | ca. 40 | mmol $m^{-2}$ $yr^{-1}$ |

## Forest Growth

An increased production of tree biomass leads to an equivalent $H^+$ production in the rooting zone due to base cation accumulation in biomass. Nilsson et al. (1982) found highly varying acidification rates that were dependent on tree species and yield class. For the stand growth of beech and spruce, mean $H^+$ production rates are 30–130 mmol $m^{-2}$ $yr^{-1}$ during the rotation period. Acidification is highest (up to 200 mmol $m^{-2}$ $yr^{-1}$), very early in the rotation, when the annual increment is maximal (Bredemeier 1987; Kreutzer 1988). The rate becomes very low in old, poorly stocked and slowly growing stands. For young stands in Denmark, Holstener-Jørgensen et al. (1988) found a clear relationship between wood increment and pH reduction in the rooted soil. In their study, an increase of basal area increment of 1 $m^2$ $ha^{-1}$ caused a drop in pH of 0.24 during the first 20 years after planting. Management can increase growth by choosing fast-growing species. Furthermore, silvicultural measures that change microclimatic conditions, especially thinning, have a positive effect on growth. The better tree growth, which has been observed in German forests in recent decades, can be partly attributed to a gradually improved thinning regime (Kenk et al. 1991).

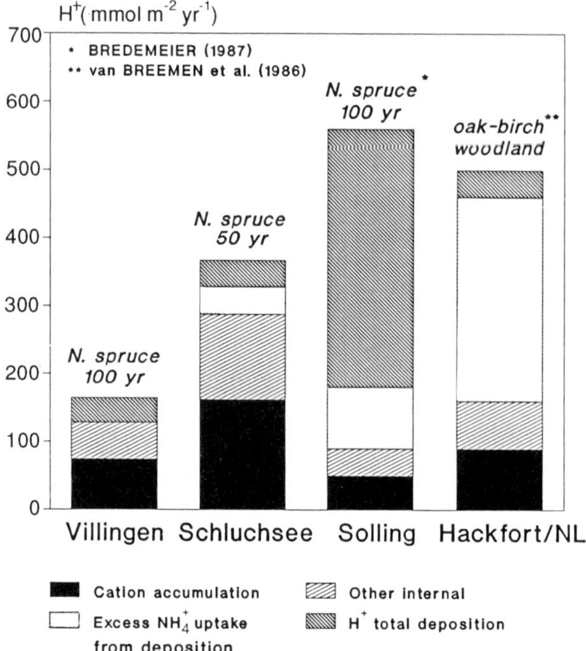


**Figure 6.5** Internal vs. external proton load of selected forest ecosystems in central Europe (cation accumulation covers incorporation of excess base cations in stand increment and accumulation of forest floor; other internal covers mainly anion production resulting from excess mineralization; excess $NH_4^+$ uptake relates to $H^+$ release associated with uptake of atmospheric $NH_4^+$ exceeding uptake of atmospheric $NO_3^-$).

## Biomass Harvesting

Long-rotation forestry, as normally practiced in central Europe, primarily harvests bole wood with bark. In cases where fully mechanized harvesting is employed, crown components (branches, twigs, foliage) are also removed from the site. Intensified harvesting, including other tree components (crown, coarse roots), leads to a gradual acidification of the rooted soil, because silicate weathering at most sites cannot compensate for these losses (Figure 6.4). Even though intensified types of harvesting are not commonly used in central European forestry, the numbers illustrate the detrimental effect of the intensive utilization of forest biomass in former times (Table 6.1).

## Choice of Tree Species

Replacement of deciduous forests by conifer plantations (cf. Figure 6.4: conversion of beech into spruce) may have a serious acidifying effect on soils and waters because:

1. In conifer stands the temporary accumulation of forest floor is more pronounced than in deciduous or mixed stands. This is due to the fact that litter produced by conifers is more resistant to decomposition. $H^+$ production originating from forest floor accumulation can reach 100 mmol $m^{-2}$ $yr^{-1}$ in young dense coniferous stands (Nilsson et al. 1982; Kreutzer 1988). If raw humus is formed, podzolization is favored. Under lateral drainage conditions, this process may also influence surface water by an increased transport of organic acids. The acidification rate resulting from accumulation of a humus layer is compensated for later in the rotation, when decomposition of organic matter accelerates under the better light and temperature regime. In managed forests, however, such a discoupling of the acidification and deacidification phases does not often allow full compensation within the rotation (Ulrich and Matzner 1986). This is due to excess nitrification, which leads to drainage losses of base cations mainly during clear-cutting and in newly established stands.
2. After conversion of deep-rooting beech stands into shallow-rooting spruce plantations, a spatial and temporal discoupling of nutrient uptake and mineralization evolves. Especially in the mineral soil, spruce utilizes only a part of the mineralized nutrient ions. Kreutzer (1989) reported excess nitrification from spruce plantations in Bavaria, where spruce was in the third rotation after beech or mixed deciduous stands. The same pattern of net anion production in deeper humic horizons of the mineral soil, as a result of tree species conversion, was also found in the Black Forest (Feger 1993). The spruce ecosystem "Schluchsee" (Figure 6.5) represents a typical site of excess mineralization in the mineral soil. In addition to $NO_3^-$, considerable amounts of $SO_4^{2-}$ are also released. These internally produced anions play a key role in the transport of acidity from the soil into surface waters, especially at sites where atmospheric deposition is relatively low.
3. The growth rate of conifers (in central Europe, mainly the Norwegian spruce: *Picea abies* Karst.) is normally higher than the growth rate of deciduous trees.

The planting of $N_2$-fixing trees (e.g., various species of *Alnus*), which can cause an excess nitrification and subsequent leaching of $NO_3^-$ (Van Miegroet and Cole 1984), is only locally important. Since such species are typically chosen for sites influenced by groundwater, the acidifying effect of $NO_3^-$ leaching is commonly compensated for through denitrification processes (Kreutzer 1988).

**Silvicultural System**

Plantation forestry, with its spatial separation of age classes, bears a significant risk for discoupling processes and can result in net acidification. This is mostly pronounced in the clear-felling system. Many studies have reported excess nitrification and cation leaching after clear-cutting, especially if the ground vegetation is suppressed. The intensity and frequency of thinning has a major influence on microbial transformations. Marginal thinning can augment the accumulation of organic matter on the forest floor, which, in turn, will increase podzolization processes. In contrast, if dense stands with thick layers of surface humus are strongly thinned, excess mineralization is likely to occur. A decrease in the O horizon through microbial decomposition is intrinsically a deacidifying process (cf. Figure 6.1). Nevertheless, if the release of ions is not temporarily met by plant uptake or storage in the soil and, as a result, excess nitrification takes place, net acidification of the soil and leaching of $NO_3^-$ results. Furthermore, clear-cutting and strong thinning in stands can cause an increase in lateral water flow through changes in the soil structure and soil-water regime. The combination of excess nitrification, mobilization of organic acids, and the reduced buffering of acids in the subsoil and bedrock may considerably enhance surface water acidification. However, there are few data on the hydrochemical effects of various silvicultural systems that are currently applied in central European forestry practice. Here, a set of watershed experiments is needed.

## Afforestation

The conversion of arable land or formerly unmanaged land (e.g., heathland) into forests can acidify the soil by forest floor accumulation and incorporation of base cations in the aggrading stand biomass (cf. Harriman et al., this volume). Furthermore, these changes may favor a spatial discoupling of mineralization and plant uptake (changes in depth and intensity of the rooting zone), which would lead to excess mineralization in extensively rooted humus-rich mineral soil horizons (cf. review by Lindner 1992).

## Fertilization

Acid-forming fertilizers are not commonly used in central European forestry. Nitrogen fertilization ceased in the 1960s for various reasons (high costs, increasing atmospheric N deposition, potential $NO_3^-$ pollution of drinking water sources). Recently, the application of Ca/Mg-limestone materials and/or readily soluble Mg salts, mainly in the form of sulfate, has become widespread because a considerable part of the recent forest dieback is due to nutritional disturbances. Even though the application of limestone material means an introduction of alkalinity to the soil system, negative effects have been observed in terms of a stimulation of nitrification processes and subsequent leaching of $NO_3^-$, which initially is not always accompanied by base cations (Kreutzer et al. 1991). Furthermore, the potential of organic acid mobilization from organic horizons exists. The anions released may acidify deeper seepage and surface water, at least in an initial phase. The application of neutral salts (e.g., $MgSO_4$) potentially increases the transport of acidity from solid soil to the hydrosphere by introducing the mobile $SO_4^{2-}$ anion (Feger et al. 1991).

## INTERACTIONS BETWEEN DEPOSITION AND SILVICULTURAL MEASURES

In addition to influencing the acidification of the solid soil and the internal production of mobile anions, silvicultural practices also affect the interactions between forest stands and atmospheric deposition.

Both the rates and the chemical composition of deposition depend on the surface properties of the canopy. Choice of tree species as well as the density and age structure of stands are crucial in this respect. In the German Solling region, the rates of $H^+$ and $SO_4^{2-}$ deposition under mature spruce are several times higher than that under beech of the same age (Ulrich et al. 1979). This is caused by the higher filtering area of the spruce stand and by the fact that hardwoods are leafless during the winter season, when rates of interception-deposition in spruce are at a maximum. As a result, spatial and temporal variations in canopy properties due to growth and management must be considered when current deposition rates are extrapolated.

A strong $H^+$ deposition may acidify the top soil to such a degree that the decomposition of forest floor material is strongly impeded. It has therefore been hypothesized that the accumulation of the forest floor persists beyond the rotation representing a strong internal acidification source (Ulrich et al. 1979). However, there is no clear evidence for such a mechanism. In contrast, there are numerous reports that reveal a general change of O horizons from raw humus types into moder/mull types during recent decades (cf. Rehfuess 1990). This change parallels a reduction in the C/N ratio and is obviously due to an increased N deposition rate. Such changes in humus types can explain the fact that increased N deposition does not generally result in an increase in $NO_3^-$ leaching.

Atmospheric N deposition is crucial to the acid-base budget of forests (Van Breemen et al. 1983). In particular, $NH_4^+$ deposition primarily originating from agricultural sources has a strong acidifying effect. Root uptake of $NH_4^+$ from deposition results in an equivalent release of $H^+$ to the soil. Because tree growth is stimulated by an improved N supply (cf. Kenk et al. 1991), more base cations are taken up, thereby increasing the $H^+$ production rate and base cation depletion in the soil. Under the situation of an extremely high N deposition, i.e., more than 500 mmol $m^{-2}$ $yr^{-1}$ (e.g., in parts of the Netherlands [cf. Figure 6.5] and in northern Germany), N supply can no longer be retained by biological mechanisms in the forest ecosystem (stand growth, ground flora, soil microbes). As a result, excess nitrification takes place and acts as an additional source for the acidification of soils and waters. Nitrogen deposition may also stimulate the mineralization of soil organic matter and thereby enhance the internal production and subsequent leaching of $NO_3^-$. Nevertheless, the stimulation of decomposition of organic matter must not necessarily be followed by excess nitrification. At many sites, N input is totally tied up by the transformation of biologically inactive O layers (raw humus) to more active humus forms (mull, moder) with a lower C/N ratio. The beneficial effect of N deposition, however, is restricted to N-limited forest ecosystems, in particular those that were depleted in N (and other nutrients) by adverse management practices in the past (Kreutzer 1989). There is no question that the ability of these systems to retain excess N from deposition is time-limited.

## SUMMARY AND CONCLUSIONS

In many forest sites in central Europe, acidic and base-poor soils are a precondition for freshwater acidification. However, deep-reaching acidity of soil profiles is not a new phenomenon and cannot exclusively be attributed to the elevated rates of acid deposition during the recent decades. Under the impact of a heavy load of atmospheric deposition ($H^+$, $NH_4^+$) soils can further acidify. Thus, atmospheric deposition adds to the process of natural and human-made acidification. Deposition of acids to forests in central Europe shows a marked regional-scale variation (Göttlein and Kreutzer 1991). Many soils were acidified naturally and/or anthropogenically long before acid rain began. The long-term intensive utilization of forests has significantly enhanced the natural soil acidity. As a result, acidic fresh waters occurred also naturally, with organic acids representing the main anions. Lateral water and element transport in the upper soil plays a key role, since under such conditions the buffering capacity of subsoil horizons and the bedrock is minimal.

With respect to freshwater acidification, the critical aspect of atmospheric deposition in these highly acidified soil systems is not the further acidification of the solid soil, but primarily the marked increase of mobile anions of strong mineral acids, mainly $NO_3^-$ and $SO_4^{2-}$, which offer an efficient transport mechanism for acidity from the soil to streams and lakes. Since deposition of N and S varies markedly on a regional scale, both in absolute rates and chemical composition (Göttlein and Kreutzer 1991), and the retention/release of anions is highly controlled by internal processes, the separation of effects for specific sites is difficult.

Modern long-rotation forestry can enhance soil acidification through intensified whole-tree harvesting as well as in a spatial and temporal discoupling of biological accumulation and mineralization of elements. The latter process favors the internal production of mobile anions, especially of $NO_3^-$. Excess mineralization can be minimized by avoiding forest floor accumulation and drastic silvicultural measures (e.g., clear-cutting, thinning). A tighter element cycling can be achieved by establishing vertically structured stands with a high percentage of deep-rooting species. In this respect, deciduous trees that produce easily decomposable litter should be favored. Liming can be a suitable measure to decrease biologically inactive forest floor layers. On the other hand, it is difficult to increase the base saturation of the mineral soil efficiently through liming. To avoid triggering off an excess mineralization of N, the application of limestone should be restricted to sites characterized by accumulation of inactive raw humus.

## REFERENCES


Bredemeier, M. 1987. Stoffbilanzen, interne Protonenproduktion und Gesamtsäurebelastung des Bodens in verschiedenen Waldökosystemen Norddeutschlands. Berichte des Forschungszentrums Waldökosysteme/Waldsterben Göttingen Reihe A **33,** 183 pp.

Delmas, R.J., and G. Gravenhorst. 1983. Background precipitation acidity. In: Acid Deposition. Proceedings of a Workshop Held in Berlin, Sept. 9, 1982, ed. S. Beilke and A.J. Elshout, pp. 82–107. Dordrecht: Reidel.

Ebermayer, E. 1876. Die gesamte Lehre von der Waldstreu. Berlin: Springer.

Feger, K.H. 1989. Hydrologische und chemische Wechselwirkungsprozesse in tieferen Bodenhorizonten und im Gestein in ihrer Bedeutung für den Chemismus von Waldgewässern. *DVWK-Mitteilungen* **17**:185–204.

Feger, K.H. 1993. Bedeutung von ökosysteminternen Umsätzen und Nutzungseingriffen für den Stoffhaushalt von Waldlandschaften. In: Freiburger Bodenkundl. Abh., vol. 31. Freiburg i.Br.: Selbstverlag Institut für Bodenkunde und Waldernährungslehre, 243 pp.

Feger, K.H., and G. Brahmer. 1987. Biogeochemical and hydrological processes controlling water chemistry in the Black Forest (West Germany). In: Proceedings of the Intl. Symposium on Acidification and Water Pathways, Bolkesjø, Norway, May 4–8, 1987, vol. 2, pp. 23–32.

Feger, K.H., H.W. Zoettl, and G. Brahmer. 1991. Assessment of the ecological effects of forest fertilization using an experimental watershed approach. *Fertilizer Res.* **27**:49–61.

Fiedler, H.J., and H. Hofmann. 1985. Ältere und neuere Messungen zur Bodenazidität in Fichtenbeständen des Erzgebirges. Berichte Jahrestagung Agrarwiss. Ges. der DDR, Nov. 20, 1985, pp. 64–97. Dresden.

Fiedler, H.J., and W. Katzschner. 1989. Zur Relation zwischen basischen Kationen und Anionen in Waldgewässern der Mittelgebirge. *Hercynia N.F. Leipzig* **26**:94–101.

Förster, H. 1988. Bodenkundliche und hydrologisch-hydrochemische Untersuchungen in ausgewählten Hochlagengebieten des Inneren Bayerischen Waldes. Dissertation Ludwig-Maximilians-Universität München, 265 pp.

Glatzel, G. 1991. The impact of historic land use and modern forestry on nutrient relations of central European forest ecosystems. *Fertilizer Res.* **27**:1–8.

Göttlein, A., and K. Kreutzer. 1991. Der Standort Höglwald im Vergleich zu anderen ökologischen Fallstudien. *Forstw. Forschungen* **39**:22–29.

Hofmann, W., and H.J. Fiedler. 1986. Charakterisierung von Zersatzzone und Schuttdecken in mitteleuropäischen Gebirgsböden. *Chemie Erde* **45**:23–57.

Holstener-Jørgensen, H., M. Krag, and H.C. Olsen. 1988. The influence of 12 tree species on the acidification of the upper soil horizons. *Det forstlige forsgsvæsen i Danmark* **42**:15–25.

Kenk, G., H. Spiecker, and G. Diener. 1991. Referenzdaten zum Waldwachstum. *KfK/PEF-Berichte* **82,** 59 pp.

Kreutzer, K. 1988. The impact of forest management practices on the soil acidification in established forests. In: Effects of Land Use in Catchments on the Acidity and Ecology of Natural Surface Waters, ed. H. Barth, pp. 75–90. Commission of the European Communities, Air Pollution Report 13. Brussels: E. Guyot SA.

Kreutzer, K. 1989. Änderungen im Stickstoffhaushalt der Wälder und die dadurch verursachten Auswirkungen auf die Qualität des Sickerwassers. *DVWK-Mitteilungen* **17**:121–132.

Kreutzer, K., A. Göttlein, and P. Pröbstle. 1991. Auswirkungen von saurer Beregnung auf den Bodenchemismus in einem Fichtenaltbestand (*Picea abies* [L.] Karst.). *Forstw. Forschungen* **39**:174–186.

Lindner, M. 1992. Ökologische Auswirkungen von Erstaufforstungen: Stand der Forschung und Folgerungen für die waldbauliche Praxis. *Forstarchiv* **63**:143–148.

Nilsson, S.I., H.G. Miller, and J.D. Miller. 1982. Forest growth as a possible cause of soil and water acidification: An examination of concepts. *Oikos* **39**:40–49.

Rehfuess, K.E. 1990. Waldböden. Entwicklung, Eigenschaften und Nutzung. 2nd Ed. Hamburg: P. Parey, 294 pp.

Reuss, J.O., and D.W. Johnson. 1985. Effect of soil processes on the acidification of water by acid deposition. *J. Environ. Qual.* **14**:26–31.

Stahr, K. 1979. Die Bedeutung periglazialer Deckschichten für Bodenbildung und Standortseigenschaften im Südschwarzwald. In: Freiburger Bodenkundl. Abh., vol. 9. Freiburg i.Br.: Selbstverlag Institut für Bodenkunde und Waldernährungslehre, 273 pp.

Ulrich, B., und V. Malessa. 1987. Tiefengradienten der Bodenversauerung. *Z. Pflanzenernähr. Bodenk.* **152**:81–84.

Ulrich, B., and E. Matzner. 1986. Anthropogenic and natural acidification in terestrial ecosystems. *Experientia* **42**:344–350.

Ulrich, B., R. Mayer, and P.K. Khanna. 1979. Deposition von Luftverunreinigungen und ihre Auswirkungen in Waldökosysteme im Solling. In: Schriften aus der Niedersächs. Forstl. Fak. d. Univ. Göttingen u. d. Niedersächs. Forstl. Versuchsanstalt, vol. 58. Frankfurt: J.D. Sauerländer.

Van Breemen, N., P.H.B. de Visser, and J.J.M. van Grinsven. 1986. Nutrient and proton budgets in four soil-vegetation systems underlain by Pleistocene alluvial deposits. *J. Geol. Soc. London* **143**:569–666.

Van Breemen, N., J. Mulder, and C.T. Driscoll. 1983. Acidification and alkalinization of soils. *Plant and Soil* **75**:283–308.

Van Miegroet, H., and D.W. Cole. 1984. The impact of nitrification on soil acidification and cation leaching in a Red Alder Ecosystem. *J. Environ. Qual.* **13**:586–590.

Wieting, J. 1986. Gewässerversauerung durch Luftschadstoffe in der Bundesrepublik Deutschland. *Wasserwirtschaft* **76**:58–62

Zeitvogel, W., and K.H. Feger. 1990. Pollenanalytische und nutzungsgeschichtliche Untersuchungen zur Rekonstruktion des historischen Verlaufs der Boden- und Gewässerversauerung im Nordschwarzwald. *AFJZ* **161**:136–144.

Zoettl, H.W., K.H. Feger, und G. Brahmer. 1985. Chemismus von Schwarzwaldgewässern während der Schneeschmelze 1984. *Die Naturwissenschaften* **72**:268–270.

# 7

# Influence of Management Practices in Catchments On Freshwater Acidification: Afforestation in the United Kingdom and North America

R. HARRIMAN[1], G.E. LIKENS[2], H. HULTBERG[3], and C. NEAL[4]

[1]Freshwater Fisheries Laboratory, Faskally, Pitlochry, Perthshire, Scotland, PH16 5LB, U.K.

[2]Institute of Ecosystem Studies, New York Botanical Garden, Mary Flagler Cary Arboretum, Box AB, Millbrook, NY 12545–0129, U.S.A.

[3]Swedish Environmental Research Institute (IVL), P.O. Box 47086, 40258 Gothenburg, Sweden

[4]Institute of Hydrology MacLean Building, Wallingford, Oxfordshire OX10 8BB, U.K.

## ABSTRACT

The recent increase in the development of intensively managed forests has coincided with the discovery of widespread surface water acidification resulting from the deposition of acidic pollutants. In impacted regions the association between surface water acidification and conifer afforestation has been attributed to increased interception of sulfur and nitrogen compounds from the atmosphere. An implicit consequence of this link is that afforestation causes minimal acidification to surface waters in pristine areas. However, other soil and surface water acidifying processes have also been identified (e.g., base cation uptake and removal) which act independently of pollutant inputs. In this chapter, the acidifying potential of forests during three distinct phases of the forest cycle are evaluated: (a) ploughing/planting phase, (b) aggrading phase, and (c) clearfelling phase. While catchment-to-catchment variations inevitably exist, the general pattern of response is similar. Future studies should attempt to quantify the effects of repeated cycles of managed forests both in terms of water quality and sustainability of production. Such studies should be located in both impacted and pristine areas with special emphasis on base cation and nitrogen cycling.

*Acidification of Freshwater Ecosystems: Implications for the Future*
Edited by C.E.W. Steinberg and R.F. Wright 

## INTRODUCTION

Forests are an integral part of the earth's natural ecosystem. Recent concerns about extensive deforestation have resulted in increased national programs of forest planting to replace these diminishing resources. Many of these programs have been designed to produce trees that grow quickly, thus providing an adequate financial return. For example, with a beneficial climate associated with adequate water supplies, pine and Eucalyptus species can be harvested within 4–15 years (e.g., Brazil and South Africa). Here we are particularly concerned with plantation development in Europe and North America, where harvesting occurs over a 40- to 60-year time scale. This often involves planting exotic conifers at a high density and is supported, where necessary, by artificial fertilization programs. The major planting programs in the U.K. (mainly Scotland and Wales) commenced in the 1950s and have continued on a patchwork basis to the present day. Concern over the impacts of forestry practices during the ploughing/planting, aggrading, and clearfelling stages was initially directed towards water losses (e.g., Hoover, 1944, in North America and Calder and Newson, 1979, in the U.K.).

Only during the past 15 years has the impact of afforestation on surface water acidification been highlighted (Harriman and Morrison 1982; Stoner et al. 1984). In these early papers, the major point of debate was the extent to which natural and anthropogenic factors determined the acidification of surface waters, i.e., do forests acidify *surface waters* in both polluted and unpolluted areas? To provide a more detailed assessment of this problem, we have attempted to identify and quantify the potential acidifying mechanisms during three phases of forest growth: (a) ploughing and planting phase, (b) aggrading phase, and (c) clearfelling phase.

The aim of this approach is to provide an assessment of the integrated effects from disturbance through forest development. For some factors it is difficult to estimate the acidifying potential. However, we hope that these areas of uncertainty will provide focal points for discussion and future research.

## ACIDIFICATION CONCEPTS

The history of surface water acidification and the underlying concepts have been reported in great detail over the past decades. Here we simply highlight the fundamental mechanisms of surface water acidification and direct the reader to the literature pertaining to each aspect.

Soil acidification is seen as an increase in the store of $H^+$ in the soil. Soils that are naturally acidic from biologically generated $H^+$ (carbonic and organic acids) become more acidic because of acid deposition through exchange of base cations from soils, i.e., exchangeable acidity increases and exchangeable bases decrease. Soils with medium- to high-base saturation respond to increased concentrations of mobile acid anions by increased base cation leaching (soil acidification), whereas $H^+$ and $Al^{n+}$ are mainly, but not exclusively, leached from soils into surface waters where base saturation is exceptionally low. This response is generally termed the *mobile anion* effect (Johnson and Cole 1980).

The mechanism by which increased strong acid anion concentrations can acidify surface waters, with little or no change in soil acidity, has been eloquently described by Reuss and Johnson (1986). These authors showed that an alkalinity depression in the soil solution is translated into a pH reduction in drainage water after $CO_2$ degassing.

Soil acidification can also occur via plant uptake and biomass accumulation/removal of base cations. However, the link between this process and surface water acidification is not yet established. This emphasizes the need to make a clear distinction between soil and surface water acidification. Here we focus on *surface water* acidification.

Perhaps the most controversial explanation of the effects of acidic inputs to catchments is the anion substitution hypothesis (Krug and Frink 1983). The key element of this hypothesis is that low pH, sulfate-dominated clear waters in polluted areas were originally organic acid-dominated colored waters of similar pH. Because of the scarcity of historical data, this hypothesis is difficult to test. Paleolimnological studies, however, have indicated large pH shifts in some clearwater lakes (Battarbee et al. 1985), suggesting that the role of this ion-substitution mechanism is of minor importance on a regional scale. Experimental tests with colored stream water have also failed to provide support for this hypothesis (Hedin et al. 1990). What is clear is that the biologically important substances, such as aluminium (Al), are now present in ionic and potentially toxic forms compared with the relatively nontoxic species found in highly colored waters.

Because of charge balance requirements in surface waters, the most commonly used indicator of acidification has been acid-neutralizing capacity (ANC), which can be represented as:

$$\text{ANC} = [\text{Ca}^{2+} + \text{Mg}^{2+} + \text{Na}^{+} + \text{K}^{+}] - [\text{Cl}^{-} + \text{SO}_4{}^{2-} + \text{NO}_3{}^{-}] = \text{C}_B - \text{C}_A, \qquad (7.1)$$

where $C_B$ is the equivalent sum of base cations, $C_A$ is the equivalent sum of strong acid anions, and units are equivalents per liter (Reuss and Johnson 1986).

ANC can also be represented in an acid-base context by the traditional alkalinity formulation:

$$\text{ANC} = [\text{HCO}_3]^{-} + [\text{A}^{-}] - [\text{H}^{+}] - [\Sigma\text{Al}^{+}], \qquad (7.2)$$

where $A^-$ represents organic anions.

Both of these formulations are acceptable; however, care must be taken to ensure that the ANC value obtained by acid-base titration (Equation 7.2) is the same as that obtained by the charge balance methods (Equation 7.1). An evaluation of the ANC concept, including appropriate titration methodologies, has been reported by Cantrell et al. (1990).

Equation 7.1 is a useful starting point from which to address the question of forestry effects on surface water acidification, because any reduction in $C_B$ will reduce ANC if $C_A$ remains constant or increases. Any increase in $C_A$ will also reduce ANC if $C_B$ remains constant or decreases.

There are four major factors that may influence the $C_B$–$C_A$ relationship during the three phases of forest growth:

1. hydrological and physical changes;
2. base cation changes (uptake, weathering, and ion exchange);
3. canopy leaching and interception of acidic pollutants by the forest canopy; and
4. organic acid changes.

## Hydrological Impacts

### *Ploughing/Planting*

In managed catchments, the effects of ploughing are visually manifest as surface soil disturbance and sediment release into streams. This may have a potentially damaging short-term effect on stream biota but is not considered to be an acidifying process.

The major objective of ploughing is to improve drainage around rooting systems, which implies a modification of the flow pathways and a possible lowering of the water table. If more water is diverted through the lower mineral horizons, then an increase in ANC, base cations and, to a lesser extent, nitrate would be expected (Likens 1992). However, if stormflow peaks increase, with more water emerging from surface organic horizons, then ANC would probably decrease due to a lower $C_B$ value.

Results of various studies are somewhat contradictory. Robinson (1986) found that peak flows increased after ploughing while other studies have shown the opposite effect (e.g., Heikurainen et al. 1978). These differences are probably related to soil drainage characteristics: increased peaks result from ploughing of poorly drained soils whereas for well-drained soils, where ploughing lowers the water table, a reduction in peak flows may occur. The consequences of physical disruption of surface soil layers may be most pronounced on macropore ("pipe") flow through these horizons. Significant changes in the timing, amount, and quality of drainage water run-off may occur.

Because $C_B$ decreases relative to $C_A$ with increasing flow in upland catchments, the concern about ploughing is usually one of increasing acidity (decreasing ANC) if peak flows increase. Recent studies of pH/flow relationships have shown that the pH/flow curve reaches a plateau well before peak flows are reached (Harriman et al. 1990). Thus acidity, pH, and ANC remain relatively constant over a wide range of stormflow conditions, and increased peak flows will probably not change the minimum pH significantly.

Lowering of the water table may also increase the potential for oxidation and mineralization of organic N and S compounds if surface soil horizons are continually drying and wetting in response to weather conditions. Most long-term catchment studies show little change in water quality during and after the ploughing phase; however, the majority of these studies have been carried out in high rainfall areas. Nevertheless, even these upland sites may show elevated $NO_3$ and $SO_4$ peaks during the first storms after warm/dry summers. Two- to threefold increases in sulfate have been measured during such conditions resulting in severe pH depressions (Harriman 1988). Such occurrences are not necessarily associated with ploughing activities.

*Aggrading Forests*

Until recently, the understanding of interception losses by aggrading forests in the U.K. was based on comparative studies of water budgets in forest and moorland catchments in Wales (Calder and Newson 1979). Additional interception losses were estimated at around 2% for every 10% forest cover in the catchment. Recent studies at Balquhidder, in central Scotland, have indicated that longer grasses and various moorland vegetation types may intercept more water at exposed, high elevation sites, thus reducing the interception differences between forest and moorland sites (Hall and Harding 1993). Therefore, regional differences in interception estimates probably depend both on input (snow, rain, etc.) and on vegetation composition prior to and after afforestation. Transpiration losses may also reduce the total water yield from forested catchments; however, most broadleaf and conifer trees can control these losses during dry periods (Hall and Roberts 1990).

*Clearfelling Effects on Water Yields*

Intuitively, the removal of trees should reduce interception and transpiration, thus increasing total water yield. Yet with increased radiation to soil surfaces and a generally large brash component remaining in the catchment, the hydrological response may be complex, especially where old roots provide channels to lower soil horizons. Even in catchments showing simple vegetation responses to clearfelling, the detection of small hydrological changes may be extremely difficult. Nevertheless, at Hubbard Brook, the annual variations in precipitation and run-off in catchment W2 prior to felling were extremely variable while evapotranspiration remained relatively constant. After felling, the average annual run-off at W2 increased by 30%.

Other long-term studies in the northern and central hardwood forests of U.S.A. show a significant relationship between first year increase in water yield and percent basal area cut in the catchment (Figure 7.1; Hornbeck et al. 1992). The major contribution to increased yield is the elevated peak discharge and stormflow volume during the growing season (Douglass and Swank 1972). The extent to which the increased water yield is sustained is dependent on numerous factors, including rainfall pattern; however, the typical scenario is one of sharply declining yield after cutting,

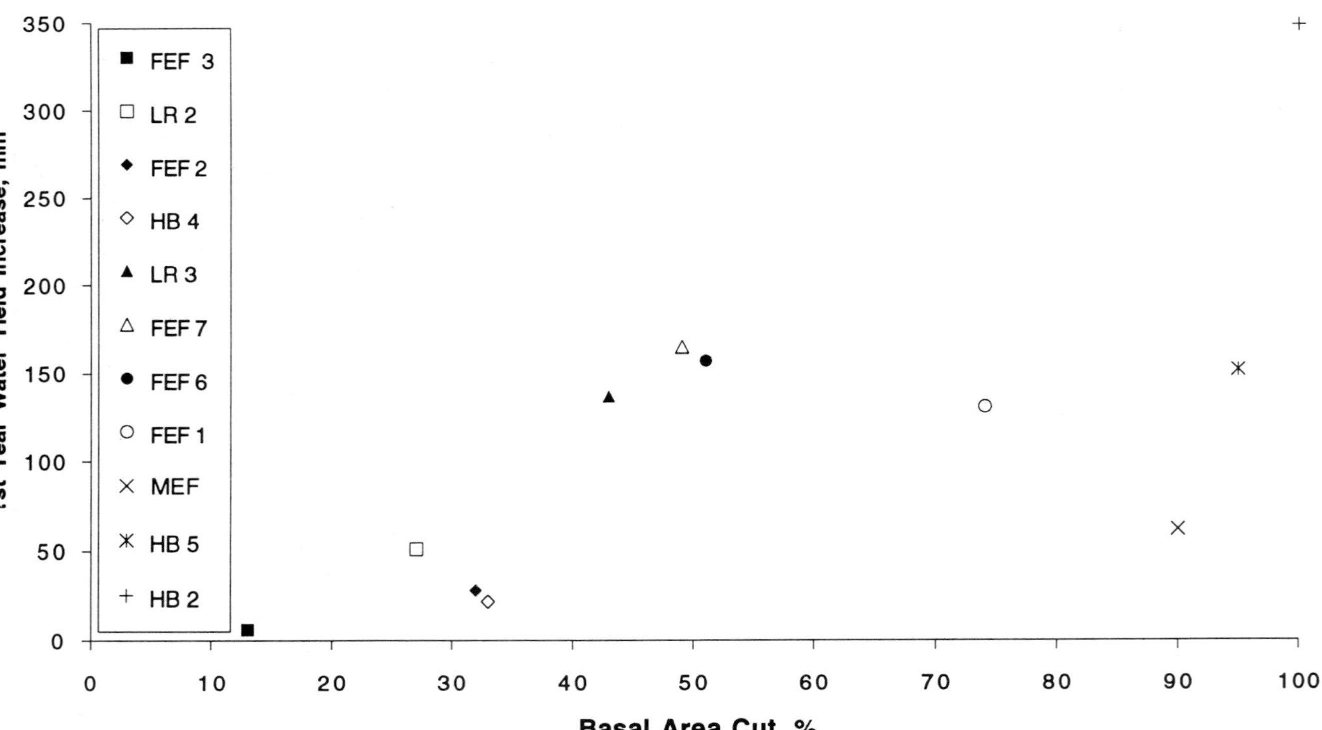


**Figure 7.1** Relationship between first year water yield and % forest basal area cut for selected sites in northern U.S.A.: FEF = Fernow Experimental Forest; LR = Leading Ridge; MEF = Marcell Experimental Forest; HB = Hubbard Brook. From Hornbeck et al. (1992).

followed by a slow return to baseline levels over a period of 5–10 years. The greatest rate of decline in yield occurs in catchments with both extensive natural vegetation regrowth and a change from hardwood to softwood species.

Although all stages of forest planting, growth, and felling are likely to change hydrologic flow paths, intensity of peak flows, and total water yield, the impacts of these effects on surface water acidification are likely to be small. While major concerns associated with the ploughing phase are linked to peak flow increases, the felling phase may cause increases in summer flows and associated acid-base chemistry (Neal et al. 1992). Run-off will probably reach a minimum level as trees reach maturity. At this stage, the concentration effect will be at its greatest, as will the potential for acidification via ion exchange processes in the soil.

## Base Cation Changes: Uptake, Weathering, and Ion Exchange

Base cation uptake with subsequent removal at harvesting represents a loss from the soil system and is a soil acidification process that may result in a *reduction* in base cation run-off into surface waters. By comparison, acid deposition results in *increased* strong acid anion run-off associated with increased base cation run-off.

### *Ploughing/Planting Phase*

For most catchments, the ploughing/planting stages are unlikely to affect base cation run-off except in cases where ploughing breaks through impervious iron pan, allowing drainage through base cation-rich mineral horizons. Few studies have measured base cation changes prior to and after drainage; however, for the few unpublished studies available, base cation levels remained relatively constant.

### *Aggrading Phase*

Probably the most controversial aspect of aggrading managed forests is whether base cation uptake by trees causes surface water acidification. This fundamental acidification question must be addressed as it will determine whether managed forests can acidify surface waters irrespective of pollutant inputs. To acidify surface waters the mix of base cations (Ca, Mg, K, Na) and acid cations ($H^+$, $Al^{n+}$) must change in favor of the acid cations *without* an equivalent change in anion composition. In natural forest ecosystems where pollutant loadings are minimal, there is no evidence of recent surface water acidification from paleoecological studies (e.g., Kreiser et al. 1990). However, on a longer time scale (centuries), past periods of surface water acidification may have coincided with major land disturbance (Rosenqvist 1990). Few studies report data from managed forests in pristine areas, but Nilsson et al. (1982) argued that although net acidification of soils is the likely outcome of base cation uptake, this was unlikely to lead to surface water acidification in the absence of increased mobile anion inputs. Soil acidification resulting from base cation uptake is a generally accepted concept; Nilsson et al. (1982) showed how rates of acidification are a function of forest growth (yield class), stocking and thinning policy, and nitrogen uptake pathways. Net acidification rates peaked between 20–30 years after planting for all year classes. In sensitive catchments, base cation run-off is often equivalent to estimated base cation uptake by trees, i.e., 0.5 Keq $ha^{-1}$ $yr^{-1}$ base cation uptake ≡ 50 μeq $L^{-1}$ stream concentration in a catchment with 1000 mm $yr^{-1}$ run-off. If trees obtain a significant base cation contribution from atmospheric sources, then the rate of soil acidification may be reduced.

Recent concern about forest health and sustainability (Sverdrup and Warfvinge 1990) has centered on the base cation depletion concept, especially where demands for increased productivity require a shorter forest growth cycle. These authors suggest that repeated harvesting and replanting will increase soil acidification (ultimately causing forest dieback) and acidify surface waters.

### *Clearfelling Phase*

Maximum soil acidification rates occur in managed forest catchments with high yield crops. Upon clearfelling, the base cation uptake will fall dramatically and deacidification (increased $C_B$ and ANC) should result from mineralization of humus/brash. Unfortunately, other complex processes also come into play at this stage, which result in major changes in the composition of run-off water. A major change associated with clearfelling is the disruption of the nitrogen cycle, which causes large increases in nitrate run-off (Likens et al. 1970; Stevens and Hornung 1987). The question, in terms of surface water acidification, is how much of the nitrate release is associated with base cation leaching. To date, clearfelling studies indicate a variable pH and base cation response to increased nitrate leaching. While the pattern of nitrate run-off is generally similar, some studies indicate little or no changes in run-off pH after harvesting (e.g., Martin et al. 1986), while others show significant pH depressions associated with increased Al concentrations (Neal et al. 1992).

In most cases a proportion of the nitrate release is associated with increased base cation run-off and a portion with increased run-off of acid cations. Thus harvesting may increase base cation run-off *and* acidify surface waters. However, in some instances base cation levels may decline, even during baseflow conditions (Neal et al. 1992). An important research need is to determine the extent to which acidified soils and/or high N inputs influence the acidification potential of nitrate released after harvesting.

The amplitude and duration of the nitrate peak, as with the water yield increase, is linked to the extent of basal area removed and to the relative distribution of forest undergrowth prior to and after felling (Martin and Pierce 1980). These authors also found a significant relationship between stream nitrate and calcium levels following clearfelling and strip-cutting in New Hampshire catchments. Progressive strip-cutting over a 3-year period caused the least disturbance in terms of base cation leaching into streams. An exception to this general trend was reported by Nicolson (1975), who found that nitrate levels declined after clearfelling in subcatchments of the Experimental Lakes Area (ELA). Maximum nitrate release occurs when all nonforest vegetation growth is suppressed, as shown at Hubbard Brook. Minimum nitrate release occurs where extensive vegetation growth occurs under old, slow-growing, thinly spread forests (Figure 7.2). Under these circumstances, clearfelling may result in a smaller nitrate release than the reduction in sulfate caused by lower interception and dry depositon of S. The net response in this case would be a small increase in pH and a fall in base cation levels (Hultberg 1991).

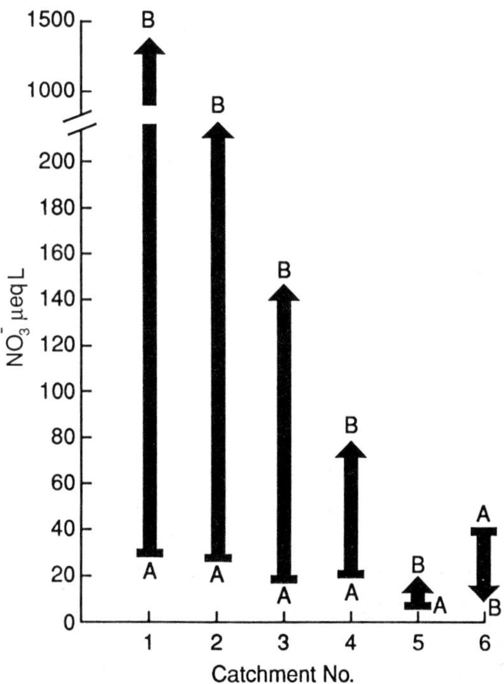


**Figure 7.2** Variations in peak nitrate leaching from selected catchments following deforestation. Catchment 1 (C1) = W2 Hubbard Brook, U.S.A.: complete deforestation; hardwood forest; no timber removed; ground vegetation supressed by herbicide. C2 = Plynlimon, Wales: 50% clearfelling; spruce forest plantation; limited ground vegetation prior to felling. C3 = Loch Ard, Scotland: spruce forest plantation: 50% clearfelling; little or no ground vegetation prior to felling. C4 = Balquhidder, Scotland: 50% clearfelling; thinned spruce/pine plantation; extensive ground vegetation prior to felling. C5 = Gårdsjön subcatchment, Sweden: 50% clearfelling; old natural softwood forest; low productivity with extensive ground vegetation prior to felling. C6 = Erzgebirge Mountain Area, Czech Republic: deforestation of upper spruce catchment due to forest dieback, lower catchment of mature beech; ground vegetation increasing, upper catchment replanted with hardwoods. A = pre-felling nitrate levels; B = peak nitrate level after felling.

Attempts to clarify base cation leaching processes in soils after harvesting have also proved difficult. Johnson et al. (1988) could find no significant relationship between base cation removal by whole tree harvesting and the supply of exchangeable base cations in soils. This applied to both deciduous and conifer forests. Subsequently, Johnson et al. (1991) found a variable response of pH and exchangeable base cations in different soil horizons. In terms of potential surface water acidification, the most interesting finding was the decrease in the ratio of exchangeable base cations to exchangeable $H^+$ and $Al^+$ in the soil surface horizons. This may be the net result of increased anion run-off associated with base cations.

## Canopy Effects

### *Aggrading Forests*

The extent to which the forest canopy affects surface water acidification depends on a multitude of factors but essentially two major processes can be considered. First, the interception of airborne substances, both natural (seasalts, volcanic emissions) and anthropogenic (acidic pollutants), increases the loading to the catchment. This process has been identified as the most important factor in increasing stream and lake acidification. Harriman and Morrison (1982) suggested that pollutant scavenging by the closed forest canopy was the major cause of increased acidity and aluminum levels in streams draining forested catchments compared with nonforested catchments in Scotland. Similar results were obtained for Welsh (Stoner and Gee 1985) and Swedish (Hultberg and Grenfelt 1986) catchments. In all these studies, S deposition was relatively high (> 300 meq $m^{-2}$ $yr^{-1}$). The quantity of extra amounts of intercepted airborne substances depends on canopy, catchment, and meteorological characteristics.

Conifers generally intercept aerosols, gases, and mist/cloud components more effectively than deciduous foliage (e.g., Lindberg et al. 1986). However, even within conifer species there may be differences in cycling and collecting efficiencies (Miller et al. 1991). Forests planted on upland sites in high rainfall areas can intercept cloud droplets, and this process has been shown to make an important contribution to the increased S and N inputs to catchments (Fowler et al. 1989; Weathers et al. 1988). Fowler et al. (1989) modeled the effects of elevation, aspect, rainfall, etc., on S and N loadings to Keilder forest in northern England and predicted a 30% increase in S inputs and 90% increase in nitrogen inputs due to afforestation (Figure 7.3). Further studies have indicated that these estimates are probably conservative because $SO_2$ inputs, major ion content of cloud water, and frequency of cloud cover were probably underestimated. This emphasizes the need to estimate the proportion of total forest at different altitudes in each catchment. Comparative studies in catchments subject to similar wet S deposition show a significant relationship between percentage forest cover and S input (e.g., Hultberg and Grenfelt 1992; Figure 7.4). These authors suggest that throughfall and run-off monitoring will provide a reasonably good estimate of total S and Cl deposition to conifer stands. A similar case has also been made for deciduous stands (Likens et al. 1990).

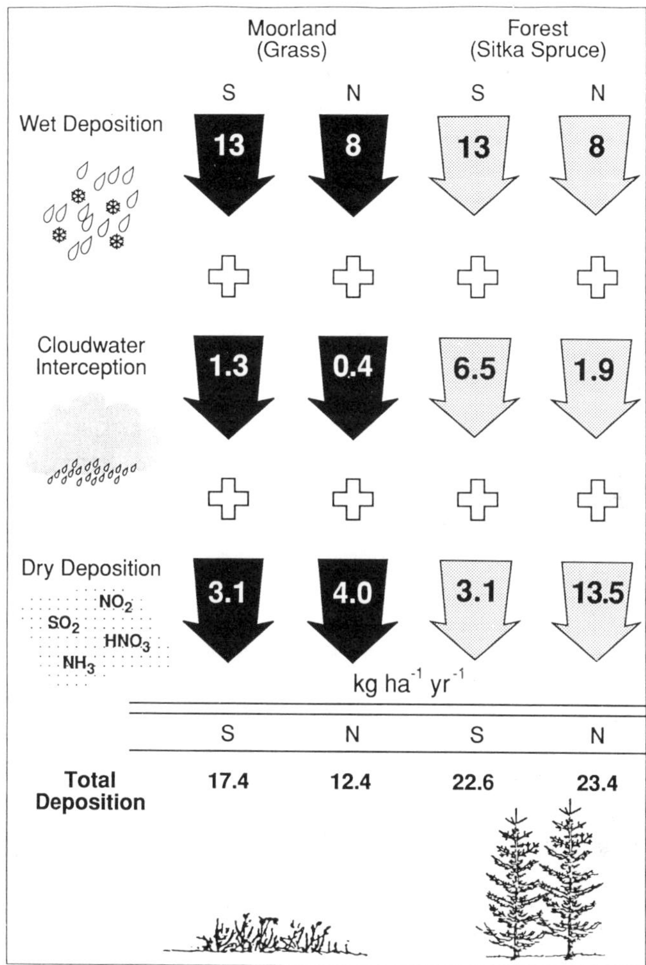


**Figure 7.3** Comparison of estimated inputs of sulfur and nitrogen to forested and nonforested catchments at Kielder, northern England, via different deposition processes (1 kg S $ha^{-1}$ $yr^{-1}$ = 6 meq $m^{-2}$ $yr^{-1}$; 1 kg N $ha^{-1}$ $yr^{-1}$ = 7 meq $m^{-2}$ $yr^{-1}$). Adapted from Fowler et al. (1989).

The net effect of these collecting mechanisms is to increase S and seasalt components in stream and lakewater. Although N compounds are intercepted to a greater extent than S and seasalts, they generally are almost completely removed via biologically mediated processes in vegetation and soils. Consequently in most sensitive upland catchments, the acidifying potential of N inputs, measured as $NO_3$ leaching in surface waters, is relatively low (Henriksen 1988).

Exceptions to this general picture have been reported for forested upland catchments in Wales, where elevated nitrate levels were detected in throughfall, soil water,

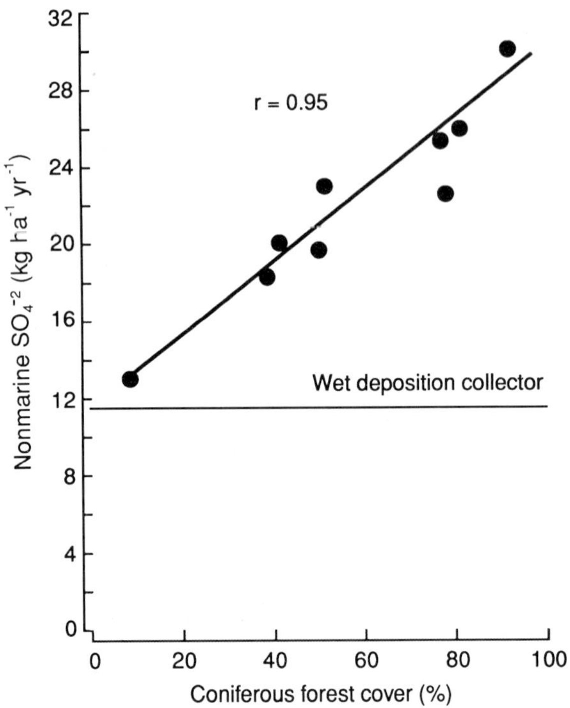


**Figure 7.4** Relationship between sulfate output and percentage forest cover in various subcatchments of Lake Gårdsjön. Adapted from Hultberg and Grenfelt (1992) (1 kg $SO_4^{-2}$ $ha^{-1}$ $yr^{-1}$ = 6 meq $m^{-2}$ $yr^{-1}$).

and drainage streams (Stevens et al. 1993). These catchments are similar in that (a) they contain mature conifers, which intercept air pollutants at a significant rate; (b) they are at the stage where the requirement for atmospherically derived N is relatively low; and (c) the forest floor has a limited cover of natural vegetation. These catchments could be considered as N-saturated.

An empirical relationship has been established between catchment sensitivity and pH for forested and nonforested catchments in high S deposition areas in Scotland (Harriman et al. 1987; Figure 7.5). For unforested sites, acidification below pH 5.0 occurs in catchments with base cation levels < 60–80 $\mu eqL^{-1}$, whereas mature forested ( 50% cover) catchments are acidified to pH < 5 at base cation levels < 120–140 $\mu eqL^{-1}$ (excess Ca + Mg).

In pristine or semi-pristine sites, forest canopies predominantly intercept seasalts, assuming other natural atmospheric sources (e.g., volcanic emissions) are insignificant.

The second major process is associated with chemical interactions within the forest canopy. The effects of these canopy interactions on surface water acidification are

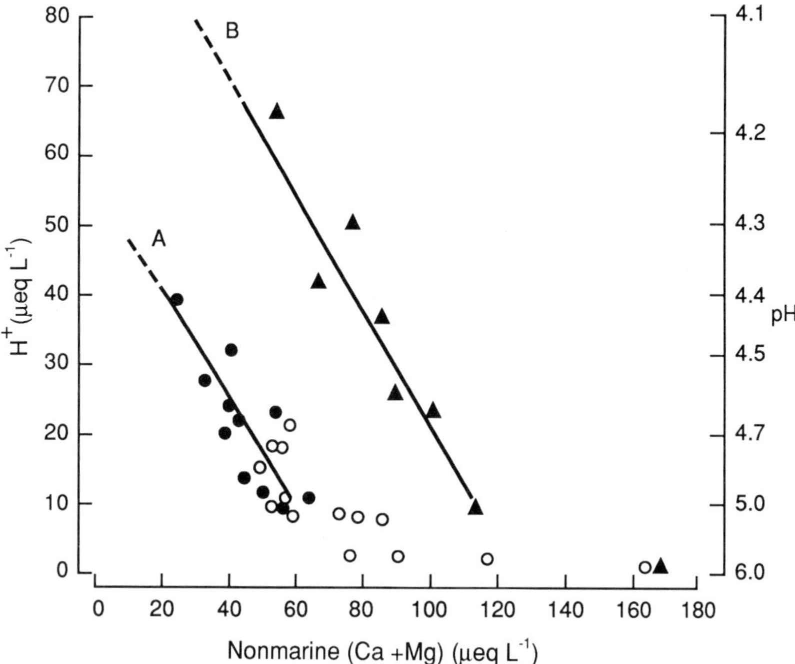


**Figure 7.5** Relationship between $H^+$ and nonmarine (Ca + Mg) for surface waters draining nonforested (O), young forest (< 15 yr; ●), and semi-mature forest (> 15 yr; ▲) catchments. From Harriman et al. (1987).

difficult to quantify because increases or decreases in ANC, measured in foliar leaching, represent only a section of the elemental cycle in the forest ecosystem. Exchange of H for base cations on foliar surfaces may reduce the pH of throughfall; however, the reverse process may take place at a different point in the cycle (e.g., root exchange zone).

## *Clearfelling Phase*

Given that the interception of acidic pollutants by the forest canopy is the single most important factor in increasing surface water acidification, the removal of the trees should increase ANC and reduce sulfate levels in run-off. However, for the first few years after felling, the flush of nitrate in many catchments obscures any changes in pH and base cations resulting from a reduction in sulfate and seasalt components in run-off. Most felling studies show an immediate reduction in seasalt components in run-off after felling while reductions in S are more variable (Neal et al. 1992; Harriman, pers. comm.). Hultberg and Grenfelt (1992) found an almost immediate decline of S in run-off from a felled catchment while Reynolds et al. (1992) showed a slow and relatively small sulfate decline, even when $NO_3$ levels decreased back to pre-felling values. Data from severely impacted areas in the Krusne hory mountains of the Czech Republic point to a significant decline in nitrate, sulfate, and base cation levels in drainage streams during the past decade (Cerny 1992). In this instance, the forests in the catchment were removed because of severe damage due to air pollution rather than being systematically felled. The net effect, however, was essentially the same, i.e., removal of the interception mechanism for wet and dry deposition. Renewal of the forest ecosystem by replanting in the 1980s has probably contributed to the continuing nitrate decline, but the subsequent regrowth may eventually reverse this trend as interception processes gradually increase S and N inputs. Clearly, the time scale and amplitude of response to reduced S inputs will vary from catchment to catchment. Factors that influence these responses include soil adsorption/desorption properties and the proportion of strongly bound sulfur that has accumulated in the soil. These responses may also be influenced by the short-term acidification of soils induced by nitrification following harvesting. Any reduction in sulfate run-off after felling may therefore be due both to reduced inputs (removal of canopy interception) and increased sulfate adsorption caused by enhanced soil acidification (Mitchell et al. 1989).

## Organic Acids

Few studies have been published on the accumulation and release of organic substances during the forest cycle, although the factors affecting the abundance and cycling of organic matter are currently receiving greater attention (Schindler et al. 1992). Recent studies associated with the NAPAP program in U.S.A. have attempted to evaluate the role of organic acids in influencing the pH and ANC of surface waters. In general, if organic acids have acid functional groups with $pK_a < 4$, they will behave like strong acids and may reduce run-off pH. The extent of this pH reduction in surface waters and methods of calculating organic anion contributions as a function of dissolved organic carbon (DOC) and pH have been reported in numerous studies (e.g., Driscoll et al. 1989; Oliver et al. 1983). Between pH 4–5, organic anions contribute approximately 5–8 $\mu eql^{-1}$ per mg DOC to the ANC value. Increasing DOC may therefore reduce pH below 5.0 even when ANC is positive.

### *Ploughing/Planting Phase*

In streams draining sensitive upland catchments, the contribution of organic anions usually increases with flow reflecting an increased proportion of run-off through organic surface soils. Yet under circumstances where wetland or bog areas are found in upper regions of the catchment, the flow/DOC relationship may be reversed. Two situations arise during and after ploughing, which may increase or decrease organic anion contributions to surface waters:

1. If water is channelled into more mineral soils via drainage ditches, then DOC levels may decrease due to microbial activity in subsurface soil horizons. DOC generally decreases with increasing soil depth (McDowell and Likens 1988).
2. Alternatively, the drying of surface organic horizons may result in the decomposition of existing organic matter, thus increasing DOC concentrations in run-off.

We have insufficient data to draw any firm conclusions on the relative importance of each mechianism.

*Aggrading Forests*

Litter accumulation, particularly in managed conifer plantations, increases with forest age up to some equilibrium level prior to felling, but the effects of this accumulation on DOC levels are difficult to quantify. Factors such as change in light, temperature, and soil moisture conditions will probably be more important in determining season-to-season and year-to-year variations of DOC in run-off.

*Clearfelling Phase*

Recent data indicate that increased levels of DOC occur during the first few years after felling (Neal et al. 1992; Harriman, unpublished data). On a seasonal basis the changes are variable and relatively small ($< 2$ mg $L^{-1}$); they appear, however, to be correlated with increased levels of organic aluminum.

## Modeling the Effects of Afforestation

In recent years, modeling studies have made major contributions to national research programs and international conferences on acid rain (e.g., NAPAP, SWAP, Glasgow Conference on Acidic Deposition). The MAGIC model, in particular, has been used to assess long-term chemical changes in stream water due to forest growth ( Neal et al. 1986; Cosby et al. 1990). The MAGIC model is optimized using site-specific parameters based on present-day data. The key driving variable is the input of S (and seasalts) which, via exchange, dissolution and adsorption processes, determines the ultimate soil solution and water quality. Most of the processes discussed previously are incorporated in the MAGIC simulations. The central controlling variable is the pool of exchangeable base cations in the soil. Jenkins et al. (1990) used MAGIC to predict that trees growing in an unpolluted region on sensitive soils would slightly reduce ANC in run-off water. By comparison, they predicted that run-off from extensively forested catchments in polluted regions would show a significant reduction in ANC.

These modeling studies again emphasize the need to clarify the role of base cation uptake and leaching in determining rates of surface water acidification and subsequent recovery. The consequences of repeated forest planting and harvesting must be addressed both in terms of sustainable forest production and likely deterioration in water quality. Both field and modeled data should be simultaneously evaluated to provide a continuing validation of the key model components which determine surface water acidification.

## CONCLUSIONS

We have considered a number of potentially acidifying processes that may influence surface water acidification during different stages of the forest cycle. Despite the considerable literature on soil acidification via base cation uptake (capacity effect), we are still uncertain as to whether this process causes extensive surface water acidification in the absence of increases in strong acid anions ($C_A$). This uncertainty stems from the lack of field data that links changes in base cation run-off to forest growth in unpolluted regions.

While other acidifying processes have been identified (i.e., increased $NO_3$ and organic acid leaching at harvesting), we feel that these are relatively small or transient compared with the cumulative effect of interception by the forest canopy. This process produces an increase in acid anion concentrations in soil solution and surface waters (intensity effect), resulting in reduced ANC levels and increased Al leaching.

While assessing the available information on forest effects, we identified several questions and areas of uncertainty:

1. Will ploughing of already acidified catchments be potentially more acidifying than in pristine areas?
2. Will repeated cycles of managed forests ultimately reduce base cation run-off into surface waters? Is enough information available on base cation cycling to be confident about inputs for modeling purposes?
3. What are the chemical and biological characteristics of catchments that determine the quantity and extent of nitrate leaching during and after clearfelling? Can we predict the likely increase in $H^+$ and Al when nitrate levels increase?
4. What can planting, growth, and removal of soft and hardwood forests in impacted and pristine areas, and the consequent effects on water quality tell us about acidification and recovery?

Ultimately the effects of disturbance resulting from forest development will be more accurately quantified, but we believe that managed conifer plantations located in areas of high acid inputs represent the scenario for the greatest reduction in the ANC of surface waters.

## REFERENCES


Battarbee, R.W., R.D. Flower, A.C. Stevenson, and B. Rippey. 1985. Lake acidification in Galloway: A paleoecological test of competing hypotheses. *Nature* **314**:350–352.

Calder, I.R., and M.D. Newson. 1979. Land use and upland water resources in Britain: A strategic look. *Water Resour. Bull.* **15**:1628–1639.

Cantrell, K.J., S.M. Serkiz, and E.M. Perdue. 1990. Evaluation of acid-neutralizing capacity data for solutions containing natural organic acids. *Geochim. Cosmochim. Acta* **54**:1247–1254.

Cerny, J. 1992. Long-term changes of run-off chemistry in an area affected by forest dieback. Proc. of Intl. Conf. Headwater Control II: Environmental Regeneration in Headwaters. Prague.

Cosby, B.J., A. Jenkins, R.C. Ferrier, J.D. Miller, and T.A.B. Walker. 1990. Modeling stream acidification in afforested catchments: Long-term reconstructions of 2 sites in central Scotland. *J Hydrol.* **120**:143–162.

Douglass, J.E., and W.T. Swank. 1972. Streamflow modification through management of eastern forests. Washington, D.C.: U.S. Dept. Agric. Forest Service Res. Paper SE–94, 15 pp.

Driscoll, C.T., R.D. Fuller, and W.D. Schecher. 1989. The role of organic acids in the acidification of surface waters in the eastern U.S. *Water, Air, Soil Pollut.* **43**:21–40.

Fowler, D., J.N. Cape, and M.H. Unsworth. 1989. Deposition of atmospheric pollutants on forests. *Phil. Trans. R. Soc. Lond. B* **324**:247–265.

Hall, R.L., and R.J. Harding. 1993. The water use of the Balquhidder catchments: A process approach. *J. Hydrol.* **145:**285–314.

Hall, R.L., and J. Roberts. 1990. Hydrological aspects of new broadleaf plantations. First annual report on the hydrological impacts of hardwood plantation on lowland Britain. Wallingford, U.K.: Inst. of Hydrology.

Harriman, R. 1988. Acid raid: The scientific background. In: Acid Rain: A Perspective, ed. J. Word and D. Browning, pp. 20–35. London: The Institute of Energy.

Harriman, R., E. Gillespie, D. King, A.W. Watt, A.E.G. Christie, A.A. Cowan, and T. Edwards. 1990. Short-term ionic responses as indicators of hydrochemical processes in the Allt a Mharcaidh catchment, Western Cairngorms, Scotland. *J. Hydrol.* **116**:267–285.

Harriman, R., and B.R.S. Morrison. 1982. Ecology of streams draining forested and nonforested catchments in an area of central Scotland subject to acid precipitation. *Hydrobiologia* **88**:251–263.

Harriman, R., B.R.S. Morrison, L.A. Caines, P. Collen, and A.W. Watt. 1987. Long-term changes in fish populations of acid streams and lochs in Galloway , southwest Scotland. *Water, Air, Soil Pollut.* **32**:89–112.

Hedin, L.O., G.E. Likens, K.M. Postek, and C.T. Driscoll. 1990. A field experiment to test whether organic acids buffer acid deposition. *Nature* **345**:798–800.

Heikurainen, L., K. Kenthamies, and J. Laine. 1978. The environmental effects of forest drainage. *SVO* **29**:49–58.

Henriksen, A. 1988. Critical loads of nitrogen in surface waters. In: Critical Loads for Nitrogen and Sulphur, ed. J. Nilsson and P. Grenfelt, vol 15, pp. 385–412. Skokloster Workshop Nordic Council of Ministers, Report 1988. Stockholm: Nord.

Hoover, M.D. 1944. Effects of the removal of forest vegetation on water yields. *Trans. Amer. Geophys. Union.* **Part 6**:969–977.

Hornbeck, J.W., M.B. Adams, E.S. Corbett, E.S. Verry, and J.A. Lynch. 1992. Long-term impacts of forest treatments on water yield: A summary for northeastern United States. *J. Hydrol.*, in press.

Hultberg, H. 1991. Biochemical cycling of sulphur and its effects on surface water chemistry in an acid catchment in SW Sweden. The first European Symposium on Terrestrial Ecosystems. Florence, Italy. May 1991. Brussels: BT/CEC.

Hultberg, H., and P. Grenfelt. 1986. Gardsjon Project: Lake acidification, chemistry in catchment run-off, Lake liming and microcatchment manipulations. *Water, Air, Soil Pollut.* **30**:31–46.

Hultberg, H., and P. Grenfelt. 1992. Sulphur and sea-salt deposition as reflected by throughfall and run-off chemistry in forested catchments. *Environ. Pollut.* **75**:215–222.

Jenkins, A., B.J. Cosby, R.C. Ferrier, T.A.B. Walker, and J.D. Miller. 1990. Modeling stream acidification in afforested catchments: An assessment of the relative effects of acid deposition and afforestation. *J. Hydrol.* **120**:163–181.

Johnson, C.E., A.H. Johnson, and T.G. Siccama. 1991. Whole-tree clear-cutting effects on exchangeable cations and soil acidity. *Soil. Sci. Soc. Am. J.* **55**:502–508.

Johnson, D.W., and D.W. Cole. 1980. Anion mobility: Relevance to nutrient transport from forest ecosystems. *Environ. Intl.* **3**:79–90.

Johnson, W.W., J.M. Kelly, W.T. Swank, D.W. Cole, H. Van Miegroet, J.W. Hornbeck, R.W. Pierce, and D. Van Lear. 1988. The effects of leaching and whole tree harvesting on cation budgets in several forests. *J. Environ. Qual.* **17,3**:418–424.

Kreiser, A.M., P.G. Appleby, J. Natkanski, B. Rippey, and R.W. Battarbee. 1990. Afforestation and lake acidification: A comparison of 4 sites in Scotland. *Phil. Trans. R. Soc. Lond. B* **327**:377–383.

Krug, E.C., and C.R. Frink. 1983. Acid rain on acid soil: A new perspective. *Science* **221**:520–525.

Likens, G.E. 1992. The Ecosystem Approach: Its Use and Abuse. Oldendorf/Luke, Germany: Ecology Inst., 167 pp.

Likens, G.E., F.H. Borman, L.D. Hedin, C.T. Driscoll, and J.S. Eaton. 1990. Dry deposition of sulphur: A 23-year record for the Hubbard Brook forest ecosystem. *Tellus* **42B**:319–329.

Likens, G.E., F.H. Borman, N.M. Johnson, D.W. Fisher, and R.C. Pierce. 1970. Effects of forest cutting and herbicide treatment on nutrient budgets in the Hubbard Brook watershed. *Ecol. Monogr.* **40**:23–47.

Lindberg, S.E., G.M. Lovett, D.R. Richter, and D.W. Johnson. 1986. Atmospheric deposition and canopy interaction of major ions in a forest. *Science* **231**:141–145.

Martin, C.W., and R.S. Pierce. 1980. Clearcutting patterns affect nitrate and calcium in streams of New Hampshire. *J. Forestry* **78(5)**:268–272.

Martin, C.W., R.S. Pierce, G.E. Likens, and F. Bormann. 1986. Clearcutting effects on stream chemistry in the White Mountains of New Hampshire. U.S.D.A. Forest Service Research Paper NE–579, 12 pp.

McDowell, W.H., and G.E. Likens. 1988. Origin composition and flux of dissolved organic carbon in the Hubbard Brook Valley. *Ecol Monogr.* **58**:177–195.

Miller, J.D., H.A. Anderson, J.M. Cooper, R.C. Ferrier, and M. Stewart. 1991. Evidence for enhanced atmospheric sulphate interception by Sitka spruce from evaluation of some Scottish catchment study data, 1991. *Sci. Total Environ.* **103**:37–46.

Mitchell, M.J., C.T. Driscoll, R.D. Fuller, M.B. David, and G.E. Likens. 1989. Effect of whole tree harvesting on the sulphur dynamics of a forest soil. *Soil Sci. Soc. Am. J.* **53**:933–940.

Neal, C., C.J. Smith, and S. Hill. 1992. Forestry impact on upland water quality. Institute of Hydrology Report 114/S/W. Wallingord, Oxfordshire, U.K.

Neal, C., P. Whitehead, R. Neale, and J. Cosby. 1986. Modeling the effects of acidic deposition and conifer afforestation on stream acidity in the British uplands. *J. Hydrol.* **86**:15–26

Nicoloson, J.A. 1975. Water quality and clearcutting in a boreal forest ecosystem. Can. Hydrol. Symp. 75 Proc. NRCC No. 15195, pp. 734–738.

Nilsson, I.S., H.G. Miller, and J.D. Miller. 1982. Forest growth as a possible cause of soil and water acidification: An examination of the concepts. *Oikos* **39**:40–49.

Oliver, B.G., E.M. Thurman, and R.L. Malcolm. 1983. The contribution of humic substances to the acidity of coloured natural waters. *Geochim. Cosmochim. Acta* **47**:2031–2035.

Reuss, J.W., and D.W. Johnson. 1986. Acid Deposition and Acidification of Soils and Waters. Ecological Series, vol. 59. New York: Springer.

Reynolds, B., P.A. Stevens, J.K. Adamson, S. Hughes, and J.D. Roberts. 1992. Effects of clearfelling on streams and soil water aluminium chemistry in 3 U.K. forests. *Environ Pollut.* **77**, in press.

Robinson, M. 1986. Changes in catchment run-off following drainage and afforestation. *J. Hydrol.* **86**:71–84.

Rosenqvist, I.Th. 1990. Pre-industrial acid-water periods in Norway. In: Surface Water Acidification Program, ed. J.B. Mason, pp. 315–320. Cambridge Univ. Press.

Schindler, D.W., S.E. Bayley, P.J. Curtis, B.R. Parker, M.P Stainton, and C.A. Kelly. 1992. Natural and man-caused factors affecting the abundance and cycling of dissolved organic substances in precambrian shield lakes. *Hydrobiologia* **229**:1–21.

Stevens, P.A., and M. Hornung. 1987. Nitrate leaching from a felled sitka spruce plantation in Beddgelert Forest, north Wales. *Soil Use Manag.* **4**:3–8.

Stevens, P.A., D.A. Norris, T.H. Sparks, and A.L. Hodgson. 1993. Nitrate leaching losses from Sitka spruce forest and moorland in Wales. *Water, Air, Soil Pollut.*, in press.

Stoner, J.H., and A.S Gee. 1985. Effects of forestry on water quality and fish in Welsh rivers and lakes. *J. Inst. Water Eng. Sci.* **39**:27–45.

Stoner, J.H, A.S. Gee, and K.R. Wade. 1984. The effects of acidification on the ecology of streams in the upper Tywi catchment in West Wales. *Environ. Pollut., Series A* **35**:125–157.

Sverdrup, H., and P. Warfvinge. 1990. The role of weathering and forestry in determining the acidity of lakes in Sweden. *Water, Air, Soil Pollut.* **52**:71–78.

Weathers, K.C., G.E. Likens, F.H. Bormann, S.H. Bicknell, B.T. Bormann, B.C. Daube, Jr., J.S. Eaton, J.N. Galloway, W.C. Keene, K.D. Kimball, W.H. McDowell, T.G. Siccama, D. Smiley, and R. Tarrant. 1988. Cloud water chemistry from ten sites in North America. *Environ. Sci. Tech.* **22**:1018–1026.

# 8

# Role of Organic Acids in Acidification of Fresh Waters

H.F. HEMOND
Massachusetts Institute of Technology, Bldg. 48, Room 419,
Cambridge, MA 02139 U.S.A.

## ABSTRACT

Organic acids are important agents of freshwater acidity that must be considered in any assessment of watershed acidification. As little as a few mmol $L^{-1}$ of natural dissolved organic acids (OA) can acidify a dilute water. Extensive surveys during the past decade suggest that a significant fraction, perhaps as much as one-third, of acidic waters in temperate zones are acidic because of OA. Although complexity and variability hinder their precise measurement and characterization, natural OA have a substantially strong acidic character, produce on the order of 5–10 meq $H^+$ $mg^{-1}$ organic carbon, lower the pH of natural waters, and decrease the measured Gran acid-neutralizing capacity (ANC) of water. Organic acids also increase the acid-base buffer capacity of water, rendering the pH less sensitive to changes in mineral acid content. Organic acids also complex metals, alter the light regime, participate in aquatic photochemistry, and serve as a microbial substrate. Wetlands and organic-rich riparian areas are important sources. The degree to which OA production is affected by acid deposition is poorly known. The biochemical pathways of OA production and consumption, as well as their paths of physical transport to fresh waters, remain poorly understood.

## INTRODUCTION

That natural OA can contribute to the acidity of fresh waters has become a widely accepted fact during the last decade. However, the importance of OA relative to other sources of acidity, such as acidic deposition, remains an issue of uncertainty. The quantitative role of OA relative to acids originating from anthropogenic emissions of $SO_2$ and $NO_x$ is of economic and political as well as scientific interest. Temporal trends are also important; it is of particular interest to determine whether natural waters have been *acidified* over the time span of industrialization as a result of natural processes or anthropogenic pollution of the environment.

*Acidification of Freshwater Ecosystems: Implications for the Future*
Edited by C.E.W. Steinberg and R.F. Wright © 1994 John Wiley & Sons Ltd.

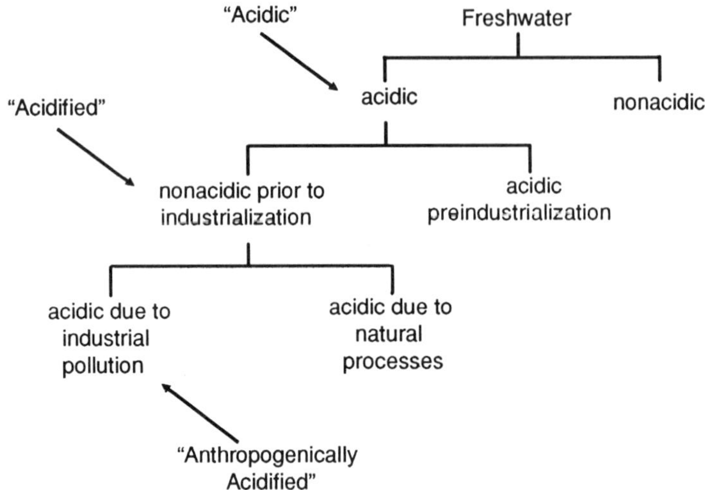


**Figure 8.1** A hierarchy of acidic waters. Those of concern to regulatory authorities are those that have been "anthropogenically acidified." Organic acids may result in naturally acidic lakes and may also influence the course of "anthropogenic acidification."

In this chapter, I define an *acidic* natural water as one whose pH, at a given partial pressure of $CO_2$ ($PCO_2$), is lower than that of pure water equilibrated with the same $PCO_2$. For distilled water in equilibrium with the atmosphere, this corresponds to a pH of about 5.6.

An *acidified* water is a natural water whose pH, titration ANC, or charge balance alkalinity (defined as the equivalent sum of base cations, $C_B$, minus the equivalent sum of strong acid anoins, $C_A$) has decreased over time, due to either anthropogenic disturbance or natural processes. Waters of special concern to air quality regulators are those that are both acidic and have been acidified by anthropogenic acid deposition (Figure 8.1). A better understanding of OA will help to place natural waters into this hierarchy more accurately.

Here I focus on the presently observable acid-base status of waters and will not attempt to determine whether any given *acidic* water has also been *acidified*. Temporal trends of acidity in waters influenced by OA can still only be addressed in a limited manner, such as through paleolimnological records (e.g., Battarbee 1989; Charles et al. 1989) and hindcasting models (Schindler et al. 1989).

The discussion of natural water acidity is complicated by its *episodic* nature. The acidity of many streams varies with season and storm events. Many low pH episodes associated with high streamflows are partly due to OA. Since many studies do not address temporal variability, and several extensive data sets (e.g., the ELS and ALSC in North America; Linthurst et al. 1986; ALSC 1990) are based on one or two samplings in a calendar year, generally during good weather, it will be necessary to temper this discussion with the knowledge that episodic processes are underrepresented in the literature.

## pH VERSUS ANC AS THE MEASURE OF ACIDITY

Both ANC and alkalinity may be better indicators of a water's acidity than pH. An acidic natural water has a Gran ANC less than zero. ANC is a more stable parameter than pH (invariant, for example, with changes in dissolved $CO_2$). Since ANC varies linearly with strong acid addition, it provides a more sensitive measure of acidification. If ANC is known and a suitable constraint on $CO_2$ content is invoked, pH can be calculated (e.g., Stumm and Morgan 1970). Unfortunately, this is not the case in soft waters containing significant amounts of natural OA (Hemond 1990a).

On a time scale of minutes to hours, pH may vary (especially in higher pH systems and in productive but poorly buffered waters) as $CO_2$ is produced, consumed, or lost from the system. Equally important, two waters of equal measured pH or equal Gran ANC but different OA content may respond very differently to a perturbation, such as the addition of a mineral acid, since natural OA substantially increase the buffer capacity.

Nevertheless, pH is the most important single measure of acidity, and buffer capacity must also be considered in weighing the ecological consequences. At least three of the four quantities—pH, $PCO_2$ or total carbonate content ($C_T$), alkalinity (defined as $C_B$–$C_A$), and OA concentration—must be known to specify the acid-base status of a water. Aluminum must be specified in cases where it is important to the ionic charge balance.

## NATURE AND DISTRIBUTION OF ORGANIC ACIDS

Natural OA in fresh waters originate from the degradation of biomass in the upland catchment, wetlands, near-stream riparian zones, the water column, and sediments. The terrestrial sources dominate in streams and lakes that have a short hydraulic residence time. Although soil porewaters in upper soil horizons often contain very high (order of 100 mg C $L^{-1}$) levels of OA (Cronan 1990), most of this OA does not reach surface waters but is sorbed onto mineral soil horizons and is ultimately biodegraded *in situ*. I have argued that wetlands and organic-rich riparian zones contribute most of the OA found in small streams and lakes of temperate zones (Hemond 1990a). At times of high water tables, a rapid lateral transport of DOC-rich water from upper soil horizons to surface water bodies can be an important source. In-lake sources are likely to dominate in very small basins or catchments lacking soils and vegetation (e.g., some alpine and arctic settings; Baron et al. 1991).

Because natural OA constitute a complex mixture, they are characterized in terms of operationally chosen, observable properties rather than chemical structure. From an acidification standpoint, the OA of most interest are those having low to modest chemical and biological reactivity; highly reactive OA, such as fatty acids, are rapidly consumed by the microbial community. Glaze et al. (1990) presented a conceptualization of the relationship between reactivity and concentration of OA in fresh water,

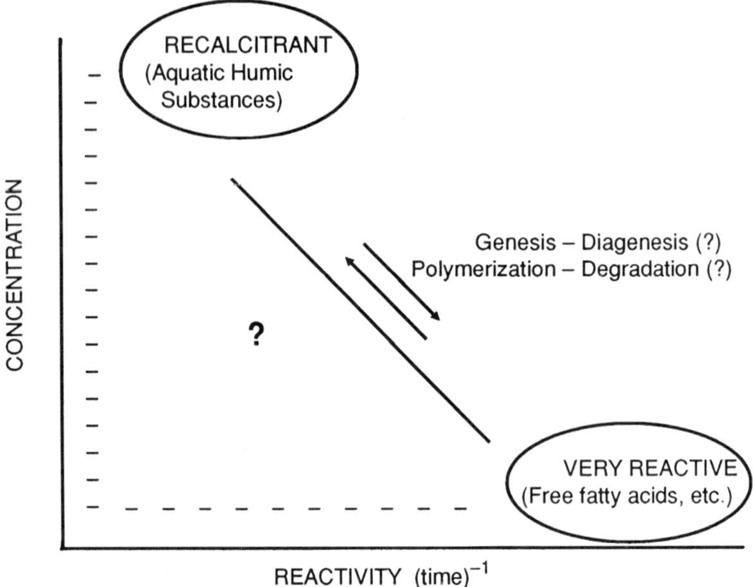


**Figure 8.2** Conceptual diagram presented in Glaze et al. 1990, as suggested by Maybeck. Recalcitrant organic acids dominate in most waters by virtue of their longer lifetimes.

with the largest fraction of the OA pool represented by "recalcitrant" OA (mostly aquatic humic substances; see Figure 8.2).

Aquatic humic substances are not usually defined in terms of reactivity but rather in terms of operational separation techniques, such as extractability from soils, solubility at various pHs, or retention on certain synthetic resins (Gjessing 1976; Thurman 1985). Nonetheless, natural OA can be sufficiently biologically and/or photochemically reactive to undergo significant chemical evolution in waters of sufficiently long residence time, as has been observed by McKnight et al. (1988) in post-eruption Spirit Lake and by Clair et al. (1989) in microcosm experiments. Photochemical and biological processes act ultimately as sinks for OA; in addition, flocculation with metals can remove OA from natural waters (Weilenman et al. 1989) and may increase in importance in acidic lakes.

## ACIDITY OF ORGANIC ACIDS

Historically, confusion about the acidity of natural OA has resulted from (a) attempts to describe their acid-base behavior with a single pKa (as though they were a single low-molecular-weight compound rather than a complex mixture, with multiple acidic functionalities on some molecules), and (b) the assumption that OA are weak acids, whereas a portion of the acidity is actually quite strong, with some ionization occurring at pHs as low as 3 or even lower (e.g., Brakke et al. 1987).

Models capable of representing the acid-base chemistry of OA fall typically into one of three categories:

1. purely empirical models, in which an equation fitted to acidimetric titration data is employed to represent $H^+$ released by the OA as a function of pH;
2. models in which the chemical mass action relationship is applied to a finite set or a continuous spectrum of simple (often but not necessarily monoprotic) acids, each having a different pKa and representing a certain proportion of the total concentration;
3. "polymer" models, in which the observed range in the apparent pKa is modeled as being due, at least in part, to electrostatic effects. Acidic groups on polymeric humic acid molecules ionize, leaving the molecules with negative charges, which affect the apparent pKa of other acidic groups on the molecules. Both site heterogeneity and electrostatic effects can be combined into a single model (e.g., Tipping et al. 1988).

Perdue (1990) provides a review of many existing models. Although they differ in concept and purposes, and attempt widely varying degrees of realism at the molecular level, any of these types of models is capable of reasonably simulating the contribution of a given OA mixture to the acidity of water. A complication, however, is that qualitative differences in OA occur from location to location, as well as over time at fixed locations (e.g., Kramer et al. 1990). Ideally, therefore, the model would have somewhat different parameters for each different water sample. Acidimetric titration data from which to extract parameters are not available for many waters, thus precluding a rigorous site-specific estimate of the OA contribution to buffer capacity. The contribution that OA make to $H^+$ concentration can be obtained without reference to the nature of the OA mixture from the "anion deficit," which is the sum of measured cation concentrations minus measured anion concentrations, usually expressed in μeq $L^{-1}$. Because the anion deficit is often a small difference between two larger quantities, its determination requires careful analytical chemistry, and it is subject to significant statistical error, especially in waters where OA concentrations are low and ionic strength is high. Another source of error can arise if a significant fraction of the metal ions (e.g., $Ca^{+2}$) is organically complexed, causing their contribution to charge to be overestimated by atomic absorption measurements. The contribution of OA to the $H^+$ concentration may also be estimated from dissolved organic carbon (DOC) or total organic carbon (TOC) data, but this assumes an acid-base model for OA.

Despite these limitations, a clear pattern has now emerged in the literature, showing important contributions of OA to acidity in many fresh waters. Typically, "anion deficit" is taken as a direct measure of organic anion concentration, although in some cases conclusions about the $H^+$ contributed from OA are based on DOC/TOC data in

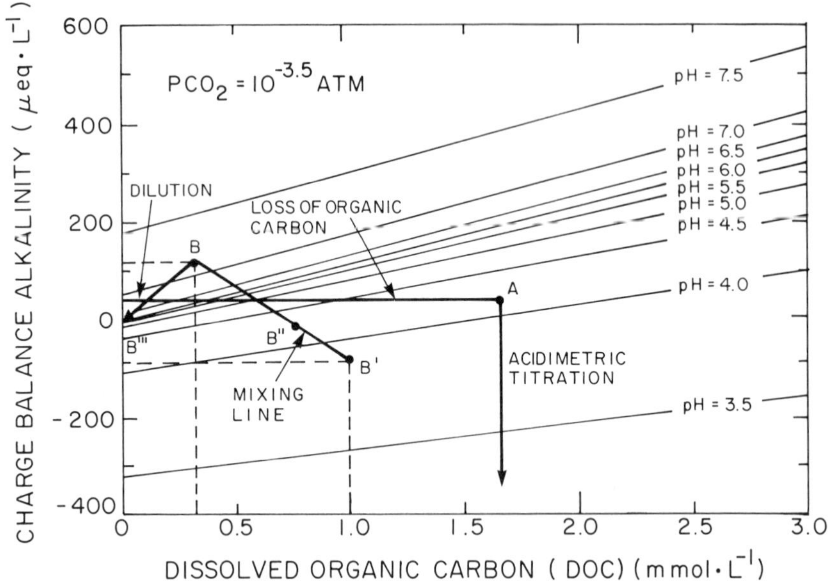


**Figure 8.3** CBALK-pH-DOC plot for fixed $PCO_2$, or $C_T$, provides graphical solutions to a variety of chemical problems in waters containing natural organic acids.

conjunction with models for organic acidity. Quantitative analyses of the contribution of OA to *buffer capacity* of natural waters are few. Some rough estimates can be made using a simple empirical representation of OA ionization (Oliver 1983) and the pH-OA/DOC-charge balance alkalinity nomograph approach of Hemond (1990a; Figure 8.3). Similar results would be obtained using, for example, the empirical nomographs produced by Munson et al. (1990), based on a statistical analysis of data from over 1,400 Adirondack lakes.

## Observations of OA in Fresh Waters

Although the dominating role of OA in peatlands has long been hypothesized, and later demonstrated by Hemond (1980) and Gorham et al. (1985), OA evidently were not widely thought to be especially important in the acidity of surface waters, such as streams and lakes, until quite recently. Studies of watershed acidity during the past decade in Europe and North America have now revealed a significant role for OA in surface water acidity. For example, Wilkinson et al. (1992) report an average 70–90 $\mu eq\ L^{-1}$ of organic anion concentration ($Org^-$) in non-bog Canadian Shield Lakes studies (although mean DOC was reported as varying from only 0.4–0.5 mmol $L^{-1}$, less than would be expected in view of an anion deficit of this magnitude). Campbell et al. (1992) report that episodic acidification of three Quebec North Shore rivers is due primarily to OA, with average "anion deficits" rising from around 35 μeq $L^{-1}$ to the range of 70–100 μeq $L^{-1}$ during high flows at snowmelt (causing pH drops from about 7 to about 5). Kerekes et al. (1986) report that OA are responsible for the acidity of a substantial portion of lakes surveyed in Nova Scotia.

In the U.S., NAPAP (1990) concluded that 22% of all acidic lakes (all larger than 4 ha in area) and 27% of all acidic streams surveyed were dominated by OA. The organically acidic waters were commonly associated with organic soils or wetlands in their watersheds, and had a DOC level > 0.8 mmol $L^{-1}$ (Baker et al. 1991). An even larger fraction of acidic lakes sampled in the Adirondacks region of New York State was found to be organically acidic: more than 38% of the lakes had a pH < 5 due to OA (ALSC 1990). The ALSC survey included lakes as small as 0.5 ha in surface area, which may account for the larger average organic influence. Thus an average charge density was about 5 μeq $mg^{-1}$ organic carbon, and pH depression due to OA was 0.5 to 2.5 pH units. Again, a positive relationship between OA concentrations and wetlands within the watersheds was observed. Organically acidified lakes in Florida have been documented by Pollman and Canfield (1986).

In Europe, numerous instances of organic acidity are now reported in the literature. For example, in Finland, Kortelainen and Mannio (1988) reported that TOC is a better predictor of pH than is nonmarine $SO_4^{-2}$ concentration in rivers draining to the Baltic; mean measured TOC exceeded 0.8 mmol $L^{-1}$, and 70% of the variance in TOC could be explained by watershed area and extent of peatlands in the drainage area. Kämäri et al. (1991) reported organic anions ($Org^-$) to be the major anion in small lakes in southern Finland; 91% of lakes studied had TOC > 0.4 mmol $L^{-1}$, with a median of 1.0 mmol $L^{-1}$.

At the Svartberget catchment in northern Sweden, OA resulted in episodic depressions of stream pH from 7.5 to 4 during periods of high flow (Bishop 1991). A study of 90 headwater streams in the area showed that in the most acidic quartile there was a negative correlation between sulfate and acidity, but a positive correlation between TOC and acidity (Bishop 1991).

In the "Reversing Acidification in Norway" (RAIN) experiment, where atmospheric acid inputs were excluded from a small catchment at Risdalsheia, the resultant decline of sulfate in runoff was partially offset by increases in ionized OA, with little concomitant change in pH (Wright et al. 1988).

In Scotland, most of the acidic lakes north of Great Glen fault are reported to be "naturally acidified," presumably by OA (Acidification in Scotland 1988). In Lake Huzenbach, a small dystrophic (2 ha) cirque lake in the Black Forest region of Germany, Thies (1990) reported anion deficits as high as 223 μeq $L^{-1}$, which is sufficient in most analyses to more than account for measured $H^+$ concentration.

Moore (1989) discussed DOC patterns in 8 catchments in Maimai, New Zealand. Here, levels of DOC ranged from 0.4–1.0 mmol $L^{-1}$, peaking during storms. Based on flowpath studies, organic debris in the channel appeared to be an important DOC source. Although acid-base chemistry was not reported, it seems likely that OA are important.

Not all studies conclude that OA are important to the acidity of natural waters; for example, Henriksen et al. (1988) reported the influence of OA to be very small in high-elevation lakes in southern Norway. Nevertheless, in general, organic acidity has a major influence on the acid-base status of fresh waters. However, it is not known what percentage of natural waters are now acidic due to OA. Confounding factors include the choice of waters sampled (e.g., what minimum lake size and stream order are included in the analysis), the degree to which episodic events are considered, and analytical uncertainties. The NAPAP percentages of organically acidic waters are probably of the right general magnitude, at least for small temperate zone lakes and headwater streams in the United States.

## THE SIGNIFICANCE OF ORGANIC ACIDIFICATION

### Buffer Capacity

Waters that are acidic due to the presence of OA have higher buffer capacities than do waters of equal pH constituted from strong mineral acids, such as $HNO_3$ and $H_2SO_4$. An example is shown in Figure 8.4, derived using the nomograph of Figure 8.3. Two waters are considered, both at pH 5.5. One contains no OA, whereas the other contains 0.5 mmol $L^{-1}$ of DOC. To move from pH 5.5 to pH 5.0, the water containing no OA requires the addition of 8 μeq $L^{-1}$ of strong acid, while the OA water requires over 60% more 13 μeq $L^{-1}$ strong acid to cause the same pH change!

The enhanced buffer capacity of the OA waters means that, assuming OA concentrations do not change, their pH will not be as sensitive as the non-OA waters to either increased or to decreased concentrations of strong mineral acid. Therefore, while the OA waters would be expected to undergo less additional acidification from *increased* atmospheric deposition than would the waters without OA, they would also respond to a lesser degree to *decreases* in atmospheric acidic deposition, or to remedial efforts such as liming.

### Sulfate/OA Substitution

Strong acid anions, such as sulfate, may replace $Org^-$ under increasing atmospheric loading of acid (Krug and Frink 1983). This could result from changes in the quantity or the quality (e.g., the pKa distribution) of OA produced within an ecosystem. The net effect would be an additional "buffering" of strong acid input.

Several hypotheses, generally relating to microbial OA production or to OA solubility, have been proposed in support of such a substitution process. The data are too sparse, and the temporal and spatial variability of OA concentrations too great to allow general conclusions to be drawn. The artificially acidified half of Little Rock Lake in Wisconsin, U.S.A., lost 0.1 mmol $L^{-1}$ of DOC following acidification; DOC loss on acidification was also observed in lake 302 at the Experimental Lakes Area

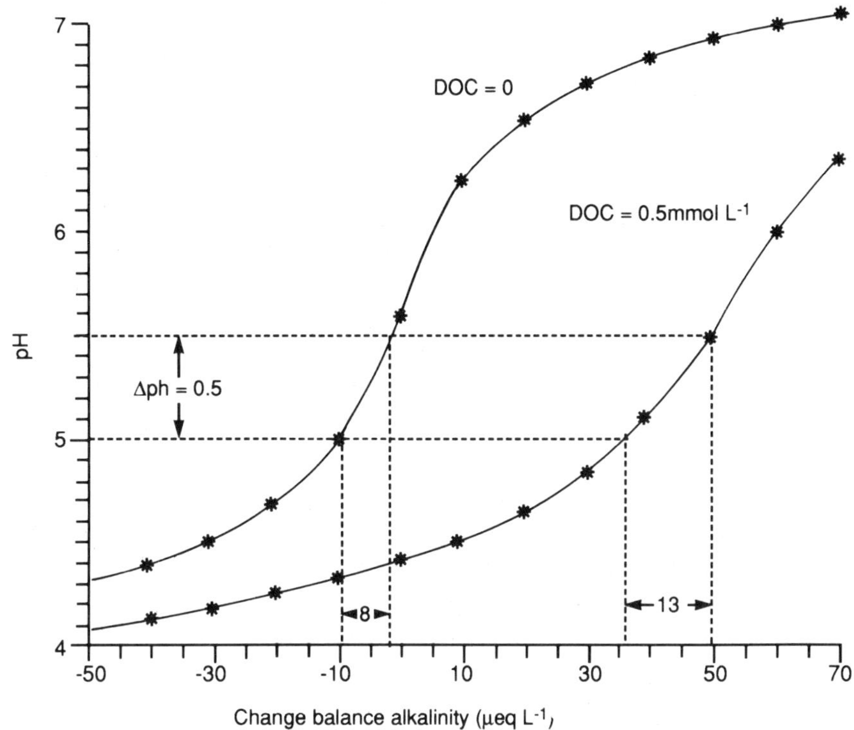


**Figure 8.4** Titration curves of two waters: one without organic acid, the other containing 72 mmol $L^{-1}$ of organic acid conforming to the Oliver (1983) model. Of particular note is the large difference in pH sensitivity to strong acid or base additions (the horizontal axis ANC = $C_B$–$C_A$).

(ELA), in northwest Ontario, Canada. However, lake 223 at the ELA showed no change in DOC upon acidification (Schindler et al. 1991). Hedin et al. (1990) reported no DOC response to $H_2SO_4$ amendments at Hubbard Brook in New Hampshire, U.S.A. Lowered mineral acid inputs at Risdalsheia catchment in southern Norway were associated with a rising trend in $Org^-$ concentrations (Wright et al. 1988).

A second, indirect link between OA and mineral acids may exist via wetland/peatland processes. Wetlands provide areas in which sulfate reduction and possible subsequent long-term sulfur storage can occur (Hemond 1990b). Thus, increased atmospheric sulfate deposition into a system dominated by high OA levels may not translate into a commensurate rise in sulfate concentration in the surface water. The acidification of colored waters in southern Sweden demonstrates, however, that sufficient atmospheric inputs of mineral acid can overwhelm natural organic buffering systems.

### Other Ecosystem Effects

In addition to affecting acidity, OA can alter the bioavailability and toxicity of metals. The most prominent role in this regard is the complexation of aluminum (e.g., Driscoll et al. 1980); inorganic Al is toxic to biota, especially when calcium concentrations are low. Complexation of aluminum by OA seems to decrease its toxicity to fish, although complexation may render aluminum more toxic to some invertebrates. OA, because of their finite chemical and/or biological reactivity, can also constitute a significant oxygen sink in some waters (e.g., dystrophic lakes). Because of their light absorptivity, they may alter light (and heat) distributions in lakes. They also serve as chromophores, capturing light energy and coupling it to subsequent chemical reactions.

## PRIORITIES FOR STUDY

The key questions about organic acidification concern the production, transport, and fate of OA in ecosystems. Although their chemical properties are only imperfectly described, and much work remains to be done, existing empirical models seem adequate for the estimation of OA contributions to past, present, and future acidity, *if* the concentration of OA in the water of interest is known.

At present, it is not known how the quantity and quality of OA changes with landscape evolution, land-use change, or changes in anthropogenic acidic deposition. Answers to such questions are needed, especially for determining critical acid deposition loads in areas where natural DOC levels are generally high in fresh waters. The ability to assess change is needed to progress from an understanding of the OA role in acidity to an understanding of the OA role in *acidification.* A framework for assessing OA change would seem to require, as a prerequisite, a better understanding of where, and by what specific processes and pathways, the OA in fresh waters are formed and transported. This then can provide a basis for prediction of change in each process, and the resulting OA, in response to altered atmospheric inputs of $NO_3^-$ or $SO_4^{-2}$, to changes in forestry or agricultural operations, to successional processes, or to changes in climate.

## ACKNOWLEDGEMENT

I thank D. Schindler, K. Bishop, T. Sullivan, and D. Brakke for constructive reviews of earlier drafts of this paper, and the support of USGS (Grant No. BCS–8906032), and the U.S. Department of Energy (Grant No. DE–FG02–29ER30196). This manuscript has not been subjected to agency review, and no official endorsement is implied.

## REFERENCES

Acidification in Scotland. 1988. Proc. Symp. Scottish Development Department. Edinburgh, U.K., November 1988.

ALSC. 1990. Adirondack Lake Survey: An Interpretive Analysis of Fish Communities and Water Chemistry, 1984–87. J.P. Baker and S.A. Gherini (coordinators), Adirondack Lakes Survey Corporation Interpretive Report, Ray Brook, NY.

Baker, L.A., A.T. Herliny, P.R. Kaufmann, and J.M. Eilers. 1991. Acidic lakes and streams in the United States: The role of acidic deposition. *Science* **252**:1151–1154.

Baron, J., D. McKnight, and A.S. Denning. 1991. Sources of dissolved and particulate organic material in Loch Vale watershed, Rocky Mountain National Park, Colorado, U.S.A. *Biogeochem.* **15(2)**:89–110.

Battarbee, R.W., A.C. Stevenson, B. Rippey, C. Fletcher, J. Natkanski, M. Wik, and R.J. Flower. 1989. Causes of lake acidification in Galloway, south-west Scotland: A paleological evaluation of the relative roles of atmospheric contamination and catchment change for two acidified sites with non-afforested catchments. *J. Ecol.* **77**:651–672.

Bishop, K., 1991. Episodic increases in stream acidity, catchment flow pathways, and hydrograph separation. Ph. D. Thesis. Cambridge, MA: Jesus College.

Brakke, D.F., A. Henriksen, and S.A. Norton. 1987. The relative importance of acidity sources for humic lakes in Norway. *Nature* **329**:432–434.

Campbell, P.G.C., H.J. Hansen, B. Dubreuil, and W.O. Nelson. 1992. Geochemistry of Quebec North Shore salmon rivers during snowmelt: Organic acid pulse and aluminum mobilization. *Can. J. Fish. Aquat. Sci.*, in press.

Charles, D.F., R.W. Battarbee, I. Renberg, and J.P. Smol. 1989. Paleolimnological analysis of lake acidification trends in North America and Europe using diatoms and chrysophytes. In: Acid Precipitation, ed. D.C. Adriano and M. Havas, pp. 207–276. New York: Springer.

Clair, T.A., F. Barlocher, P. Brassard, and J.R. Kramer. 1989. Chemical and microbial diagenesis of humic matter in fresh waters. *Water, Air, Soil Pollut.* **46**:205–211.

Cronan, C.S. 1990. Patterns of organic acids transport from forested watersheds to aquatic ecosystems. In: Organic Acids in Aquatic Ecosystems, ed. E.M. Perdue and E.T. Gjessing, pp. 245–260. Dahlem Workshop Report LS 48. Chichester: Wiley.

Driscoll, C.T., J.P. Baker, J.J. Bisogni, and C.L. Schefield. 1980. Aluminum speciation and its effect on fish in dilute acidified waters. *Nature* **284**:161–164.

Gjessing, E.T. 1976. Physical and Chemical Characteristics of Aquatic Humus. Ann Arbor: Ann Arbor Science Publ., 120 pp.

Glaze, W.H., et al. 1990. What is the composition of organic acids and how are they characterized? In: Organic Acids in Aquatic Ecosystems, ed. E.M. Perdue and E.T. Gjessing, pp. 75–95. Dahlem Workshop Report LS 48. Chichester: Wiley.

Gorham, E., J. Ford, M. Santelman, and S. Eisenreich. 1985. Chemistry of bog waters. In: Chemistry of Lakes, ed. W. Stumm, pp. 339–363. New York: Springer.

Hedin, L.O., G.E. Likens, K.M. Postek, and C.T. Driscoll. 1990. A field experiment to test whether organic acids buffer acid deposition. *Nature* **345**:798–800.

Hemond, H.F. 1980. Biogeochemistry of Thoreau's Bog, Concord, Massachusetts. *Ecol. Monogr.* **50(4)**:507–526.

Hemond, H.F. 1990a. Acid-neutralizing capacity, alkalinity, and acid-base status of natural waters containing organic acids. *Environ. Sci. Technol.* **24(10)**:1486–1489.

Hemond, H.F. 1990b. Wetlands as the source of dissolved organic carbon to surface waters. In: Organic Acids in Aquatic Ecosystems, ed. E.M. Perdue and E.T. Gjessing, pp. 301–313. Dahlem Workshop Report LS 48. Chichester: Wiley.

Henriksen, A., D.F. Brakke, and S.A. Norton. 1988. Total organic carbon concentrations in acidic lakes in Southern Norway. *Environ. Sci. Technol.* **22**:1103–1105.

Kämäri, J., M. Forsius, P. Kortelainen, J. Mannio, and M. Verta. 1991. Finnish lake survey: Present status of acidification. *Ambio* **20**:23–27.

Kerekes, J., S. Beauchamp, T. Tordon, and T. Pollock. 1986. Sources of sulphate and acidity in wetlands and lakes in Nova Scotia. *Water, Air, Soil Pollut.* **31**:207–314.

Kortelainen, P., and J. Mannio. 1988. Natural and anthropogenic acidity sources for Finnish lakes. *Water, Air, Soil Pollut.* **42**:341–352.

Kramer, J.R., P. Brassard, P. Collins, T.A. Clair, and P. Takats. 1990. Variability of organic acids in watersheds. In: Organic Acids in Aquatic Ecosystems, ed. E.M. Perdue and E.T. Gjessing, pp. 127–139. Dahlem Workshop Report LS 48. Chichester: Wiley.

Krug, E.C., and C.R. Frink. 1983. Acid rain on acid soil: A new perspective. *Science* **221**:520–525.

Linthurst, R.A., D.H. Landers, J.M. Eilers, D.F. Brakke, W.S. Overton, E.P. Meier, and R.E. Crowe. 1986. Characteristics of lakes in the eastern United States, vol. 1: Population descriptions and physicochemical relationships. EPA 1600/4–86/007a. U.S. Environmental Protection Agency, 275 pp.

McKnight, D.M., K.A. Thorn, and R.L. Wershaw. 1988. Rapid changes in dissolved humic substances in Spirit Lake and South Fork Castle Lake, Washington. *Limnol. Oceanogr.* **33(6)**:1527–1541.

Moore, T.R. 1989. Dynamics of dissolved organic carbon in forested and disturbed catchments, Westland, New Zealand. 1. Maimai. *Water Resour. Res.* **25(6)**:1321–1330.

Munson, R.K., S.A. Gherini, C.T. Driscoll, R.M. Newton, and C.L. Schofield. 1990. Interpretive analysis of ALSC physical-chemical data. NAPAP 1990 Intl. Conf. on Acidic Deposition: State of Science and Technology.

NAPAP (U.S. National Acid Precipitation Assessment Program). 1990. 1990 Integrated Assessment Report (November 1991). Washington, D.C.: NAPAP.

Oliver, B.G., E.M. Thurman, and R.L. Malcolm. 1983. The contribution of humic substances to the acidity of colored natural waters. *Geochim. Cosmochim. Acta* **47**:2031–2035.

Perdue, E.M., 1990. Modeling the acid-base chemistry of organic acids in laboratory experimental and fresh waters. In: Organic Acids in Aquatic Ecosystems, ed. E.M. Perdue and E.T. Gjessing, pp. 111–126. Dahlem Workshop Report LS 48. Chichester: Wiley.

Pollman, C.D., and D.E. Canfield. 1986. Florida: A case study of hydrologic and biogeochemical controls on seepage lake chemistry. In: Acid Deposition and Aquatic Ecosystems, ed. D.F. Charles. New York: Springer.

Schindler, D.W., S.E.M. Kasian, and R.H. Hesslein. 1989. Biological impoverishment in lakes of the midwestern and northeastern United States from acid rain. *Environ. Sci. Technol.* **23(5)**:578–580.

Schindler, D.W., T.M. Frost, K.H. Mills, P.S.S. Chang, I.J. Davies, L. Findlay, D.F. Malley, J.A. Shearer, M.A. Turner, P.J. Garrison, C.J. Watras, K. Webster, J.M. Gunn, P.L. Brezonik, and W.A. Swenson. 1991. Comparisons between atmospherically-acidified lakes during stress and recovery. *Proc. Roy. Soc. Edinburgh* **97**:193–226.

Stumm, W., and J.J. Morgan. 1970. Aquatic Chemistry. New York: Wiley.

Thies, H. 1990. Acidification studies at northern Black Forest cirque lakes. In: Hydrology in Mountainous Regions. I. Hydrological Measurements; The Water Cycle. IAHS Publ. No. 193, pp. 511–515.

Thurman, E.M. 1985. Organic Geochemistry of Natural Waters. Dordrecht: Nijhoff/Junk.

Tipping, E., C.A. Backes, and M.A. Hurley. 1988. The complexation of protons, aluminium and calcium by aquatic humic substances: A model incorporating binding-site heterogeneity and macroionic effects. *Water Res.* **22(4)**:597–611.

Weilenman, N., C.R. O'Melia, and W. Stumm. 1989. Particle transport in lakes: Models and measurements. *Limnol. Oceanogr.* **34**:1–18.

Wilkinson, K.J., H.G. Jones, P.G.C. Campbell, and M. Lachance. 1992. Estimating organic acid contributions to surface water acidity in Quebec (Canada). *Water, Air, Soil Pollut.* **61**:57–74.

Wright, R.F., E. Lotse, and A. Semb. 1988. Reversibility of acidification shown by whole-catchment experiments. *Nature* **334**:670–675.

Standing, left to right:
Don Charles, Heinz Zoettl, Bridget Emmett, Alvaro Ovalle, Ron Harriman, Harry Hemond

Seated, left to right:
Karl-Heinz Feger, Dieter Leßmann, Helga Van Miegroet, Hans Hultberg

# 9

# Group Report: Can We Differentiate between Natural and Anthropogenic Acidification?

B. EMMETT, Rapporteur

D. CHARLES, K.H. FEGER, R. HARRIMAN,
H.F. HEMOND, H. HULTBERG, D. LEßMANN,
A. OVALLE, H. VAN MIEGROET, H.W. ZOETTL

## INTRODUCTION

Differentiation between anthropogenic and natural factors in the acidification of fresh waters is difficult since these two factors frequently co-occur. Anthropogenic factors include changes in chemical inputs, the physical structure of a system, and watershed characteristics (e.g., the composition of terrestrial vegetation), and thus may include land-use change, increases in acidic deposition, and global climate change. Anthropogenic acidification is often characterized by sulfate-dominated waters that are frequently rich in inorganic aluminum, which is known to have adverse effects on freshwater biota. Naturally acidic waters are generally associated with high concentrations of dissolved organic acids where specialized biological communities have developed, although acidification due to the natural release of mineral acids may also occur in small geographical regions, e.g., where acidic volcanic lakes are common (Yoshimura 1933), in the spontaneous burning of pyrite shales (Havas and Hutchinson 1983), and in fresh waters draining soils that contain sulfur-bearing minerals (Paces 1985; Cook et al. 1993).

It is important to differentiate between anthropogenic and natural acidification. If the cause is anthropogenic, changes to reduce deposition or to mitigate effects using land-management practices may be required. If water is found to be naturally acidic, it may be desirable not to interfere. For example, if liming is carried out to restore

*Acidification of Freshwater Ecosystems: Implications for the Future*
Edited by C.E.W. Steinberg and R.F. Wright © 1994 John Wiley & Sons Ltd.

natural biological communities, but the higher pH is not maintained, species may be lost and the community structure and productivity reduced upon return to acidic conditions if liming is not continued. More importantly, it is desirable to quantify the importance of atmospheric inputs to acidification, where both types of acidification are occurring, to quantify the cost of remedial measures correctly.

Various techniques have been employed to separate anthropogenic and natural factors. These include paleoecological studies, the use of long-term chemical records, manipulation experiments, and simulation modeling. The review of these studies in the various background papers and, in particular, the paleoecological research provides evidence that it is possible to differentiate between natural and anthropogenic acidification (e.g., Battarbee and Charles, this volume). Discussion was therefore directed towards differentiating between land-use change and acidic deposition and, in particular, the relative importance of forest management versus acid deposition on freshwater acidification (Harriman et al., this volume). During these discussions, various research needs were identified relating to processes involved in acidification. In this chapter, the background to these discussions and research needs are described in detail. We also briefly addressed (a) the use of biota to distinguish between natural and anthropogenic acidification, (b) the impact on global warming on both natural and anthropogenically derived acidification, and (c) how to transfer knowledge from temperate studies to tropical zones. Finally, the group provided some ideas on how methods for differentiation between natural and anthropogenic acidification can be improved.

### Definitions of Terms

In several of the background papers, slightly differing definitions of "acidification" and "indicators of acidification" were used (e.g., Harriman et al., Hemond, and Van Miegroet, all this volume). For discussion purposes, we agreed to define acidification of fresh waters as an increase in hydrogen ion activity, whether caused by anthropogenic factors or natural factors. We also recognized that a reduction in the acid-neutralizing capacity (ANC) is a widely accepted definition of acidification. For a discussion on the use of these two definitions see Hemond (this volume) and Van Miegroet (this volume).

## RATES OF ACIDIFICATION AND THE ROLE OF NATURAL AND ANTHROPOGENIC FACTORS

The rate of acidification has increased dramatically over the last 150 years in lakes located in the base-poor glaciated region of Europe and North America. Such waters are generally poorly buffered and have a low ionic strength (Battarbee and Charles, this volume). Prior to the industrial revolution, acidification occurred at a slow rate,

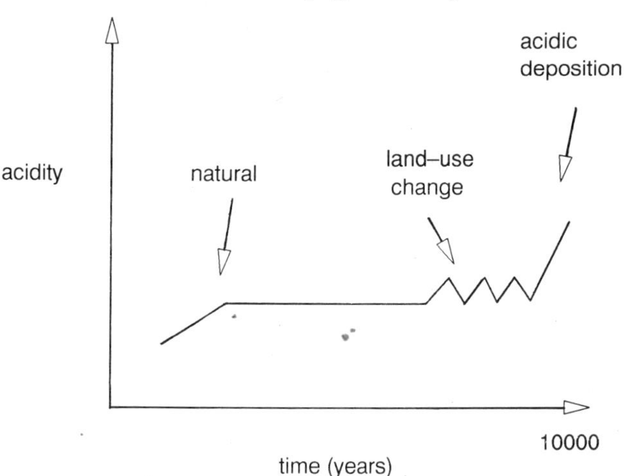


**Figure 9.1** Generalized trend in acidity of currently clearwater, low pH glacially formed lakes in the north temperate zone, from a region receiving acid deposition and influenced by anthropogenic land-use change (based on paleolimnological data).

with acidity fluctuating in response to human activity, such as forest removal and burning, etc. Increased sulfur deposition since industrialization, approximately 150 years ago, coincided with the rapid acidification inferred from paleoecological records (Figure 9.1). Changes in land use, however, may also have occurred during this period and must, therefore, be considered to have had an influence on the historical record (Battarbee and Charles, this volume; Feger, this volume).

The magnitude of the gradual pH change due to natural factors depends on the bedrock and resistance to weathering and production of organic matter. A decline of 0.5 to 1.0 pH unit can take hundreds to thousands of years (Battarbee and Charles, this volume). This natural acidification is primarily a consequence of the loss of base cations during soil formation and the development of an organic horizon (Feger, this volume). In particular, an increase in the production of organic-rich wetlands can modify the rate of pH change. The gradual decline in pH over many centuries contrasts sharply with the relatively rapid increase in acidity over the last 150 years caused by increased atmospheric deposition of sulfur. This has resulted in large-scale increases in freshwater sulfate levels without an equivalent increase in base cation concentrations, causing acidification in sensitive streams and lakes in North and Central Europe and North America. Although interest has turned to the importance of nitrate in the acidification of fresh waters more recently, sulfate deposition remains the most important factor in the rapid acidification of fresh waters since industrialization.

## WHAT IS THE RELATIVE IMPORTANCE OF FOREST MANAGEMENT PRACTICES VERSUS ATMOSPHERIC DEPOSITION IN THE ACIDIFICATION OF SURFACE WATERS?

Land-use change includes a wide range of activities, e.g., forest management, wetland drainage, urbanization, mining, agricultural conversion, etc. Changes in land use have the potential to change the acidity of soils and waters and have been suggested to be the primary cause of acidification in some regions. Mechanisms include (a) removal of base cations in biomass harvesting, (b) the alteration of water pathways, particularly the formation and destruction of macropores, thereby causing changes in soil water contact time, (c) the release of base cations following burning, (d) changes in weathering rates, (e) changes in mineralization of organic material and litter material, and (f) the increased capture of pollutants due to changes in vegetation type. The importance of these factors for surface water acidification is discussed by Feger (this volume) and Harriman et al. (this volume). We focused on the interaction of forest management and acid deposition on acidification of fresh waters.

There is much debate on the importance of forest management in the acidification of fresh waters in the absence of acid deposition (see Battarbee and Charles, this volume). According to some theoretical analyses and field observations (e.g., Feger, this volume), forest management could significantly reduce ANC of the soil through uptake and removal of base cations. However, there is no paleoecological or other evidence to suggest that forest management alone has caused a long-term decrease in lake water pH (Battarbee and Charles, this volume). For example, the invasion by spruce, in the absence of acidic deposition, does not result in the acidification of lake water (e.g., Renberg et al. 1990).

There is evidence, however, that management can change the sensitivity of a site to acid deposition. For example, changes in the species composition can alter efficiency of capture in acid deposition through impaction (e.g., afforestation with conifers in the U.K.; Harriman et al., this volume), and silvicultural practices such as removal of biomass over a prolonged period in central Europe can increase site sensitivity (Feger, this volume; Harriman et al., this volume). This evidence suggests there is a potential for managing soils and vegetation to minimize surface water acidification in areas where acidic deposition is important, e.g., through choice of species, age structure, and a thinning regime. There is a need to link such forest management studies with stream and lake studies.

From these empirical studies, our conclusion is that forest management is not sufficient nor is it necessary to have caused the recent rapid acidification observed in fresh waters in sensitive areas (Figure 9.1). Acid deposition, however, is necessary.

### Research Needs

1. Is base cation uptake by vegetation a significant factor in surface water acidification on a regional basis?

2. Are the sources of base cations for alkalinity generation in surface waters similar to those taken up by roots, and what is the role of hydrological pathways?
3. Can vegetation change or management (other than afforestation) cause chronic (i.e., more than 5 years) surface water acidification in the absence of acidic deposition?

## PROCESSES INVOLVED IN ACIDIFICATION

To evaluate the relative importance of natural and anthropogenic factors, it is necessary to understand the various acidification processes. The effects of anthropogenic impacts, such as land use and increases in acid deposition, can then be better quantified.

For freshwater acidification to occur, a source of mobile anions and an acidic soil is required (Feger, this volume; Van Breemen et al. 1983; Reuss and Johnson 1986). These anions may be generated within the soil or may enter the system from the atmosphere. If anion concentrations increase with no concurrent increase in base cations, freshwater acidification will result. In acidic soils this will be accompanied by an increase in inorganic aluminum concentration in fresh waters. The pH response of surface waters to strong acid anion inputs will primarily be dependent upon the anion retention capacity (e.g., sulfate adsorption) and the base saturation of the soil (Reuss et al. 1987). The impact of incoming anions is also dependent upon the deposition of base cations (e.g., Brydges and Summers 1989; Driscoll et al. 1989). An input of acidity is not required for acidification of fresh waters, since the input of a neutral salt may also transfer acidity from the soil exchange complex to the surface waters (e.g., during seasalt episodes; Hendershot et al. 1991). However, Hendershot et al. (1991) have emphasized that long-term acidification of the soil-water system, as a whole, arises from the input of acidity ($H^+$ or compounds that release $H^+$).

Both natural and anthropogenic factors can result in chronic (steady-state) acidification, and episodic (short-term) acidification (Figure 9.2). Natural episodic acidification may occur following various events, such as heavy rainfall, snowmelt, and seasalt events. Climatic changes that effect the drying of organic horizons and vegetation disturbance, such as windthrow or fire, may also result in a temporary increase in acidity of surface waters in base-poor regions (e.g., Bayley et al. 1993). The primary processes that can produce natural episodic acidification are dilution of base cations, nitrification, organic acid production, and inputs of neutral salts displacing hydrogen ions through ion exchange (Wigington et al. 1992). Anthropogenic factors can, in addition, promote episodic acidification if the transfer of acidic deposition, in either precipitation or snow to the streams or lakes, occurs without significant interaction with the watershed soil (e.g., Christophersen et al. 1984), through the conditioning of a catchment (Wigington et al. 1992) or disturbance of the watershed. Wetlands may act as temporary sinks for some of these anions but may release them during drawdown, resulting in an aggravation of the original stream acid

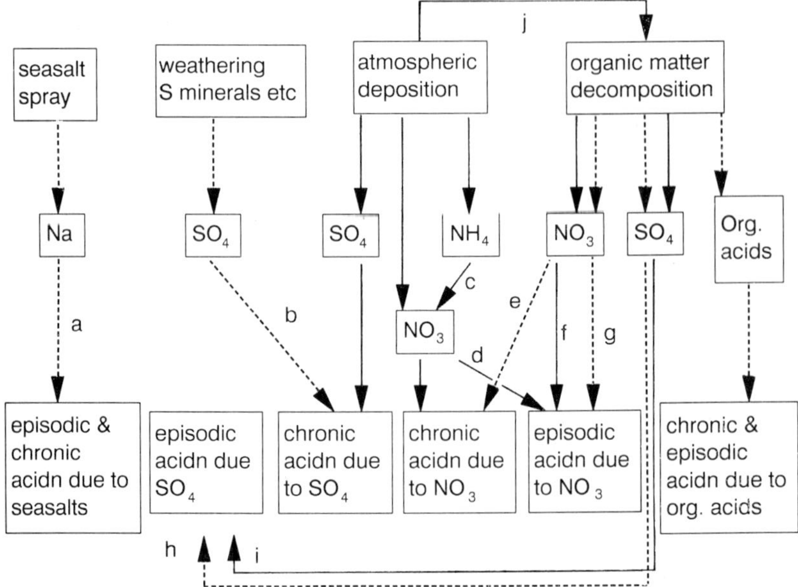


**Figure 9.2** Schematic representation of how several important S, N and organic acid biogeochemical processes, in combination with ion exchange, control chronic and episodic acidification of surface waters and how they can be affected by natural and anthropogenic factors, with - - → representing "natural" pathways and → representing anthropogenically influenced pathways. The impact of these processes on the pH response of fresh waters will depend on the deposition of base cations and the soil base saturation and anion retention capacity (e.g., sulfate adsorption). Letter codes indicate: (a) cation exchange; (b) regionally limited to areas with sulfur-bearing minerals; (c) stimulation of the nitrification process by increase in ammonium inputs; (d) snowmelt or heavy rainfall; (e) effect on nitrogen accumulation by N-fixers and increases in nitrification rate; (f) vegetation disturbance (e.g., clear-cutting); (g) temporal variation of nitrification due to climatic conditions (e.g., drought and drying of organic horizons) or vegetation disturbance (e.g., insect infestation); (h) temporary variation in organic sulfate mineralization due to environmental conditions; (i) increase in sulfate mineralization due to disturbance, e.g., drainage; (j) impact of acidic deposition on organic matter decomposition.

event. Both chronic and episodic acidification may have significant and different impacts on stream and lake biota (Ormerod and Jenkins, this volume).

Various areas of uncertainty remain concerning the role and relative importance of various pathways in the acidification of fresh waters. These areas are:

1. the role of organic acids in acidification, specifically the pathways by which they are produced and leached from the soil;
2. the roles of the nitrogen and certain parts of the sulfur cycles in the acidification process (both episodic and long term);
3. the importance of acidic episodes due to seasalts in combination with acidic deposition.

Hydrological pathways influence the transfer of anions and cations from the soil bedrock system to fresh water and control the associated process of buffering (cation exchange and weathering). Thus, surface waters with similar soil and bedrock properties, which receive comparable atmospheric inputs, may have considerably different pHs due to differences in hydrological pathways (Havas et al. 1984). The role of these pathways requires further investigation and is discussed more fully by Landers et al. (this volume). Acidification due to an increase in sulfur deposition is well documented, relative to the role of organic acids and nitrogen, and is therefore only briefly outlined below. Readers are referred to Van Miegroet (this volume) for more information.

## Organic Acids

The terms *natural organic acids*, *dissolved organic carbon* (DOC), and *aquatic humic substances* differ in their definition but are often used interchangeably. In fresh waters, a large fraction of DOC usually qualifies as humic material and is typically acidic. The term DOC was used to define this type of organic matter for the purpose of our group discussions.

In natural waters, DOC can be an important source of acidity, depending on the concentrations of both base cations and DOC. DOC has both strong and weak acid properties and can contribute up to 10 μeq $H^+$ per mg DOC (Brakke et al. 1987; Kortelainen and Mannio 1988). Consequently, the pH of some humic acid-rich bogs can be less than 4, due entirely to DOC contributions to acidity (Hemond, this volume). DOC has also been implicated in contributing to acidic episodes if water flows through upper organic horizons during rain-induced episodes in peaty catchments (e.g., Edwards et al. 1986; also reviewed in Davies et al. 1992).

DOC concentrations are high in the surface waters of many regions, where their contribution to acidity can be significant. In other areas, clearwater systems predominate and DOC concentrations are very low, and DOC is not an important contributor to acidity. Such areas are found in the mountainous areas of the western U.S.A. (Eilers et al. 1989) and lakes in southern Norway, especially at higher elevations (Henriksen et al. 1988).

Although DOC is known to be produced naturally by several processes, and the humic fraction is known to arise during the decomposition of organic matter, there is a pressing need for detailed information on the chemical pathway of DOC production. DOC represents only a small fraction of the soil organic carbon available, and little is known about the controls on the amount and character of the DOC produced. The information available on temporal and spatial variations in organic acid and DOC at the ecosystem level was recently reviewed by Mulholland et al. (1990). Some general relationships between DOC concentrations and environmental conditions in forest ecosystems have been identified; however, these are not universal (Cronan 1990). Factors identified as influencing DOC concentrations include species composition of vegetation, season, chemistry of canopy throughfall, and drainage. The properties of the DOC in terms of acidity, metal complexing ability, and reactivity to microbial degradation, however, are poorly known. Further information on structural incorporation of pollutants into DOC is required, as are details on transport, transformation, and breakdown of DOC in the soil. All of these aspects can influence the role of DOC in contributing to acidity and in binding toxic metals.

Similarly, we have little information on the relative importance of processes removing DOC from stream or lake environments (Brakke et al., this volume). Quantitative data are lacking on the biotic and abiotic processes that control DOC removal. Biotic utilization, sorption, particle formation followed by settling and photooxidation may all operate (Meyer 1990).

The changes in soil hydraulic properties, water fluxes, rooting development, and organic matter quality during land-use change can have a significant effect on both the quantity and quality (i.e., acidic properties) of DOC released to surface waters. The effects of disturbance appear variable although clear-cutting appears generally to increase concentrations of DOC (Telang et al. 1981; Sollins and McCorison 1981). An increase in the amount of wetland within a catchment would also be expected to increase the amount of DOC released, while contraction of wetlands in response to climatic warming might substantially reduce the quantity and alter the quality of DOC released to surface waters.

The impact of acid deposition on DOC production and release is unclear. The interaction with the changes in the terrestrial nitrogen cycle due to increased nitrogen deposition may have a major influence on DOC release. This may occur through changes in the C:N ratio of soil organic matter, changes in rates of decomposition, and through incorporation of N into the structure of DOC produced. Sulfur may also be incorporated into DOC structure. In general, the limited data available suggest that soil and surface water DOC concentrations are more likely to decrease than increase with higher deposition of strong acidic anions. The concentration of DOC across a gradient of increasing sulfur deposition in the northeastern and midwestern sections of the U.S.A. was found to decrease in surface waters (Sullivan et al. 1988). In addition, paleoecological reconstructions suggest some cases where DOC declined (Steinberg 1991) and others where the change was small or showed no consistent pattern (Cumming et al. 1992). Even if the change in DOC is small, acid deposition could still result in significant impacts on acid-base qualities and metal-complexing abilities of the DOC. In the acid-exclusion experiment of the Norwegian RAIN project, an increase in organic anion concentration was observed with no change in total organic carbon (TOC) concentration recorded (Wright 1989). Further analyses of lake sediment cores to resolve questions of change in DOC over time are warranted, and careful attention should be given to any changes in land use that occurred while the watersheds were studied.

The interaction of DOC and strong acidic anions is especially significant in areas containing large numbers of humic lakes such as Finland (Kortelainen and Mannio 1988), portions of the northeastern and upper midwestern U.S.A. (Brakke et al. 1988; Eilers et al. 1988), and eastern Canada (Gorham et al. 1986). Understanding the response of such systems to acidic inputs is essential in modeling critical loads. Incorporating DOC into critical load calculations has proven difficult; hence, DOC effects are typically ignored. The lack of information calls for expanded research on DOC as it relates to acidification of fresh waters, ranging from comparative surveys, mechanistic experiments and whole-lake experiments, to whole-systems manipulations such as the HUMEX project in Norway (Gjessing 1993).

In some cases, it is likely that strong acid deposition is additive with natural sources of acidity (Brakke et al. 1987; Kortelainen and Mannio 1988). In addition, as protonation of DOC occurs during acidification, the DOC may be less effective in binding aluminum, thus producing toxic conditions for fish populations. Biota may respond to long-term changes in DOC; however, the importance of these changes are relatively small compared to the large changes that can be caused by mineral acids.

## Nitrogen

The future role of nitrate in the acidification of waters and soil is a major research area. In natural systems, nitrate leaching is typically low ($< 5\ \mu eq\ L^{-1}$) and thus will have minimal impact on acidification of fresh waters. Elevated nitrate leaching can occur, for example, in N-fixing systems (such as alder), where enhanced nitrate leaching is accompanied by the mobilization of aluminum (Van Miegroet and Cole 1988; Reuss 1989). There is now evidence that retention of nitrate is also reduced in some areas with high atmospheric acid loadings, e.g., in some areas of central Europe (Henriksen and Brakke 1988), the Netherlands (Van Dijk et al. 1992), in the eastern U.S.A. (Johnson et al. 1991; Stoddard 1993; Murdoch and Stoddard 1993), and in areas where tree decline is evident (Hauhs and Wright 1986). A survey of 1005 lakes in Norway in 1974–1975 and again in 1986 indicated that the contribution of nitrate to acidification of lakes had increased in some ecosystems over the ten-year period (Henriksen and Brakke 1988). This increase could not be explained by an increase in nitrogen deposition during this period but appeared to be caused by a reduction in the retention of the incoming nitrogen, with an increasing proportion of nitrogen deposition lost in runoff. Variability in retention of nitrogen indicates that the relative importance of key nitrogen processes varies dramatically among different ecosystems and, in particular, where the vegetation structure is modified (Kreutzer 1989; Feger 1992).

The importance of nitrate in episodic acidification pulses appears to be most common following increased nitrogen loading or anthropogenic disturbance of site physical properties and/or vegetation composition (e.g., liming, draining, forestry operations, and forest decline). Exceptions may include elevated nitrate concentrations following insect infestation of a site, windthrow, and drought in wetlands. Short-term nitrate pulses are frequently associated with heavy rain and snowmelt; however, sulfate appears to be relatively more important in episodic events in Europe, Canada, and some parts of the U.S.A., with nitrate making a greater contribution in the northeastern U.S.A. (Wigington et al. 1992). Nitrate concentrations tend to increase during the first rise in streamflow, following dry periods that have occurred especially when biological uptake is low. By contrast, elevated nitrate concentrations may persist for several years, following an extended period of drought (Reynolds et al. 1992). High concentrations following snowmelt may be a consequence of release from the melting pack, but the contribution of the "event" water is highly variable depending on the changes in water flowpaths (Davies et al. 1992). In some cases, continued mineralization and nitrification in the snow-covered mineral soil and a flushing out of this nitrate prior to tree uptake contributes to the elevated concentrations during snowmelt (e.g., Foster et al. 1989). Short-term episodic acidification, due to heavy rain or snowmelt, may promote a biological response different to the one caused by chronic acidification due to long-term nitrate leaching in response to drought, clear-cutting, etc.

Land-use and management practices can have a significant effect on nitrate release to fresh waters. Elevated nitrate concentrations over several years will occur when either plant uptake or soil immobilization capacity is temporarily reduced, for example, following clear-cutting. The magnitude and duration of the pulse are dependent upon the nitrogen status of the site prior to clear-cutting, the harvesting intensity, the control of ground vegetation following felling, and the response of the soil nitrogen transformations, particularly nitrification and denitrification, following felling (e.g., Likens et al. 1970; Vitousek et al. 1982; Miller and Newton 1983; Emmett et al. 1991). In areas unaffected by acid deposition, nitrate concentrations may decrease or remain unchanged following clear-cutting (e.g., Nicholson 1975). The impact of clear-cutting in sensitive areas with low acidic deposition is not well documented. Wetland drainage, burning, and a variety of other management practices may also result in short-term increases in nitrate concentrations (Feger, this volume; Harriman, this volume) although responses are variable (e.g., Wright 1976; Bayley et al. 1993).

Chronic leakage of nitrate can occur if uptake is consistently lower than inputs plus the internal production in the soil. This may occur as a result of changes in the soil moisture regime, microclimatic conditions, and organic matter quality and can be influenced by, for example, vegetation composition and stand age (Feger, this volume). Streamwater nitrate concentrations can be reduced during the early growth phase of a forest stand relative to non-afforested catchments, but are elevated in catchments dominated by mature stands (Stevens et al. 1993). The increase in nitrate leaching in mature stands may be due to reduced uptake of nitrogen as the stand ages, increased capture of atmospheric nitrogen by the canopy, increased production in the soil, or reduced retention by the microbial population. Nitrate leakage or increased availability in the soil have been reported to be associated with old-growth forest (Leak and Martin 1975) although this has not always been observed (Davidson et al. 1992).

If nitrogen is deposited to a system at a rate in excess of the rate of retention (biotic and abiotic), then nitrogen (probably as nitrate) breakthrough will result. The maximum rate of nitrogen immobilization in the mineral soil or the conditions under which this nitrogen may be remobilized are unknown. Aber et al. (1989) have described the various processes and stages involved in this breakthrough of nitrate following increases in atmospheric nitrogen deposition. If nitrogen inputs are in excess of microbial and vegetation uptake capacity, litter decomposition and soil nitrogen mineralization may increase, resulting in a stimulation of nitrification (e.g., McNulty et al. 1990). There may be either an increase in ammonium concentrations in the soil or, if nitrification is active or nitrate inputs high, nitrate will be leached. Denitrification may remove some nitrogen and thus reduce the nitrate available for leaching. Denitrification accounted for 250 meq N $m^{-2}$ $yr^{-1}$ in a study carried out in the Netherlands, with throughfall input of nitrogen at 320 meq N $m^{-2}$ $yr^{-1}$ (Tietema and Verstraten 1991). This amount will be highly variable depending on various soil parameters such as nitrate availability, soil moisture, and carbon availability (Firestone and Davidson 1989). The relationship between nitrogen inputs, denitrification rates, and the ratio of $N_2O$ to $N_2$ efflux has not been extensively studied. A positive relationship between the amount and relative availability of nitrogen and nitrous oxide fluxes has been observed for tropical forest soils, where nitrous oxide fluxes are much greater than from temperate forest soils (Matson et al. 1989). The dominance of $N_2O$ production in acidic soils relative to $N_2$ production, the increase in this ratio with increased nitrate availability (Firestone and Davidson 1989), and the negative feedback to methane consumption (Melillo et al. 1989) may all result in increases in concentrations of radiatively active (greenhouse) gases at some sites where a high nitrogen loading is occurring.

An excess of nitrogen availability may have various consequences for the vegetation and may eventually result in elevated streamwater nitrate concentrations. Impacts may include direct toxicity to individual species, soil-mediated effects upon plant species, increased susceptibility to secondary stress factors (e.g., frost), and changes in competitive relationships between plants, resulting in loss of diversity (Bobbink et al. 1992).

Nitrate leaching greater than that observed in pristine sites has been found to occur in European coniferous forests at loadings greater than 65 meq N $m^{-2}$ $yr^{-1}$ (Dise, submitted). Yet there is conflicting evidence on the relationship between nitrate inputs and nitrate leaching beyond this threshold in Europe and the U.S.A. (Dise and Wright, submitted; Van Miegroet et al. 1992). This variability may be due to the different ratio of nitrogen species in the deposition (ammonium will be retained to a greater extent than nitrate) or to the changes in soil nitrogen processes (particularly nitrification and denitrification) at different sites. Previous land-management practices (e.g., changes in tree species, excess of biomass removal fertilization; see Feger, this volume) may be the key factor controlling the release of nitrogen from soil pools and may thereby increase the variability between sites (Zoettl 1990). The interactions between the nitrogen cycle and other pollutants (e.g., $SO_4$ and $CO_2$) are also not clearly understood.

Within lakes and streams, nitrogen may undergo various transformations (Kelly, this volume). Biological uptake of increased nitrogen inputs may be limited due to phosphorus limitation (Hecky et al. 1988). There is an acidifying potential from increased inputs of ammonium from deposition to fresh waters, as nitrification in these systems is likely to be substrate-limited (Schindler et al. 1985). Aquatic nitrifiers, however, may be more sensitive to pH fluctuations than nitrifiers in the soil, with nitrification disrupted at pH 5.4–5.6, resulting in increases in ammonium concentrations (Rudd et al. 1988). Denitrification in lake and stream sediments (and less importantly the water column) is one of the most important mechanisms for neutralizing nitric acid. The rate of denitrification may be expected to increase with increased nitrate concentrations (e.g., Rudd et al. 1990). Yet removal of nitrate by denitrification will be slower than through algal uptake (Kelly, this volume). In a review by Tiedje (1988), nitrate concentrations were considered the most important limiting factor for denitrification in sediments, while oxygen was the most common limiting factor in the lake water column.

As with sulfate, an increase in nitrate combined with an increase in hydrogen or aluminum concentrations is considered to be acidifying, and damage to freshwater biota will occur. Where nitrate is utilized in either streams or lakes, no excess acidity will result, and net alkalinization may occur. There are no documented examples of enhanced nitrate release caused solely by anthropogenic inputs directly affecting freshwater biota. A potential future problem of increased nitrogen inputs in some aquatic ecosystems will be changes in species composition of primary producers. The interactions between eutrophication and the impacts of acidification are poorly understood.

## Sulfur

Chronic sulfate-driven acidification of fresh waters due to natural factors such as weathering/oxidation of sulfide minerals or the spontaneous burning of bituminous shales (Havas and Hutchinson 1983) is possible; however, this is generally limited to small geographical areas. Episodic events may occur following natural disturbances, such as fire or drought, but are restricted to base-poor regions (Bayley et al. 1993). High sulfate concentrations in fresh waters are usually restricted to areas of high atmospheric deposition. Management practices can have some influence on sulfate-driven acidification by affecting the mineralization of organic sulfur pools in the soil, changing water/soil contact, or by altering vegetation type, and thus efficiency of pollutant capture (see Feger et al. 1990; Harriman, this volume). The magnitude of the response to the sulfate signal, however, is partly controlled by adsorption/desorption reactions in the deeper mineral soils. This differs greatly between systems and is not sufficiently well understood and requires further investigation (Van Miegroet, this volume). Sulfate removal in sediments and lake water is reviewed by Kelly (this volume).

## Seasalts

Episodic inputs of seasalts from rainfall, or derived from melting snowpack, can result in additional short-term pH and alkalinity changes in streams and small fast-turnover lakes that are located tens of kilometers inland from the sea (e.g., Mulder et al. 1990;

Langan 1985; Neal et al. 1988; Davies et al. 1992). This pH depression results from the cation-exchange processes in soils, whereby incoming sodium ions exchange with hydrogen and aluminum and base cations that have accumulated in soils because of natural or anthropogenic factors (Wright et al. 1988). Seasalt events may be more important in forested catchments because of enhanced atmospheric scavenging of atmospheric pollutants and seasalts by the canopy and a greater capacity for ion exchange (Welsh and Burns 1987; Langan 1985; Davies et al. 1992). During episodic acidification, the lowest pH values are generally observed in waters that had the lowest pre-episode pH values. However, the magnitude of the pH depression decreases with diminishing pre-episodic pH (Wigington et al. 1992). Seasalt events may therefore be more severe (lower pH) in sites conditioned by acidic deposition. The biological consequences of these episodes are difficult to quantify because calcium and magnesium levels increase concurrently with pH decrease. Biota may be less responsive to pH depressions resulting from seasalt events as opposed to non-seasalt events.

### Research Needs

1. What are the controls on DOC quality and quantity, transport, and decomposition, and how are these affected by land-use change and acid deposition?
2. What is the range of nitrate leaching rates from catchments with varying biogeochemical properties?
3. What is the effect of increasing nitrogen inputs on nitrogen cycling in terrestrial ecosystems, both before and after the state of saturation?
4. What controls exist on these processes in the soil (nitrogen immobilization, remobilization of stored nitrogen, nitrification, and denitrification), both in natural and acidified ecosystems?
5. What controls adsorption/desorption of sulfate in deeper mineral soils (see Van Miegroet, this volume)?
6. How will a decrease in sulfur deposition and/or a change in climate affect the turnover of the soil organic sulfur pools?
7. How do biological effects of non-seasalt episodes compare in naturally acidified systems and anthropogenically acidified waters?
8. To what extent has acidic deposition reduced the buffering capacity of geological material in the hydrologically important source areas. How fast is the buffering resource being reduced in specific regions?
9. What is the importance of deposition of base cations in acidification and alkalinization of soils and waters?

## IMPACT OF GLOBAL CLIMATE CHANGE

Changes in climatic factors (e.g., rainfall regimes, temperature, $CO_2$, UV-B) may all affect freshwater acidification through effects on the production of both organic and strong mineral acids. We addressed possible consequences of changes in temperature and rainfall only; for some consequences of increased UV-B irradiation, see Landers et al. (this volume). One of the most important factors will be the change in evapotranspiration. The relationship between temperature and evapotranspiration is not linear; as warming increases, a unit change in temperature will result in a proportionally larger response in evapotranspiration.

These factors have the potential to affect significantly the N, S, C, P, and base cation cycles. Generally, the effect of evapotranspiration is greater on decomposition of organic matter than on net primary productivity. Therefore, as evapotranspiration increases, there is a tendency for an increase in the proportion of vegetation biomass to litter (Tinker and Ineson 1990). If the large stores of N, S, and C in the soil system are mineralized due to increased microbial activity, this may significantly increase the contribution of internally produced mineral acids relative to acid deposition in the absence of a concurrent increase in base cations (e.g., Van Cleve et al. 1990). However, weathering rates are temperature dependent and may increase to counterbalance this source of acidification. Lakewater nitrate concentrations may increase, in response to release of nitrogen from the soil organic pool, or be reduced, as a result of increased water residence time and thus in nitrogen retention. Denitrification in the lakewater column and sediments may also become more effective at warmer temperatures (Rudd et al. 1990).

The effect on organic acid production is unclear; drier sites are likely to release less DOC despite high mineralization rates. In boreal regions, however, increased mineralization activity (due to the longer period of microbial activity) and increased leaching may occur where permafrost areas decrease (Brinkman 1990), thus possibly increasing leaching of DOC. The impact on soils is predicted to be smallest in temperate regions, where only minor changes in rainfall totals and leaching are predicted (Brinkman 1990). Increased leaching can have a major impact on various processes, such as loss of salts and nutrients. The conditioning of a site through acid deposition and the consequent reduction in the pool of base cations may result in a greater effect on acidified fresh water in a polluted area than on pristine areas. The new CLIMEX experiment, in which $CO_2$ and temperature will be experimentally increased under the two roofed catchments at Risdalsheia, Norway, focus on the interaction between previous acid deposition loadings and global warming (Jenkins et al. 1992).

Climate change can have important effects on key species in boreal lakes and their habitats, with biota responding to both the physical and chemical changes (e.g., Schindler et al. 1990). Both acidification and climate warming and drying may cause an increase in the rates of DOC removal. In the case of climatic drying, DOC supplies are also reduced. As a result, both will increase transparency, causing an increase in the depth of the euphotic zone (Schindler et al. 1991). Changes in DOC concentrations may affect removal rate of iron (Curtis, in prep.) and thus the cycle of trace metals and nutrients that co-precipitate with iron.

**Research Needs**

1. Can systems where strong acid anion pulses have occurred as a result of natural fluctuations in climatic conditions (e.g., changes in the frequency of wetting and drying cycles) be used as an analogue of some expected aspects of global climate change?
2. To what extent can climate change alter hydrological pathways and thus surface water acidification on a regional basis?
3. How will global climate change affect the role of nitrate and other acidic anions in the acidification of fresh waters?

## TO WHAT EXTENT CAN WE APPLY KNOWLEDGE FROM THE TEMPERATE TO TROPICAL ZONES?

There is a deficiency in both long-term data and an understanding of the freshwater processes in tropical regions. This lack of information makes it impossible to assess the potential damage to freshwater biota due to acid deposition, although the potential for acidification in the Tropics has been addressed (Rodhe and Herrera 1988; Ovalle and Filho 1991).

There may be differences in the relative importance of some components of the tropical ecosystem (e.g., vegetation uptake) relative to temperate studies in the responses of the system to acid deposition. Understanding these differences may help clarify the contribution of different ecosystem components to the response to acid deposition observed in temperate systems. For example, the impact of sulfate deposition may be delayed due to the high sulfate adsorption capacity of many tropical soils (Van Breemen 1990).

As dynamic models are based on chemical and physical laws, they can be applied in the Tropics. Yet the lack of information to calibrate the models is a problem, particularly with regard to the acidic episodes that predominate in some areas due to rainfall periodicity. More biogeochemical information is needed, both on episodic and long-term variation. A successful avenue of research in temperate regions, which may also be useful in the Tropics, is the manipulation of ecosystems by dosing with mineral acids and the study of process responses. Such manipulations should include both nitrogen and sulfur inputs. In the absence of historical data, paleoecological approaches should be developed to determine the rate and factors involved in acidification of tropical lakes.

**Research Needs**

1. What is the sensitivity of the more diverse aquatic ecosystems in the Tropics to acidification? How does the diversity of aquatic assemblages affect their response to acidification?

2. What are the impacts of land use, in the presence and absence of acid deposition and natural factors affecting water quality (climate, geology, vegetation, etc.).

## THE USE OF BIOTA TO DISTINGUISH BETWEEN NATURAL AND ANTHROPOGENIC ACIDIFICATION

Fresh waters rich in DOC typically have less toxic forms of metals (e.g., aluminum) due to complexation with the DOC (McCahon and Pascoe 1989). As a result, some biota may survive in DOC-rich lakes at metal concentrations that would otherwise be toxic. Thus, two lakes with the same hydrogen ion concentration but different DOC concentrations could have different and, in many cases, quite distinctive biotic assemblages (e.g., Forman 1979). The differences in chemistry associated with natural and anthropogenically derived acidic/acidified waters, therefore, result in biological differences. There is little evidence of chronic high concentrations of the more toxic forms of inorganic aluminum in surface waters due to natural factors. Biota that are tolerant to high aluminum concentrations appear to be widespread but not abundant. Indeed, the presence of ionic aluminum is one of the key characteristics of anthropogenically acidified waters. However, there are few naturally clear acidic lakes that are not receiving acidic deposition.

A variety of biota (notably diatoms, benthic invertebrates, and fish) can be used to identify critical inorganic aluminum concentrations and pH change, although a thorough knowledge of the relationship between chemistry and biotic assemblages is required in the system in order for this to be achieved. This is reviewed in more detail by Brakke et al. (this volume). The relative importance of anthropogenically derived strong mineral acids in acidic episodes such as snowmelt can be assessed by (a) monitoring the biota during early spring and (b) assessing the recovery until late fall relative to unaffected waters. Taxa with sensitive life history stages absent following episodic events (snowmelt or rainstorms) and the remaining biocenosis are potentially good indicators of whether these events have occurred (Merrett et al. 1991; Wade et al. 1989; Raddum et al. 1988; Gebhardt et al. 1989; Matschullat 1992; Raddum and Fjellheim, this volume). If natural acidic episodes were important in a particular area prior to acidic deposition, we would perhaps expect assemblages to have become adapted to these conditions and therefore be more tolerant to recent, more severe episodes caused by acidic inputs.

### Research Needs

1. What is the impact on stream biota of naturally acidic episodes in streams resulting from natural disturbances, such as fire and drought?
2. Can we differentiate between biological responses to acidic episodes caused by natural events and those caused by acid deposition?

3. What is the impact of both DOC and mineral acid-derived acidity on primary production?
4. Will the frequency of anthropogenic-derived nitrate acidification pulses increase in the future, and, if so, how will this affect the biota?
5. How might the effects of nitrate and sulfate differ, and can paleoecological methods be used to infer previous nitrogen status of waters?
6. Does DOC mitigate the effects of acidification throughout the food chain?

## HOW CAN WE IMPROVE METHODS TO DIFFERENTIATE BETWEEN NATURAL AND ANTHROPOGENIC ACIDIFICATION?

Methods used at the present include (1) analysis of current chemical conditions; (2) the use of long-term records either in paleoecological records or long-term chemical data bases; (3) substitution of space for time (the use of current low atmospheric input sites as an analogue for preindustrialization conditions at a site with high inputs), (4) manipulation experiments, where atmospheric deposition is either increased or removed, and (5) dynamic process-oriented simulation modeling of ecosystems. These methods are discussed below and we include an outline on how they may be improved.

1. To improve our ability to differentiate between natural and anthropogenic causes of acidification using current chemical conditions requires a greater understanding of the processes. At present, adequate empirical models exist to determine how much acidity comes from organic and mineral acids in a water sample. However, these models are not process-driven; they assume that organic acids have always occurred in present-day quantities with the same functional groups.

2a. Paleoecological methods provide some of the best approaches available for differentiating the role and importance of natural and anthropogenic factors causing surface water acidification over many decades. Techniques have improved rapidly over the last 5–10 years; however, greater abilities to infer and understand changes could be obtained from additional research in the following areas:

- Combine existing diatom and chrysophyte calibration data sets for several regions in Europe and the U.S.A. to help provide a better understanding of the ecological responses to chemical and physical habitat changes. The combined data sets would include data on between 500 to 1000 lakes on each continent.
- Perform additional studies of organisms preserved in sediments to quantify their ecological characteristics as well as field studies to determine habitat distribution and laboratory culture studies to determine factors affecting growth of organisms.

- Analyze remains of several groups of organisms from existing and new sediment cores to learn how they may have responded to changes in lakewater acidity and related parameters, both as individuals and as they represent entire communities. This information can then be used to better understand species-species interactions and help improve our ability to predict how whole communities change in response to surface water acidification.
- Design studies to understand ecosystem changes resulting from nitrogen breakthrough.
- Undertake more paleoecological studies to examine the relative shifts in biota caused by changes in concentrations of organic acids as compared to those caused by changes in mineral acids.
- Integrate paleoecological approaches into experimental studies and model development to investigate specific hypotheses, e.g., those relating to the effects of forest practice on lakewater acidity, occurrence and rate of reversibility, etc.

2b. Present long-term data sets are dominated by information from catchments with a high level of disturbance and acid deposition. There is a need to monitor areas with low acidic deposition in sensitive areas. Long-term data sets will be particularly useful to assess the changing relationships in the relative importance of declining sulfate and nitrogen deposition. Long-term monitoring is vital and may be essential for reversibility and predictive studies. However, it must be recognized that these long-term studies will also show responses to other changes so interpretation can be difficult. A paucity of data and less sophisticated analytical methods are the major constraints in the use of historical chemical records. It may still be useful to continue efforts at locating and critically examining historical water quality records, although many have already been analyzed (National Research Council 1986).

3. The use of current low atmospheric input sites as analogues for preindustrialization conditions would also be a useful approach to determine the impact of land use in the absence of acid deposition on lake water chemistry, e.g., a large lake and catchment survey with full information on catchment characteristics and history.

4. Manipulations can provide very useful information, in particular, on the processes involved in determining the impacts of land use and acid deposition on surface water acidification. New studies could include land use and acid input manipulations in clean areas (especially in relation to nitrogen dynamics), e.g., manipulations to study interactions between nitrogen and sulfur deposition, their relative importance in aluminum mobilization, and how they are affected by global climate change. Other priorities could include manipulations in terrestrial and lake ecosystems and land-use manipulations (e.g., clear-cutting) in high input areas (central Europe). Manipulations involving a watershed

containing a downstream lake could provide a record of long-term changes, pre-disturbance conditions, and natural variability in the lake sediments.

5. An increase in the understanding of processes is the most urgent need for improving the predictive capability of models. At present, models can be used to differentiate between natural and anthropogenic factors in acidification; however, the lack of process information limits predictive capacity.

Research projects that incorporate several of the above approaches in an integrated design are most likely to be the most successful in answering questions regarding the distinction between anthropogenic and natural acidification.

## REFERENCES

Aber, J.D., K.J. Nadelhoffer, P. Steudler, and J.M. Melillo. 1989. Nitrogen saturation in northern forest ecosystems. *Bioscience* **39**:378–386.

Bayley, S.E., D.W. Schindler, K.G. Beaty, B.R. Barker, and M.P. Stainton.1992. Effects of multiple fires on nutrient yields in streams draining boreal forest and fen watersheds: Nitrogen and phosphorus. *Can. J. Fish. Aquat. Sci.* **49**:584–596.

Bobbink, R., D. Boxman, E. Fremstad, G. Heil, A. Houdijk, and J. Roelofs. 1992. Critical loads for nitrogen eutrophication of terrestrial and wetland ecosystems based upon changes in vegetation and fauna. In: Critical Loads for Nitrogen. Report from a Workshop Held in Lökeberg, Sweden, 6–10 April, 1992, ed. P. Grennfelt, and E. Thörnelöf, Nord 1992, vol. 41, pp. 111–161. Copenhagen: Nordic Council of Ministers.

Brakke, D.F., J.M. Eilers, and D.H. Landers. 1988. Chemical characteristics of lakes in the Northeastern U.S. *Environ. Sci. Technol.* **22**:155–163.

Brakke, D.F., A. Henriksen, and S.A. Norton. 1987. The relative importance of acidity sources for humic lakes in Norway. *Nature* **329**:432–434.

Brinkman, R. 1990. Resilience against climate change?. Soil minerals, transformations and surface properties, Eh, pH. In: Soils on a Warmer Earth, ed. H.W. Scharpenseel, M. Schomaker, and A. Ayoub, pp. 51–60. New York: Elsevier.

Brydges, T.G., and P.W. Summers. 1989. The acidifying potential of atmospheric deposition in Canada. *Water, Air, Soil Pollut.* **43**:249–263.

Charles, D.F., and S.A. Norton. 1986. Paleolimnological evidence for trends in atmospheric deposition of acids and metals. In: Acid Deposition: Long-term Trends. National Research Council, Council on Monitoring and Assessment of Trends in Acid Deposition, pp. 335–431. Washington, D.C.: National Academy Press.

Christophersen, N., S. Rustad, and H.M. Seip. 1984. Modelling streamwater chemistry with snowmelt. *Phil. Trans. Roy. Soc. Lond. B* **305**:427–439.

Cook, R.B., J.W. Elwood, R.R. Turner, M.A. Bogle, P.J. Mulholland, and A.V. Palumbo. 1993. Acid-base chemistry of high-elevation streams in the Great Smoky Mountains. *Water, Air, Soil Pollut.*, in press.

Cronan, C.S. 1990. Patterns of organic acid transport from forested watershed to aquatic ecosystems. In: Organic Acids in Aquatic Ecosystems, ed. E.M. Perdue and E.T. Gjessing, pp. 245–260. Dahlem Workshop Report LS 48. Chichester: Wiley.

Cumming, B.F., J.P. Smol, J.C. Kingston, D.F. Charles, H.J.B. Birks, K.E. Camburn, S.S. Dixit, A.J. Uutala, and A.R. Selle. 1992. How much acidification has occurred in Adirondack region lakes (New York, U.S.A.) since preindustrial times?. *Can. J. Fish. Aquat. Sci.* **49**:128–141.

Davidson, E.A., S.C. Hart, and M.K. Firestone. 1992. Internal cycling of nitrate in soil of a mature coniferous forest. *Ecology* **73**:1148–1156.

Davies, T.D., M. Tranter, P.J. Wigington, Jr., and K.N. Eshleman. 1992. "Acidic episodes" in surface waters in Europe. *J. Hydrol.* **132**:25–69.

Driscoll, C.T., G.E. Likens, L.O. Hedlin, J.S. Heaton, and F.H. Bormann. 1989. Changes in the chemistry of surface waters; 25 year-results at the Hubbard Brook Experimental Forest, NH. *Environ. Sci. Technol.* **23**:137–143.

Edwards, A.C., J. Creasey, and M.S. Cresser. 1986. Soil freezing effects on upland stream solute chemistry. *Water Resour.* **20**:831–834.

Eilers, J.M., D.F. Brakke, and D.H. Landers. 1988. Chemical characteristics of lakes in the Upper Midwest, United States. *Environ. Sci. Technol.* **22**:164–172.

Eilers, J.M., D.F. Brakke, D.H. Landers, and W.S. Overton. 1989. Natural and anthropogenic causes of lake acidification in Nova Scotia. *Environ. Monit. Assess.* **12**:3–21.

Emmett, B.A., J.M. Anderson, and M. Hornung. 1991. The controls on dissolved nitrogen losses following two intensities of harvesting in a Sitka spruce forest (N.Wales). *For. Ecol. Manag.* **41**:65–80.

Feger, K.H. 1992. Nitrogen cycling in two Norway spruce (Picea abies) ecosystems and effects of $(NH_4)_2SO_4$ addition. *Water, Air, Soil Pollut.* **61**:295:307.

Feger, K.H., G. Brahmer, and H.W. Zoettl. 1990. Element budgets of two contrasting watersheds in the Black Forest (Federal Republic of Germany). *J. Hydrol.* **116**:85–99

Firestone, M.K., and E.A. Davidson. 1989. Microbiological basis of NO and $N_2O$ production and consumption in soil. In: Exchange of Trace Gases between Terrestrial Ecosystems and the Atmosphere, ed. M.O. Andreae and D.S. Schimel, pp.7–22. Dahlem Workshop Report LS 47. Chichester: Wiley.

Forman, R.T.T. 1979. Pine Barrens: Ecosystem and Landscape. New York: Academic.

Foster, N.W., J.A. Nicholson, and P.W. Hazlett. 1989. Temporal variation in nitrate and nutrient cations in drainage waters from a deciduous forest. *J. Environ. Qual.* **18**:238–244.

Gebhardt, H., M. Linnenbach, R. Marthaler, A. Ness, and H. Segner. 1989. Die Bachforelle (*Salmo trutta f. fario*)—Ein Bioindikator für die Gewässerversauerung. *Fischökologie* **1**:1–21.

Gjessing, E.T. 1993. The HUMEX project: Experimental acidification of a catchment and its humic lake. *Environ. Intl.* **18:**535–543.

Gorham, E., J.K. Underwood, F.B. Martin, and J.G. Ogden, III. 1986. Natural and anthropogenic causes of lake acidification in Nova Scotia. *Nature* **22**:451–453.

Hauhs, M., and R.F. Wright. 1986. Regional pattern of acid deposition and forest decline across a section through Europe. *Water, Air, Soil Pollut.* **31**:463–474.

Havas, M., T.C. Hutchinson, and G.E. Likens. 1984. Red herrings in acid rain research. *Environ. Sci. Tech.* **18**:176A–186A.

Havas, M., and T.C. Hutchinson. 1983. The Smoking Hills: Natural acidification of an aquatic ecosystem. *Nature* **301**:23–27.

Hecky, R.E., and P. Kilham. 1988. Nutrient limitation of phytoplankton in fresh water and marine environments: A review of recent evidence on the effects of enrichment. *Limnol Oceangr.* **33**:796–822.

Hendershot, W.H., P. Warvinge, F. Courchesne, and H.U. Sverdrup. 1991. The mobile anion concept: Time for a reappraisal. *J. Environ. Qual.* **20**:505–509.

Henriksen, A., and D.F. Brakke. 1988. Increasing contributions of nitrogen to the acidity of surface waters in Norway. *Water, Air, Soil Pollut.* **42**:183–201.

Henriksen, A., D.F. Brakke, and S.A. Norton. 1988. Total organic carbon concentrations in acidic lakes in southern Norway. *Environ. Sci. Technol.* **22**:1103–1105.

Jenkins, A., D. Schulze, N. Van Breemen, F.I. Woodward, and R.F. Wright. 1992. CLIMEX: CLIMate change EXperiment. In: Responses of Forest Ecosystems to Environmental Changes, ed. A. Teller, P. Mathy, and J.N.R. Jeffers, pp. 359–366. New York: Elsevier.

Johnson, D.W., H. Van Miegroet, S.E. Lindberg, D.E. Todd, and R.B. Harrison. 1991. Nutrient cycling in the red spruce forests of the Great Smoky Mountains. *Can. J. For. Res.* **21**:769–787.

Kortelainen, P., and J. Mannio. 1988. Natural and anthropogenic acidity sources for Finnish lakes. *Water, Air, Soil Pollut.* **42**:341–352.

Langan, S.J. 1985. Episodic acidification of stream at Loch Dee, SW Scotland. *Trans. Roy. Soc. Edinburgh* **78**:393–397

Leake, W.B., and C.W. Martin. 1975. Relationship between stand age to streamwater nitrate in New Hampshire. *USDA For. Res. Note* NE–211.

Likens, G.E., F.H. Bormann, N.M. Johnson, D.W. Fischer, and R.S. Pierce. 1970. Effects of forest cutting and herbicide treatment on nutrient budgets in the Hubbard Brook watershed ecosystem. *Ecol. Monogr.* **40**:24–46.

Matschullat, J., H. Andreae, D. Lessmann, V. Malessa, and U. Siewers 1992. Catchment acidification: From the top down. *Environ. Pollut.* **77**:143–150.

Matson, P.A., P.M. Vitousek, and D.S. Schimel. 1989. Regional extrapolation of trace gas flux based on soils and ecosystems. In: Exchange of Trace Gases between Terrestrial Ecosystems and the Atmosphere, ed. M.O. Andreae and D.S. Schimel, pp.97–108. Dahlem Workshop Report LS 47. Chichester: Wiley.

McCahon, C.P., and D. Pascoe. 1989. Short-term experimental acidification of a welsh stream: Toxicity of different forms of aluminium at low pH to fish and invertebrates. *Arch. Environ. Contam. Toxicol.* **18**:233–242.

McNulty, S.G., J.D. Aber, T.M. McLellan, and S.M. Katt. 1990. Nitrogen cycling in high elevation forest of the Northeastern U.S.A. in relation to nitrogen deposition. *Ambio* **19**:38–40.

Mellillo, J.M., P.A. Steudler, J.D. Aber, and R.D. Bowden. 1989. Atmospheric deposition and nutrient cycling. In: Exchange of Trace Gases between Terrestrial Ecosystems and the Atmosphere, ed. M.O. Andreae and D.S. Schimel, pp. 263–280. Dahlem Workshop Report LS 47. Chichester: Wiley.

Merrett, W.J., Rutt G.P., N.S. Weattherley, S.J. Thomas, and S.J. Ormerod. 1991. The response of macroinvertebrates to low pH and increased aluminium concentrations in Welsh streams: Multiple episodes and chronic exposure. *Arch. Hydrobiol.* **121**:115–125.

Meyer, J.L. 1990. Production and utilization of dissolved organic carbon in streams and riverine ecosystems. In: Organic Acids in Aquatic Ecosystems, ed. E.M. Perdue and E.T. Gjessing, pp. 281–300. Dahlem Workshop Report LS 48. Chichester: Wiley.

Miller, J.H., and M. Newton. 1983. Nutrient loss from disturbed forest ecosystems in Oregon's Coast Range. *Agro-Ecosystems* **8**:153–167.

Mulder, J., N. Christophersen, M. Hauhs, M. Vogt, S. Andersen, and D.O. Andersen. 1990. Hydrochemical controls in the Birkenes catchment as inferred from rainstorm high in seasalts. *Water Resour. Res.* **26**:611–622.

Mulholland, P.J., C.N. Dahm, M.B. David, D.M. Di Toro, T.R. Fisher, H.F. Hemond, I. Kögel-Knaber, M.H. Meybeck, J.L. Meyer, and J.R. Sedell. 1990. What are the temporal and spatial variations of organic acids at the ecosystem level? In: Organic Acids in Aquatic Ecosystems, ed. E.M. Perdue and E.T. Gjessing, pp 315–330. Chichester: Wiley.

Murdoch, P.S., and J.L. Stoddard. 1993. The role of nitrate in the acidification of streams in the Catskill Mountains of New York. *Water Resour. Res.*, in press.

National Research Council, Committee on Monitoring and Assessment of Trends in Acid Deposition. 1986. Acid Deposition: Long-term Trends. Washington, D.C.: National Academy Press. 506 p.

Neal, C., N. Christophersen, R. Neal, C.J. Smith, P.G. Whitehead, and B. Reynolds. 1988. Chloride in precipitation and streamwater for the upland catchment of river Severn, mid–Wales; some consequences of hydrochemical models. *Hydrol. Proc.* **2**:155–165.

Nicholson, J.A. 1975. Water quality and clear-cutting in a boreal forest ecosystem. *Can. Hydrol. Symp. 75 Proc.* **NRCC No. 15195**:734–738.

Ovalle, A.R.C., and E.V. Filho. 1991. Acid precipitation in Brazil: A short review. In: Acid Deposition: Origin, Impacts, and Abatement Strategies, ed. J.W.S. Longhurst, pp. 51–59. New York: Springer.

Paces, T. 1985. Sources of acidification in central Europe estimated from elemental budgets in small basins. *Nature* **315**:31–36.

Raddum, G.G., A. Fjellheim, and T. Hestthugen. 1988. Monitoring of acidification through the use of aquatic organisms. *Verh. Int. Verein Limnol.* **23**:2292–2297.

Renberg, I., T. Korsman, and N.J. Anderson. 1990. Spruce and surface water acidification: An extended summary. *Phil. Trans. Roy. Soc. Lond. B* **327**:371–372.

Reuss, J.O. 1989. Soil-solution equilibrium in lysimeter leachates under red alder. In : Effects of Air Pollution on Western Forests, ed. R.K. Olson and A.S. Lefohn, pp. 547–559. APCA Transaction Series No. 16. Pittsburgh: Air and Waste Management Association.

Reuss, J.O., B.J. Cosby, and R.F. Wright. 1987. Chemical processes governing soil and water acidification. *Nature* **329**:27–32.

Reuss, J.O., and D.W. Johnson. 1986. Acid deposition and the acidification of soils and waters. Ecological Studies 59. New York: Springer.

Reynolds, B., B.A. Emmett, and C.Woods. 1992. Variations in streamwater nitrate concentration and nitrogen budgets over 10 years in a headwater catchment in mid-Wales. *J. Hydrol.* **136**:155–175.

Rodhe, H., and R. Herrera. 1988. Acidification in tropical countries. SCOPE 36. New York: Wiley.

Rudd, J.W.M., C.W. Kelly, D.W. Schindler, and M.A. Turner. 1988. Disruption of the nitrogen-cycle in acidified lakes. *Science* **240**:1515–1517.

Rudd, J.W.M., C.A. Kelly, D.W. Schindler, and M.A. Turner. 1990. A comparison of the acidification efficiencies of nitric and sulfuric acids by two whole-lake addition experiments. *Limnol. Oceanogr.* **35**:663–679.

Schindler, D.W., S.E. Bayley, P.J. Curtis, B.R. Parker, M.P. Stainton, and C.A. Kelly. 1991. Natural and man-caused factors affecting the abundance and cycling of dissolved organic-substances in Precambrian Shield lakes. *Hydrobiologia* **229**:1–21.

Schindler, D.W., K.G. Beaty, E.J. Fee, D.R. Cruikshank, E.R. Debruyn, D.L. Findlay, G.A. Linsey, J.A. Shearer, M.P. Stainton, and M.A. Turner. 1990. Effects of climatic warming on lakes in the Central Boreal Forest. *Science* **250**:967–970.

Schindler, D.W., M.A. Turner, and R.H. Hefflein. 1985. Acidification and alkalinization of lakes by experimental addition of nitrogen compounds. *Biogeochem.* **1**:117–133.

Sollins, P., and F.M. McCorison. 1981. Nitrogen and carbon solution chemistry of an old growth coniferous forest watershed before and after cutting. *Water Resour. Res.* **17**:1409–1418.

Steinberg, C. 1991. Fates of organic matter during natural and anthropogenic lake acidification. *Water Resour.* **25**:1453–1458.

Stevens, P.A., D.A. Norris, T.H. Sparks, and A.L. Hodgson. 1993. Nitrate leaching losses from Sitka spruce forest and moorland in Wales. *Water, Air, Soil Pollut.*, in press.

Stoddard, J.L. 1993. Long-term changes in watershed retention of nitrogen. Its causes and aquatic consequences. In: Environmental Chemistry of Lakes and Reservoirs, ed. L.A. Baker. Advances in Chemistry Series No. 237. Washington, D.C.: Am. Chemical Soc., in press.

Sullivan, T.J., C.T. Driscoll, J.M. Eilers, and D.H. Landers. 1988. Evaluation of the role of seasalt inputs in the long-term acidification of coastal New England lakes. *Environ. Sci. Technol.* **22**:185–190.

Telang, S.A., G.W. Hodgson, and B.L. Baker. 1981. Effects of clear-cutting on abundances of oxygen and organic compounds in a mountain stream of the Marmot Creek Basin. *Can. J. For. Res.* **11**:545–553.

Tiedje, J.M. 1988. Ecology of denitrification and dissimilatory nitrate reduction to ammonium. In: Biology of Anaerobic Microorganisms, ed. J.B. Zehnder, pp. 179–244. New York: Wiley.

Tietema, A., and J.M. Verstraten. 1991. Nitrogen cycling in an acid forest ecosystem in the Netherlands at increased nitrogen input: The nitrogen budget and the effects of nitrogen transformations on the proton budget. *Biogeochem.* **156**:21–46.

Tinker, P.B., and P. Ineson. 1990. Soil organic matter and biology in relation to climate change. In: Soils on a Warmer Earth, ed. H.W. Scharpenseel, M. Schomaker, and A. Ayoub, pp.71–88. New York: Elsevier.

Van Breemen, N. 1990. Impact of anthropogenic atmospheric pollution on soil, with special relevance to tropical and subtropical soils, and possible consequences of the greenhouse effect. In: Soils on a Warmer Earth, ed. H.W. Scharpenseel, M. Schomaker, and A. Ayoub, pp. 137–144. Developments in Soil Science 20. New York: Elsevier.

Van Breemen, N., J. Mulder, and C.T. Driscoll. 1983. Acidification and alkalinization of soil. *Plant and Soil* **75**:283–308.

Van Cleve, K., W.C. Oechel, and J.L. Hom. 1990. Response of black spruce (*Picea mariana*) ecosystems to soil temperature modification in interior Alaska. *Can. J. For. Res.* **20**:1530–1535.

Van Dijk, H.F.G., A.W. Boxman, and J.G.M. Roelofs. 1992. Effects of a decrease in atmospheric deposition of nitrogen and sulfur on the mineral balance and vitality of a Scots pine stand in the Netherlands. *For. Ecol. Manag.* **51**:207–215.

Van Miegroet, H., and D.W. Cole. 1988. The influence of N-fixing alder on acidification and cation leaching from a forest soil. In: Forest Site Evaluation and Long-term Productivity, ed. D.W. Cole and S.P. Gessel, pp. 113–124. Seattle: Univ. of Washington Press.

Van Miegroet, H., D.W. Cole, and N.W. Foster. 1992. Nitrogen distribution and cycling. In: Atmospheric Deposition and Forest Nutrient Cycling, ed. D.W. Johnson and S.E. Lindberg, pp. 178–196. New York: Springer.

Vitousek, P.M., J.R. Gosz, C.C. Grier, J.M. Melillo, and W.A. Reiners. 1979. A comparative analysis of potential nitrification and nitrate mobility in forest ecosystems. *Ecol. Monogr.* **52**:155–177.

Wade, K.R., S.J. Ormerod, and A.S. Gee. 1989. Classification and ordination of macroinvertebrates assemblages to predict stream acidity in upland Wales. *Hydrobiologia* **171**:59–78.

Welsh, W.T., and J.C. Burns. 1987. The Loch Dee Project: Runoff and surface water quality in an area subject to acid precipitation and afforestation in SW Scotland. *Trans. Roy. Soc. Edinburgh* **78**:249–260.

Wigington, P.J., Jr., T.D. Davies, M. Tranter, and K.N. Eshleman. 1992. Comparison of episodic acidification in Canada, Europe and United States. *Environ. Pollut.* **78**:29–35.

Wright, R.F. 1976. The impact of forest fire on the nutrient influxes to small lakes in northeastern Minnesota. *Ecology* **57**:649–663.

Wright, R.F. 1989. RAIN project: Role of organic acids in moderating pH change following reduction in acid deposition. *Water, Air, Soil Pollut.* **46**:251–259.

Wright, R.F., S.A. Norton, D.F. Brakke, and T. Frogner. 1988. Experimental verification of episodic acidification of fresh waters by seasalts. *Nature* **334**:422–424.

Yoshimura, S. 1933. Katanuma, a very stong acid-water lake on Volcano Katanuma, Miyagi prefecture, Japan. *Arch. Hydrobiol.* **26**:197–202.

Zoettl, H.W. 1990. Remarks on the effects of nitrogen deposition to forest ecosystems. *Plant and Soil* **128**:83–89.

# 10

# Effects of Acidification on Trace Metal Transport in Fresh Waters

J. VESELÝ
Czech Geological Survey, 118 21 Prague 1, Czech Republic

## ABSTRACT

Anthropogenic acidification changes the relative importance of individual factors controlling both transport and concentration of trace elements in fresh waters. The mobility of Al, Mn, Be, Cd, and Zn in catchments is increased due to the atmospheric deposition of strong acids. The concentrations of these metals in surface waters rapidly increase with a decrease in pH, especially below pH 5. The transport of Cu, Hg, and probably Pb, even in acidified waters, is predominantly associated with high molecular weight organics and biota. The changes in the transport of Cu and Hg are mostly related to changes in the concentration and transport of organic compounds that result from acidification.

## INTRODUCTION

The principal sources of trace metals in fresh waters are soils and sediments (in which trace metals were previously deposited), atmospheric deposition, and weathering. Trace metals are removed from fresh waters predominantly by sedimentation following their incorporation onto or into particles, generally through complexation with surface sites. The surfaces on which metals are adsorbed are not just hydroxylated mineral surfaces but also living organisms, debris of plankton or bacteria and macromolecules adsorbed onto inorganic particles. The concentrations of trace metals in fresh waters are a result of complex competition between processes of adsorption and/or precipitation, on one hand, and formation of soluble complexes on the other.

Acidification dramatically alters the relative contribution of individual factors influencing transport and concentrations of trace metals in fresh waters (Nelson and Campbell 1991; Nriagu 1989). Sorption of metal cations onto particles, for example, has a new and strong competitor: free $H^+$ ions. The significance and abundance of $OH^-$ and $HCO_3^-$ complexes is lower and the role of the organocomplexes changes. Another reason for this is the fact that organic carbon concentrations in water typically decrease

*Acidification of Freshwater Ecosystems: Implications for the Future*
Edited by C.E.W. Steinberg and R.F. Wright 

as a result of acidification (Schindler et al. 1992; Steinberg 1991). Bacterial processes, which take place in water and at the water-sediment interface, thereby playing a crucial role in trace metal partitioning between both phases, are also affected (Kelly, this volume). With a decrease in the pH, the relative stability of ions in a lower oxidation state is enhanced, and adsorption surfaces are more likely to be affected by adsorption of organics (Davis 1984). In an acidified environment, however, the properties of particles are also subject to change: their solubility increases as does the solubility of trace metals coming from soils, vegetation, and bedrock of the catchment.

Studies of the effects of acidification on the behavior of trace metals in fresh waters are complicated by difficulties with analytical determinations of the generally low concentrations of these elements. Thus data for a number of elements are lacking. Another difficulty lies in separating the effect of strong acids from atmospheric deposition of trace metals per se, since they typically accompany sulfur in fossil fuels and are deposited along with sulfur compounds. Finally, the extent of acidification-induced changes is related to the characteristics of the catchment, including soil permeability and chemistry, presence or absence of a wetland, hydrology, and, in the case of lakes, the renewal time and mean depth. As a result, changes in the transport and concentration of trace metals caused by acidification are rather poorly understood.

Based on our present knowledge of prevailing transport mechanisms in acidified fresh waters, trace metals can be divided into two groups (LaZerte et al. 1989; Borg and Johansson 1989; Veselý and Majer, submitted):

1. trace metals mobilized by acid deposition,
2. trace metals whose transport in an acidified environment is controlled by the abundance of high molecular weight organics and biota.

## ACIDIFICATION-RELATED MOBILIZATION OF TRACE METALS

### Aluminum

Concentrations of acid-mobilized Al ($Al_t$) increase with decreasing pH, from a pH of ca. 6.3, and can exceed 2 mg $L^{-1}$ at pH < 4.4 in terrains with high deposition of strong acids (Veselý et al. 1985). $Al_t$ includes (a) organically bound Al ($Al_o$), (b) Al present in floating suspended matter (minor in acidic waters), and (c) an ecologically dangerous form: dissolved inorganic Al ($Al_i$) (Driscoll 1985). At DOC 5 mg $L^{-1}$, most of the dissolved Al ($Al_i + Al_o$) is present in the form $Al_i$ (Helmer et al. 1990).

The concentration of $Al_i$ is the sum of $Al^{3+}$ with Al bound in inorganic complexes, predominantly as $AlF_2^+$, $Al(OH)_2^+$, and $AlSO_4^+$. The highest concentration of $Al_i$ (up to 1.5 mg $L^{-1}$ $Al_i$; Hruška and Krám 1992) can characteristically be found in waters draining from terrains with high levels of acid deposition (consequently with more intense neutralization of acids due to dissolution of hydrolytic products of Al) and a low capacity to release basic cations, mostly Ca and Mg (Cronan and Schofield 1990).

Additional factors influencing the variability of $Al_i$ concentrations include diffe[illegible] in the concentrations of fluorocomplexes, the availability of the neutralizing [illegible]lytic products of Al in soils and sediments, and possibly dissolution kinetics. It follows that there is no generally valid relationship between pH and the transport of Al in acidified fresh waters (Cronan and Schofield 1990). Similarly, the concentration of $Al^{3+}$ is not solely controlled by an equilibrium with a single mineral, such as gibbsite (Hooper and Shoemaker 1985). Even at similar pH values, individual acidified terrains differ significantly in their $Al_i$ concentrations.

Nonetheless, concentrations of $Al_i$ and $H^+$ generally increase with increasing discharge in stream water, although $Al^{3+}$ may remain constant under certain conditions or decrease with decreasing pH (Sullivan et al. 1986). The $Al_i$ concentration may gradually decrease at a roughly constant pH, as shown by manipulation studies in which streams were artificially acidified (Norton et al. 1990). These changes can be attributed to variations in the pathways of circulating waters, in the kinetics of Al dissolution (Sullivan et al. 1986), or in the depletion of Al in ion-exchange positions accompanied by a gradual dissolution of decreasing soluble products of Al hydrolysis present in stream sediments (Norton et al. 1990). In contrast to easily exchangeable cations in ion-exchange positions (e.g., coming from submerged vegetation), the ability of hydrolytic products of Al to neutralize strong acids is considerable. However, it is not necessarily unlimited since more than 500 kg (18.5 mmol $m^{-2}$) of Al is annually removed from 1 $km^2$ of the most severely polluted catchments (Hruška and Krám 1992).

Whereas the neutralization of strong acids through the dissolution of hydrolytic products of Al prevails at a pH < 5, these products (i.e., amorphous Al(oxy)hydroxide and gibbsite) generally precipitate at a pH > 5, and Al is transported into sediments. The formation and solubility of these precipitates is not only controlled by the level of $Al^{3+}$ ions and the pH values, they are also affected by temperature, the presence of organic compounds (Tipping et al. 1988), the precipitation history, and most likely by the presence of Fe. Through the accumulation of these beige-colored precipitates, at sites exhibiting a pH of 5–6 (Norton and Henriksen 1983), a barrier against acidification of water further downstream is formed. This barrier hinders the transport of dissolved forms of a number of trace elements (e.g., Zn, Mn, and Pb), which are then transported into streambed sediments along with Al (Veselý et al. 1985).

## Manganese

The most acidified areas of Bohemia release 2.7 mmol $m^{-2}$ $yr^{-1}$ of Mn annually via streamflow (Veselý and Majer, submitted). In these areas, both humus-rich and mineral forest soils contain considerably less Mn (31% and 74%, respectively) when compared to soils on a similar bedrock and at a similar altitude, but with significantly lower acid deposition (Veselý 1987). Manganese is known to be mobilized from lacustrine sediments, where its contents in the upper few centimeters of acidified lakes are much lower (Bendell-Young et al. 1989). In most acidified catchments, Mn appears to be the least retained of all metals at budgets studies (Dillon et al. 1988).

Changes in the Mn transport caused by acidification are also documented by the characteristic relationship between the median concentrations of Mn and the pH in Bohemian stream waters (Figure 10.1). The concentrations of acid-mobilized Mn increase between pH 6.8 and 4.1; the relationship is more distinct in waters from territories with higher deposition rates of strong acids (Veselý and Majer, submitted), while the concentrations of Mn in bulk precipitation are very low (Figure 10.1).

A drop in the pH values associated with acidification enhances the relative stability of $Mn^{2+}$ ions and, at the same time, suppresses the formation of $Mn^{3+}$ and $Mn^{4+}$ (oxy)hydroxides ($MnO_x$) (Bendell-Young et al. 1989; Hem 1980). The formation of $MnO_x$ in an acidified environment is likely to proceed only under special conditions, such as microbial and surface catalysis or a sudden increase in the redox potential in oligotrophic waters. These oxides, e.g., birnessite, exhibit a high sorption capacity and an affinity toward Co, Pb, Ba, and As; thus the influence of $MnO_x$ on the migration of these elements under acidified conditions may be weaker.

In Bohemian stream waters exhibiting a pH < 4, Mn concentrations are significantly lower than at pH 4.0–4.2 (Figure 10.1). A likely explanation for the apparent reversal in mobility is a substantial decrease in Mn concentrations in the most severely acidified terrains, resulting from long-term exposure to acid deposition that has depleted the available Mn in soils. Another possibility is a higher abundance of waters that are largely unaffected by contact with mineral soils in the group of the most acidic waters, with a pH of 3.8–4.0.

## Beryllium

Another element that is easily mobilized into surface waters in acidified terrains is Be (Veselý et al. 1989). Approximately ten times more Be is present in very acidic waters, compared to Cd, Pb, or Cu, and a thousand times more than Hg since Be is an extremely light metal. The changes in median concentrations of Be in Bohemian stream waters with varying pHs are similar in character to those of Mn; however, the relative increase in the Be concentrations is substantially larger (approximately 100 times between a pH of 7 and 4; see Figure 10.1 ). Low Be concentrations in precipitation suggest that the increase in the Be concentration in acidified waters is almost completely caused by mobilization of the metal within the catchment.

Although the mean Be concentration generally increases sharply with decreasing pH, the increase in Be concentrations in fresh waters with decreasing pH often appears to be relatively slight. It follows that the variability in Be concentrations in acidified waters at a similar pH is rather high. As documented by an increasing difference between the upper and lower quartiles (between which 50% of the data falls), the variability in Be concentrations increases with decreasing pH (Figure 10.1). This occurs because the Be transport by acidic fresh waters is also influenced by the $F^-$ concentration, which forms the fluorocomplex $BeF^+$. Beryllium concentrations correlate with $F^-$ concentrations, from approximately 100 $\mu g\ L^{-1}$ $F^-$ (5 $\mu mol\ L^{-1}$), in a pH range of 4.2–6.0 (Veselý et al. 1989). At a pH of < 4.2, surface waters are usually

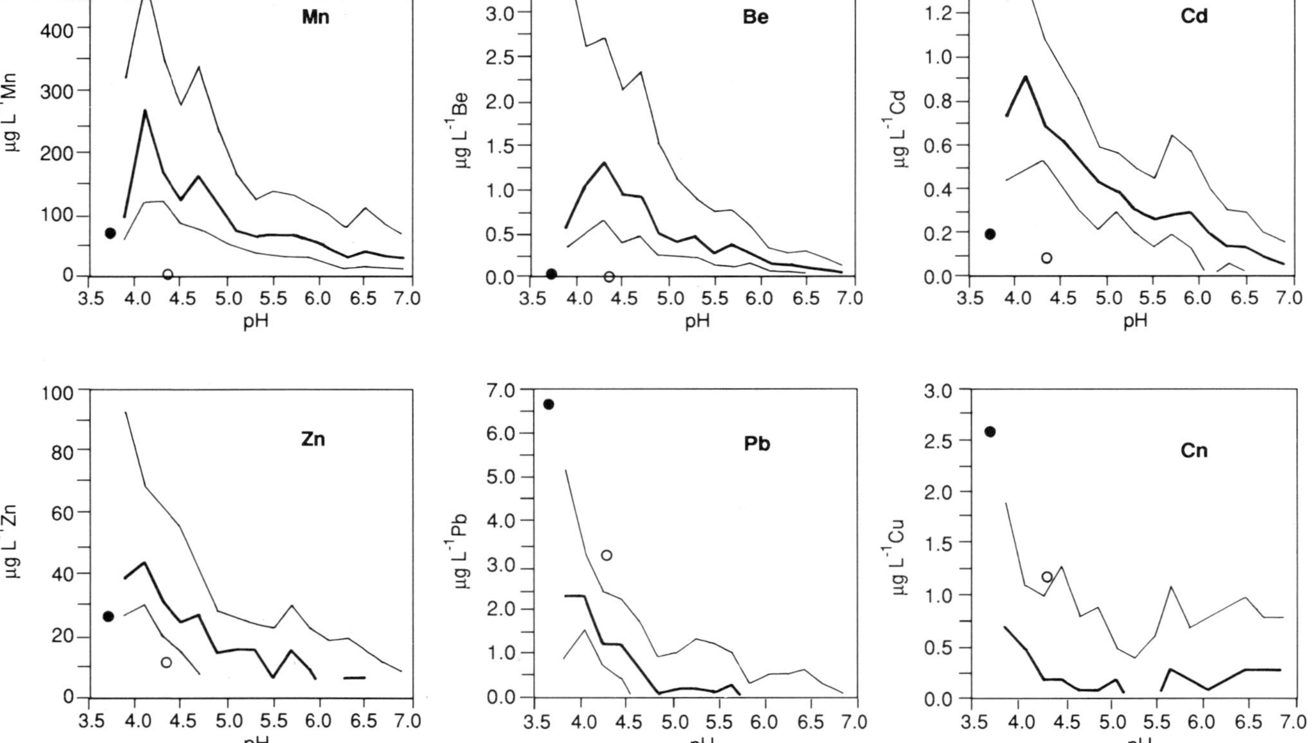


**Figure 10.1** Relationship between the mean concentrations, upper and lower quartiles of Mn, Be, Cd, Zn, Pb, and Cu and pH in Bohemian fresh waters. Values were calculated by sorting the water samples into groups with a similar pH (within 0.2 on the pH scale); only groups with n > 30 were considered. Volume-weighted average concentrations in bulk (○) and throughfall (●) in 1991 samples from the mountains in the Bohemian Forest are also shown. Conversion: Mn: 1μmol = 54.9 μg; Be: 1μmol = 9 μg; Cd: 1 μmol = 112.4 μg; Zn: 1μmol = 65.4 μg; Pb: 1μmol = 207.2 μg; Cu: 1 μmol = 63.5 μg.

depleted of free fluoride ions, due to their bonding in the stronger $AlF_2^+$ complex, and the Be mean concentration decreases (Figure 10.1).

### Cadmium and Zinc

The chemical similarity of Cd and Zn manifests itself in acidified fresh waters. Studies of the Cd concentrations and the retention of Zn in Canadian lakes (Stephenson and Mackie 1988; Dillon et al. 1988), metal fluxes to Swedish lakes (Borg and Johansson 1989; Bergkvist et al. 1989 ), and the dependence of the Cd and Zn median concentrations on the pH in Bohemian stream waters (Figure 10.1) indicate a predominantly acidification-related mobilization of Cd and Zn. Based on a comparison of the Zn and Cd mean concentrations in Bohemian stream waters and precipitation (Figure 10.1), it is apparent that atmospheric deposition is not negligible. The concentrations of these metals are increased by atmospheric deposition, especially in moderately acidified waters of heavily contaminated areas and in the case of Zn.

The seemingly erratic behavior of Cd in acidified waters is likely to be the result of bioaccumulation. Although Cd is nonessential for biota, some aquatic species bioconcentrate Cd more than a thousand times (Growder 1991). Recent empirical data indicate that pH may have a biphasic effect on Cd uptake by freshwater invertebrates. A decrease in the pH from approximately 7.0 to 5.5 increases Cd uptake, while a further decrease below pH 5.5 reduces Cd uptake by *Holopedium gibberum* (Yan et al. 1990).

Bioaccumulation could also explain the occurrence of a single, three-month long, local increase in the Cd concentration, which was observed during a 9-year monitoring study of five Bohemian lakes (pH of 4.25–5.75) situated up to 70 km apart (Veselý, unpublished results). In the lake water of the small 18 ha Černé Lake, strikingly different concentrations of acid-released Cd were found on the same day at different sampling sites (0.32 to 5.40 $\mu g\ L^{-1}$; 3 to 50 nmol $L^{-1}$). Filtered samples, however, had normal concentrations of Cd (0.30–0.35 $\mu g\ L^{-1}$; 2.7 to 3.1 nmol $L^{-1}$). The difference in Cd concentrations (in Černé Lake), with a pH of 4.5, was larger than the increase in those of Cu and Zn, elements well known for their tendency toward bioaccumulation.

## TRANSPORT CONTROL BY HIGH MOLECULAR WEIGHT ORGANIC COMPOUNDS

### Mercury

Unpolluted fresh waters contain < 0.001–0.01 $\mu g\ L^{-1}$ Hg (< 5–50 pmol $L^{-1}$) and < 0.0001 to 0.001 $\mu g\ L^{-1}$ (< 0.5 to 5 pmol $L^{-1}$) of the extremely toxic methylmercury. Fluctuations in the concentrations of organic compounds and seasonal variations are more important for the transport of both methylmercury and Hg in waters than changes in the pH (Bloom et al. 1991). The relative insensitivity of Hg concentrations toward pH is caused by a strong affinity of Hg toward most functional groups in organic molecules. Consequently, Hg concentrations are generally higher in humic waters and anoxic segments of lakes, where up to 50% Hg may be present in methylmercury species.

There is probably an inverse relationship between the methylmercury concentrations in fish and the pH values of lake water. This relationship probably does not result from a change in the specific rate of the methylmercury uptake (Bloom et al. 1991) but rather from an acidification-inflicted change in the supply of methylmercury to the ecosystem. Specifically, either bacterial processes (methylation and/or demethylation Hg) or an increase in the Hg available for methylation could be involved (Gilmour and Henry 1991).

Volatization of $Hg_0$ from water is strongly temperature-dependent and seems to be lower at a reduced water pH (Steffan et al. 1988). More Hg available for methylation might also be present in acidified waters as a result of decreased dissolved organic carbon (Winfrey and Rudd 1990). Less organic matter in water usually means a lower total Hg content; on the other hand, a higher rate of aerobic methylation of Hg is more likely in waters with less organic matter (Winfrey and Rudd 1990).

By contrast, organic compounds exhibit a stimulating effect on methylation of Hg in sediments. Anaerobic Hg methylation is associated with sulfate-reduction by sulfate-reducing bacteria, whose activity increases with increasing sulfate concentrations in water, typical of acidification. Laboratory experiments have shown, however, that a decrease in the pH of sediment porewater causes a decrease in the Hg methylation, probably due to a decrease in the available inorganic Hg at lower pH (Ramlal et al. 1985). It remains to be proven whether methylation of Hg in sediments is crucial to the increase in the methylmercury concentrations in fish during the initial stages of lake acidification.

## Copper and Lead

Several studies have documented the importance of organic matter in the transport of Cu and Pb in acidified terrains. For example, LaZerte et al. (1989) compared the concentrations of Pb and Cu in precipitation, outflow from wetland, and at the soil seep in the Plastic Lake catchment, Ontario. Although precipitation contained elevated Pb and Cu concentrations, the concentrations at the soil seep were very low. Lead and Cu from atmospheric deposition are tightly bound in the organic surface layer of the soils, and their transport to fresh waters occurs mostly in a form bound to organic substances (Borg and Johansson 1989; Kreutzer, this volume).

In contrast to acid-mobilized trace metals, mean Cu and Pb concentrations in Bohemian stream waters are lower than those in precipitation, but they are not lower in the most acidic waters ($pH < 4$), if compared to water in a pH range of 4.0–4.2 (Figure 10.1). Lead concentrations increase with decreasing pH and, if we compare stream waters with a similar pH, are higher in moderately rather than in highly contaminated Bohemian terrains. This may be caused by a higher content of organics in stream waters of moderately contaminated terrains. Contrary to Pb, the mean concentrations of Cu do not depend upon the pH between pH 4.4 and 7 (Figure 10.1).

However, a different mechanism may control Cu and Pb concentrations, especially in lakes, if the ratio of organic substances produced in the lake to organic substances transported from the catchment is high. In contrast to fulvic-humic substances transported from the catchment, organic substances produced in water are characterized by the presence of more nitrogen-containing functional groups; organic substances remobilized from lake sediments are characterized by the presence of more sulfur-containing groups. Therefore, the affinities of these substances toward trace metals are different. Substances produced in lakes contain fewer dissociable groups (mostly –COOH) and are more hydrophobic; thus their residence time in water is shorter (Buffle 1990). By contrast, organocomplexes of Al and Be are formed by bonding primarily to the –COOH and –OH functional groups of fulvic-humic substances. These tend to remain in water and be transported from the lake via outflow.

Increased sedimentation of metals with organic substances produced in lakes may result in a decrease in the metal concentration of the lake water. A high variability in the Cu and Pb concentrations observed in the surface water of the Čertovo Lake, Bohemia (Figure 10.2) can be explained in this way. A drop in the concentrations of Cu and Pb was not accompanied by a similar drop in Cd and Be. Scavenging of Cu and Pb can be enhanced by scavenging organic substances after their adsorption onto mineral particles (Davis 1984; White and Driscoll 1985). Figure 10.2 also shows a long-time decrease in the Pb lakewater concentration, which is caused by a gradual decrease in Pb in gasoline.

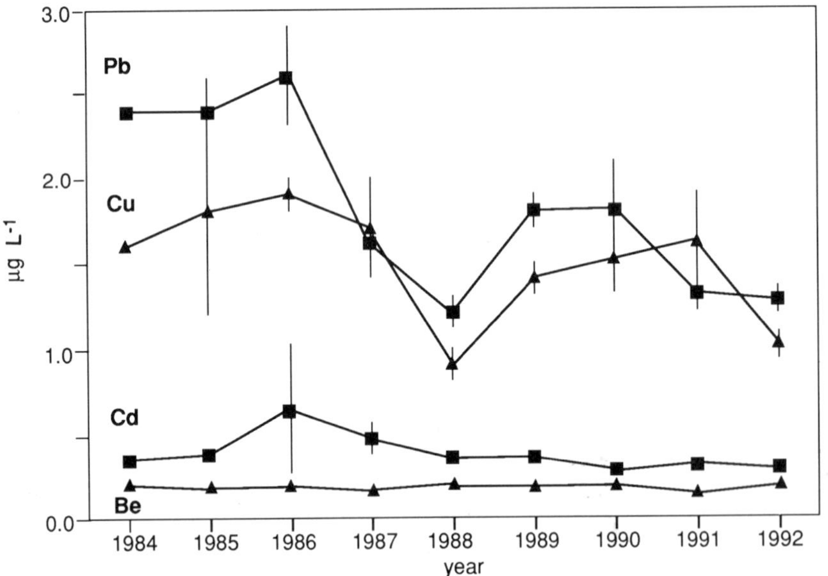


**Figure 10.2** Changes in the average summer concentrations of Pb, Cu, Cd, and Be in surface water of Čertovo Lake, Bohemia (1984–1992). pH range: 4.23–4.48; the standard deviations for Be and Cd ≤ 0.05 μg $L^{-1}$ are not given. See conversion chart in Figure 10.1.

## CONCLUSIONS

Aluminum and Mn transport increases in an acidified environment, primarily due to increased solubility and/or limited formation of hydrated oxides of the metals. These changes are enhanced by the formation of stable Al fluorocomplexes and a higher stability of $Mn^{2+}$ ions in acidified fresh waters. Less abundant trace elements (Cd, Be) do not normally form hydrolytic products, and their increased transport results from their suppressed sorption and remobilization of the elements previously sorbed on soils, sediments, and vegetation.

Long-term, intense leaching of the landscape by acidified precipitation gradually decreases the concentration of Mn and possibly other elements in upper soil horizons due to enhanced transport. A gradual decrease in their concentrations can thus be assumed to occur even in fresh waters, under identical conditions. For this reason, in Bohemian stream waters that have a pH of 3.8–4.0, compared to waters with a pH of 4.0–4.2, the mean Mn, Be, Zn, and Cd concentrations can be lower. As the lowest pH values in surface waters are close to those in throughfall samples, surface waters are usually less affected by the contact with the terrain. Their pathways are shallower and the chemical composition is more similar to throughfall precipitation than to the composition of waters with higher pH values. Thus the concentrations of mobilized elements may be lower, owing to a lower influence of the terrain. However, the simultaneous effect of depletion and suppressed influence of the terrain are most probable.

The concentration changes with variable pH are less clear in the case of trace elements, whose transport is primarily controlled by organic substances in an acidic environment. The concentrations of these trace elements are usually lowest between pH 5 and 5.8 and increase at pH < 4.4. Elements of this group (e.g., Cu, Hg) tend to form bonds with functional groups of organic molecules containing S or N, which are more abundant in substances originating in water and remobilizing from sediments. Cycling of these elements may be affected in this way, especially in lakes.

Furthermore, acidification-mobilized trace elements probably include Co and Sc, whereas the transport of Fe and Ag in an acidified environment is predominantly controlled by organic substances. The transport of ecologically important trace elements forming oxyanions (P, As, Se, and Mo) is also associated with organic substances in an acidic environment, although it is also affected by sorption to hydrated oxides, predominantly those of Fe. For example, from the drainage area of Gårdsjön Lake (Sweden), acidification resulted in a decrease in the input of P possibly due to efficient fixation of P in podzol soils (Jansson et al. 1986). The $PO_4^{3-}$ concentrations were minimal in a pH 5.2–5.8 range, and the rate constants of the $PO_4^{3-}$ uptake by microorganisms decreased in response to an acid shock in two Canadian lake waters (Nalewajko and O'Mahony 1988). Processes that decrease the pH values enhance the stability and transport of these elements in the lower oxidation state (e.g., $As^{3+}$; Xu et al. 1991) and appear to enhance the mobility and reduce availability of Se in soils (Mushak 1985).

## ACKNOWLEDGEMENTS

I gratefully acknowledge the comments made by S.A. Norton and T. Paces, and the help of V. Majer, K. Štulík, and M.Novák in preparing this manuscript.

## REFERENCES

Bendell-Young, L.I., H.H. Harvey, P.J. Dillon, and P.J. Scholer. 1989. Contrasting behavior of manganese in the surficial sediments of thirteen south-central Ontario lakes. *Sci. Total Environ.* **87–88**:129–139.

Bergkvist, B., L. Folkeson, and D. Berggren. 1989. Fluxes of Cu, Zn, Pb, Cd, Cr and Ni in temperate forest ecosystems: A literature review. *Water, Air, Soil Pollut.* **47**:217–286.

Bloom, N.S., C.J. Watras, and J.P. Hurley. 1991. Impact of acidification on the methylmercury cycle of remote seepage lakes. *Water, Air, Soil Pollut.* **56**: 477–491.

Borg, H., and K. Johansson. 1989. Metal fluxes to Swedish forest lakes. *Water, Air, Soil Pollut.* **47**: 427–440.

Buffle, J. 1990. The ecological role of aquatic organic and inorganic components, deduced from their nature, circulation and interactions. In: Metal Speciation in the Environment, ed. J.A.C. Broekaert, S. Gucer, and F. Adams, NATO ASI Ser., vol. 23, pp. 469–501. Berlin: Springer.

Cronan, C.S., and C.L. Schofield. 1990. Relationships between aqueous aluminum and acidic deposition in forested watersheds of North America and northern Europe. *Environ. Sci. Technol.* **24**:1100–1105.

Davis, J.S. 1984. Complexation of trace metals by adsorbed natural organic matter. *Geochim. Cosmochim. Acta* **48:**679–691.

Dillon, P.J., H.E. Evans, and P.J. Scholer. 1988. The effects of acidification on metal budgets of lakes and catchments. *Biogeochem.* **5**:201–220.

Driscoll, C.T. 1985. Aluminum in acidic surface waters: Chemistry, transport, and effects. *Environ. Health Persp.* **63**:93–104.

Gilmour, C.C., and E.A. Henry. 1991. Mercury methylation in aquatic systems affected by acid deposition. *Environ. Pollut.* **71**:131–169.

Growder, A. 1991. Acidification, metals and macrophytes. *Environ. Pollut.* **71**:171–203.

Helmer, E.H., N.R. Urban, and S.J. Eisenreich. 1990. Aluminum geochemistry in peatland waters. *Biogeochem.* **9**:247–276.

Hem, J.D. 1980. Redox coprecipitation mechanisms of manganese oxides. *Adv. Chem. Ser.* **189**:45–72.

Hooper, R.P., and C.A. Shoemaker. 1985. Aluminum mobilization in an acidic headwater stream: Temporal variation and mineral dissolution disequilibriums. *Science* **229**:463–465.

Hruška, J., and P. Krám. 1992. Monitoring of the Lysina catchment with high aluminum concentration in runoff. In: Proc. Environmental Regeneration in Headwaters, ed. J. Křeček and M.J. Haigh, pp. 111–116. Prague: ENCO.

Jansson, M., G. Persson, and O. Broberg. 1986. Phosphorus in acidified lakes: The example of Lake Gårdsjön, Sweden. *Hydrobiologia* **139**:81–96.

LaZerte, B., D. Evans, and P. Grauds. 1989. Deposition and transport of trace metals in an acidified catchment of central Ontario. *Sci. Total. Environ.* **87–88**:209–221.

Mushak, P. 1985. Potential impact of acid precipitation on arsenic and selenium. *Environ. Health Persp.* **63**:105–113.

Nalewajko, C., and M.A. O'Mahony. 1988. Effects of acid pH shock on phosphate concentrations and microbial phosphate uptake in an acidifying and circumneutral lake. *Can. J. Fish. Aquat. Sci.* **45**:254–260.

Nelson, W.O., and P.G.C. Campbell. 1991. The effects of acidification on the geochemistry of Al, Cd, Pb and Hg in fresh waters environments: A literature review. *Environ. Pollut.* **71**:91–130.

Norton, S.A., and A. Henriksen. 1983. The importance of $CO_2$ in evaluation of effects of acidic deposition. *Vatten* **39**:346–354.

Norton, S.A., J.S. Kahl, A. Henriksen, and R.F. Wright. 1990. Buffering of pH depressions by sediments in streams and lakes. In: Soils, Aquatic Processes, and Lake Acidification, ed. S.A. Norton, S.E. Lindberg, and A.L. Page, vol. 4, Acidic Precipitation, pp. 133–157. Berlin: Springer.

Nriagu, J.O., ed. 1989. Trace elements in lakes. Proc. Int. Conf. at McMaster Univ., Hamilton, Canada, 14–18 Aug., 1988. *Sci. Total. Environ.*, Special Issue, **87–88**.

Ramlal, P.S., J.W.M. Rudd, A. Furutani, and L. Xun. 1985. The efect of pH on methyl mercury production and decomposition in lake sediments. *J. Can. Fish. Aquat. Sci.* **42**:685–692.

Schindler, D.W., S.E. Bayley, P.J. Cuetis, B.R. Parker, M.P. Stainton, and C.A. Kelly. 1992. Natural and man-caused factors affecting the abundance and cycling of dissolved organic substances in PreCambrian Shield lakes. *Hydrobiologia* **229:**1–21.

Steffan, R.J., E.T. Korthals, and M.R. Winfrey. 1988. Effects of acidification on mercury methylation, demethylation, and volatization in sediments from an acid-susceptible lake. *Appl. Environ. Microbiol.* **54**:2003–2009.

Steinberg, C. 1991. Fate of organic matter during natural and anthropogenic lake acidification. *Water Res.* **25:**1453–1458.

Stephenson, M., and G.L. Mackie. 1988. Total cadmium concentrations in the water and lithoral sediments of central Ontario lakes. *Water, Air, Soil Pollut.* **38**:121–136.

Sullivan, T.J., N. Christophersen, I.P. Muniz, H.M. Seip, and P.D. Sullivan. 1986. Aqueous aluminium chemistry response to episodic increases in discharge. *Nature* **323**: 324–327.

Tipping, E., C. Woof, P.B. Walters, and D.M. Ohnstad. 1988. Conditions required for the precipitation of aluminium in acidic natural waters. *Water Res.* **22**:585–592.

Vesely, J. 1987. The effects of emissions on the chemical composition of the forest soils. *Lesnictví* **33:**385–398 (in Czech).

Vesely, J., P. Benes, and K. Sevcík. 1989. Occurence and speciation of beryllium in acidified fresh waters. *Water Res.* **23:**711–717.

Vesely, J., Z. Šulcek, and V. Majer. 1985. Acid-base changes in streams and their effect on the contents of some heavy metals in stream sediments. *Vest. Ústr. Úst. geol. (J. Geol. Survey Prague)* **60:**9–23.

White, J.R., and C.T. Driscoll. 1985. Lead cycling in an acidic Adirondack lake. Environ. Sci. Technol. 19:1182–1187.

Winfrey, M.R., and J.W.M. Rudd. 1990. Environmental factors affecting the formation of methylmercury in low pH lakes: A review. *Environ. Toxicol. Chem.* **9:**853–869.

Xu, H., B. Allard, and A. Grimvall. 1991. Effects of acidification and natural organic materials on the mobility of arsenic in the environment. Water, Air, Soil Pollut. 57–58:269–278.

Yan, N.D., G.L. Mackie, and P.J. Dillon. 1990. Cadmium concentrations of crustacean zooplankton of acidified and nonacidified Canadian Shield lakes. Environ. Sci. Technol. 24:1367–1372.

# 11

# Changes Caused by Acidification to the Biodiversity: Productivity and Biogeochemical Cycles of Lakes

D.W. SCHINDLER
Department of Zoology, CW-312 Biological Sciences, University of Alberta, Edmonton, Alberta, T6G 2E9, Canada

## ABSTRACT

Acid precipitation appears to disrupt the nitrogen cycle in some lakes and the phosphorus cycle in some catchments. While overall productivity of acidified lakes does not appear to decline unless phosphorus inputs decrease, production in littoral areas can decrease as the result of carbon limitation. Sediment respiration declines with acidification to pH values below 5.0.

In littoral communities, growth of filamentous algae (usually chlorophytes) in North America or *Sphagnum* in Scandinavia and central Europe changes the whole character of the littoral communities.

In general, biodiversity of all taxonomic groups declines as lakes acidify, and there is a switch to acid-tolerant taxa. In some cases, this appears to disrupt the functioning of food webs; however, case histories are too few to assess how widespread such effects might be.

## INTRODUCTION

In the early 1970s, Swedish scientists observed that acidifying lakes were clearer and had lower nutrient concentrations and standing crops than nearby circumneutral lakes (Grahn et al. 1974). Higher than normal concentrations of coarse detritus were observed on lake bottoms. These were hypothesized to result from a slowing of decomposition by acidification, causing reduced nutrient cycling. Grahn and his colleagues hypothesized that acidification was causing lakes to undergo "oligotrophication," i.e., to become less productive, support lower plant biomass, and have lower

*Acidification of Freshwater Ecosystems: Implications for the Future*
Edited by C.E.W. Steinberg and R.F. Wright 

nutrient concentrations than less acidic lakes. This concept is opposite to that of "eutrophication," which refers to the increasing productivity, plant biomass, and nutrient concentrations in lakes that receive sewage or other sources of nutrients (Wetzel 1983).

Studies in several regions of the world have examined aspects of the "oligotrophication" hypothesis (for a review, see Schindler 1988). Briefly, current evidence shows that acidification has significant effects on biodiversity in all lakes studied but that effects on biogeochemical cycles, plant productivity, and biomass vary significantly from lake to lake. In many cases, oligotrophication, in the sense originally hypothesized, does not occur. The mechanisms by which biogeochemical cycles are affected are also somewhat different than those originally envisioned by Grahn and his colleagues. Below I review evidence for oligotrophication from several regions and studies, and for changes in the productivity, biodiversity, and the biogeochemical cycles of nutrients in acidified lakes.

## EFFECTS OF ACIDIFICATION ON PHYTOPLANKTON

Early proponents of the oligotrophication hypothesis believed that acidic lakes were clearer than circumneutral ones because standing crops and production of phytoplankton declined, perhaps as a result of the disruption of key nutrient cycles. With a few exceptions noted later, no preacidification data existed that could be used to test this hypothesis on individual lakes. Whole-lake experiments the Experimental Lakes Area (ELA) and at Little Rock Lake (LRL) (Schindler 1980; Shearer et al. 1987; Schindler et al. 1991) and measurements in lakes of different acidity (Kippo-Edlund and Heitto 1990; see also reviews by Dillon et al. 1984 and Baker et al. 1990) do not show declines in phytoplankton production or biomass as lakes acidify. Lake Gårdsjön, Sweden, appears to be an exception for reasons outlined later.

At ELA, experimental acidification of lakes was accompanied by increases in phytoplankton production and standing crop, as well as in lake clarity and nutrient concentration. However, all of the changes were observed in reference lakes as well and were interpreted as being caused largely by climatic warming and drying during the years of the experiments, rather than by acidification (Schindler et al. 1990, 1993). The small increase in phytoplankton in excess of that observed in natural lakes was caused by increased subthermocline production resulting from greater water transparency (Shearer et al. 1987).

In contrast, all studies have shown that the number of phytoplankton species and various measures of diversity decline as lakes acidify (Dillon et al. 1984; Lydén and Grahn 1985; Baker et al. 1990; Kippo-Edlund and Heitto 1990; Schindler et al. 1991). Original species of algae are replaced with acid-tolerant forms, which are relatively fewer in number (Renberg et al. 1985; Kippo-Edlund and Heitto 1990).

## EFFECTS ON PERIPHYTON AND MACROPHYTE COMMUNITIES

Swedish, Finnish, and German scientists have observed that terrestrial species of mosses often invade the margins of acidifying lakes, replacing the original macrophyte assemblages (Grahn 1985; Heitto 1990; Melzer and Rothmeyer 1983). In the acidic Lake Gårdsjön, mats of mosses covered lake bottoms. Original algal species in the littoral zones (typically dominated by species of diatoms, small cyanophytes, and chlorophytes) were also replaced largely with a few species of filamentous chlorophytes or bluegreens (Almer et al. 1978; Lazarek 1985).

In North America, invasions of acidic lakes by mosses are rare, although many of the *Sphagnum* species that invaded Scandinavian and German lakes are present in eastern Canada. Dillon et al. (1984) report *Sphagnum* from a few acidic lakes in eastern Canada. Instead, the littoral zones of acidifying North American lakes typically become dominated by a few species of filamentous chlorophytes, usually of the family Zygnemataceae (reviewed by Baker et al. 1990;). Epilithic (rock-attached) waterline bands of periphyton (attached algae) form very early in the acidification process (pH > 6), making changes in periphyton production one of the lacustrine processes most sensitive to acidification (Schindler 1990; Turner et al. 1987 and unpublished). At lower pH values, the genera *Mougeotia*, *Zygnema*, and *Zygogonium* form huge metaphytic (littoral, unattached) biomasses in the warmer months of summer (Schindler 1980; Müller 1980; Stokes 1986; Howell et al. 1990). By pH values of 5.3, filamentous chlorophytes dominate littoral areas, occupying epilithic, epiphytic (attached to macrophytes), and metaphytic sites. Midsummer blooms intensify as lakes become even more acidic, although interannual variability is large (Turner et al. 1991 and unpublished).

The changes in periphyton communities were first believed to result from increased photosynthesis and decreased respiration (Hendrey et al. 1976; Almer et al. 1978; Müller 1980). Detailed investigation at ELA has shown that this hypothesis is probably incorrect and that photosynthesis declines while respiration of littoral periphyton communities increases as lakes acidify. The P/R ratio approaches 1 at pH values below 5, even in midsummer when long daylight hours assist in maintaining a positive P/R balance. At depths below 3 m, metaphyton cover is usually almost nonexistent; at all depths, biomass and production decline as day length declines in autumn. The effect is not caused by toxicity of the hydrogen ion, but by carbon limitation. At low pH, the only form of dissolved organic carbon available as a substrate for photosynthesis is dissolved $CO_2$, which is below atmospheric saturation due to depletion by photosynthesis. Turbulence also favorably affects the ability of algae to obtain $CO_2$, and the increased dominance by filamentous greens as pH declines may be because they are better adapted to maintain positive P/R balances at low $CO_2$ than the original taxa (Turner et al. 1991 and in review). Similarly, Lazarek (1985) found that epiphytic communities had low production. He hypothesized that decreased grazing and decomposition caused their accumulation. Dramatic changes in the macroinvertebrate grazing communities of the littoral zone have been shown by Raddum and Fjellheim (this volume), and paleoecological data suggest increasing abundance of littoral microcrustacea at low pH in Adirondack and German lakes (Charles et al. 1990; Steinberg et al. 1988). Nonetheless, the effects of altered grazing on periphyton caused by acidification have not been studied in detail. In summary, littoral periphyton communities appear to decline in production as lakes acidify, providing evidence for oligotrophication in the strictest sense. In contrast, the biomass of algae in littoral areas increases, reminding one of the *Cladophera* epidemics observed during the eutrophication of the lower St. Lawrence Great Lakes. The number and diversity of periphyton species clearly decline as lakes acidify.

With the exception of the increases in *Sphagnum* mentioned above, studies of macrophytes in acidifying lakes have given conflicting results, and there is no clear-cut evidence of the direction of response (Baker et al. 1990; Grahn 1985; Heitto 1990; Melzer and Rothmeyer 1983). Wile et al. (1985) show that the depth to which macrophytes grow in acidic lakes may increase due to increased transparency. It is impossible to relate the present information to the oligotrophication hypothesis in any meaningful way.

## DISRUPTION OF NUTRIENT CYCLES

While early surveys revealed that acidifying lakes had lower concentrations of phosphorus than circumneutral ones (Almer et al. 1974; Dillon et al. 1979), it now appears that the same poor soils and hardrock geology that predispose lakes to acidification yield low concentrations of both base cations and phosphorus, so that lakes and streams that are acidifying probably always had lower phosphorus concentrations than those with higher buffering capacity in the same areas. In Sweden, long-term (> 10 years) decreases in phosphorus concentrations have been observed, with concentrations by the early 1980s generally 50–70% of those observed a decade earlier (Hultberg and Andersson 1982; Jansson et al. 1986). The declines appear to result from decreased yields of phosphorus from catchments due to reaction of phosphorus with aluminum and iron oxides (Broberg 1987). Other areas with long-term data sets do not show such trends.

There is no field evidence that the internal phosphorus cycle is disrupted by acidification, either from experimentally or atmospherically acidified lakes (Dillon et al. 1984; Ogburn 1984; Schindler et al. 1991). However, Detenbeck and Brezonik (1991) found that in laboratory experiments, sorption of phosphorus by sediments increased greatly as pH declined from 6.0 to 4.5. Under normal circumstances, the sediments of softwater lakes release very little phosphorus (Schindler et al. 1977; Levine et al. 1986), so that increased sorption might not be detectable *in situ*.

Because phosphorus is the limiting nutrient for phytoplankton growth in most softwater lakes (Schindler 1977, 1978), changes in the cycles of nitrogen, carbon, or other nutrients are unlikely to cause oligotrophication.

## CHANGES IN MICROBIAL ACTIVITY

The aquatic nitrogen cycle is clearly disrupted by damage to the microbial community. In two experimentally acidified lakes at ELA, nitrification declined at pH values less than 5.4–5.6 (Rudd et al. 1988), and nitric acid began to contribute to acidification once nitrogen inputs exceeded the demands of phytoplankton (Rudd et al. 1990; Kelly et al. 1990). In Little Rock Lake, nitrogen fixation by phytoplankton ceased at pH 5.0 (Schindler et al. 1991). However, nitrification does not appear to be disrupted in all lakes, and it may be that a suitable nitrifying community can develop over time. The matter deserves further study.

In contrast to the situation with phosphorus, acidification appears to have caused increasing yields of nitrogen from terrestrial catchments to lakes, chiefly as nitrate. Increased nitrogen deposition has caused nitrogen saturation of many terrestrial ecosystems, stimulated increased nitrification in soils, and caused "forest decline" at some sites, thereby accelerating terrestrial nitrogen losses (Nilsson and Grennfelt 1988; Hornung 1993). Once nitrate input exceeds the demands of producers in receiving lakes, denitrification is the only process removing increased nitrogen inputs, and acidification by nitric acid begins to occur (Kelly et al. 1990).

Evidence from ELA suggests that acidification slows decomposition in winter, in contrast to the increases in decomposition in littoral areas during the summer, as mentioned above. Schindler et al. (1991) found that respiration under winter ice in lakes 223 and 302S decreased at pH values below 5.0. Detailed study indicated that metabolism of old carbon in sediments was unaffected by acidification of overlying water, but that decomposition of newly sedimented carbon declined at a pH of 5.0–5.25 (Kelly et al. 1984). The buildup of coarse detritus in acidified ELA lakes also confirms the early Swedish observations. It appears to be caused by the loss of several acid-sensitive benthic macroinvertebrates, such as the amphipod *Hyalella* and several insect species that perform the first step in the "processing" of organic detritus.

Acid titration eliminates ionic forms of dissolved inorganic carbon by converting it to $CO_2$, so that dissolved organic carbon concentrations at pH values below 5.0 are only 10–20 $\mu$mol $L^{-1}$, i.e., $CO_2$ in equilibrium with the atmosphere. While phytoplankton production in oligotrophic lakes does not appear to be limited by such low concentrations, periphyton appears to become carbon-limited, as mentioned above.

## DISRUPTIONS OF OTHER KEY CHEMICAL CYCLES BY ACIDIFICATION

Other chemical changes related to acidification may change phytoplankton distribution, production, and nutrient cycling. For example, in many softwater lakes, the attenuation of light is chiefly a function of dissolved organic matter (DOM) rather than phytoplankton (Schindler 1971). Acidification has been shown to cause DOM concentrations to decline in acidified lakes of several areas (reviewed by Schindler 1988;

Baker et al. 1990). As mentioned above, this can cause a deepening of the euphotic zone and increased availability of light for photosynthesis at all depths. It also exposes more sediment to warmer epilimnion water, possibly accelerating microbial activity. These mechanisms would counter declines in nutrient cycling to some degree.

At pH 4.5, the metaphytic mats mentioned above have been demonstrated to disrupt the normal annual production of alkalinity by sulfate reduction in littoral areas, accelerating the acidification process (Kelly et al., in review). Similar decreased production of alkalinity was observed in Little Rock Lake at pH 4.5, although the reasons have not been described (T. Frost, pers. comm.). Low pH values in sediments, symptomatic of declining alkalinity production, were also observed in Lake Hovvatn, Norway (Rudd et al. 1986; Kelly et al., in review).

## BIOTIC IMPOVERISHMENT CAUSED BY ACIDIFICATION

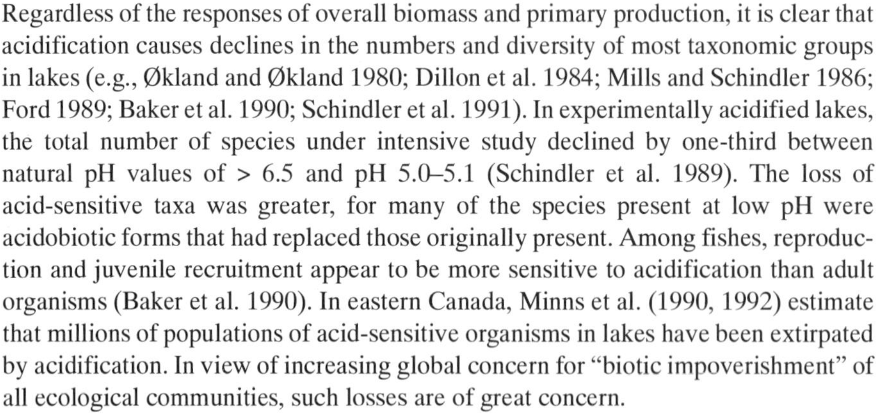

Regardless of the responses of overall biomass and primary production, it is clear that acidification causes declines in the numbers and diversity of most taxonomic groups in lakes (e.g., Økland and Økland 1980; Dillon et al. 1984; Mills and Schindler 1986; Ford 1989; Baker et al. 1990; Schindler et al. 1991). In experimentally acidified lakes, the total number of species under intensive study declined by one-third between natural pH values of > 6.5 and pH 5.0–5.1 (Schindler et al. 1989). The loss of acid-sensitive taxa was greater, for many of the species present at low pH were acidobiotic forms that had replaced those originally present. Among fishes, reproduction and juvenile recruitment appear to be more sensitive to acidification than adult organisms (Baker et al. 1990). In eastern Canada, Minns et al. (1990, 1992) estimate that millions of populations of acid-sensitive organisms in lakes have been extirpated by acidification. In view of increasing global concern for "biotic impoverishment" of all ecological communities, such losses are of great concern.

It is possible that many of the responses of species in lakes to acidification are indirect, caused by disappearance of sensitive species of predators or prey or by changes in critical habitats. At ELA, lake trout declined due to elimination of several species of acid-sensitive prey rather than to hydrogen ion toxicity, and their spawning areas became covered with filamentous green algae, thus preventing normal spawning activity (Schindler et al. 1985). Similar results of "complex interactions" have not been recorded in acidification studies, but they are well-known from intensive studies of other perturbations (Carpenter 1990), and it is possible that the paucity of intensive community studies is the reason so few have been found.

In many North American lakes, the increasing relative abundance of large phytoplankton species may, in theory, result from the disappearance of acid-sensitive species of zooplankton, which generally include larger species of *Daphnia* and calanoid copepods. In North America, small crustaceans like *Diaptomus minutus* and *Bosmina longirostris* and acidophilic rotifers like *Keratella taurocephala* generally predominate once pH values drop below 5.5 (Schindler et al. 1991). The upper size limit for food for such small grazers is generally less than 30 μmol $L^{-1}$, allowing large phytoplankton species to flourish. Still, the evidence to support such indirect effects is scanty, and it appears that either direct toxicity of the hydrogen ion or the toxicity of trace metals associated with acidic conditions (Stokes 1986) are the most important causes of changes in phytoplankton species and declines in diversity. Overall, the number of species of zooplankton also declines as lakes acidify (Baker et al. 1990; Schindler et al. 1991).

Stenson and Oscarson (1985) and Stenson and Eriksson (1989) showed that once all species of fish disappeared from Swedish lakes, large species of zooplankton became dominant because predacious insects that had previously been controlled by fish increased, depleting small zooplankton species. They also noted that littoral species became more apparent in pelagic regions, perhaps venturing farther from refugia once predators were absent. Overall, the changes in trophic interactions and size structures caused by acidification are highly variable and may depend on the sensitivity of particular key species of predators and prey to acidity.

## EVIDENCE FROM PALEOECOLOGICAL STUDIES

Paleoecological analyses of diatoms and chrysophyceans in lake sediments have been used to infer changes in lake pH over time. Strangely, only a few studies have mentioned changes in the number of taxa in fossil groups as lakes acidify, although such information must have been gathered in the process of inferring pH. It would be useful to see a straightforward account of the changes in the number of species and in community diversity that accompany acidification. Similarly, only a few studies have assessed fossils from a variety of trophic levels on the same cores (Charles et al. 1990; Steinberg et al. 1988).

## DISCUSSION

Figure 11.1 summarizes various changes to lake communities and biogeochemical processes that have been observed at one or more sites. Not all of the disruptions have been seen in any single lake; however, in closely studied systems, several processes and several aspects of community structure are disrupted, beginning at pH values as high as 6.0. In some cases, changes have been observed at only one to a few sites, or appear to occur at some sites and not others. For example, terrestrial phosphorus cycles are disrupted in catchments of southwestern Sweden, causing decreasing concentrations of this element in lakes; however, such phenomena have not been observed in other areas.

While there is evidence that some nutrient cycles are disrupted and that production of periphyton may decline with acidification, in most cases oligotrophication would not be an accurate description of the changes that acidification causes, for nutrient status and overall lake production are not noticeably affected.

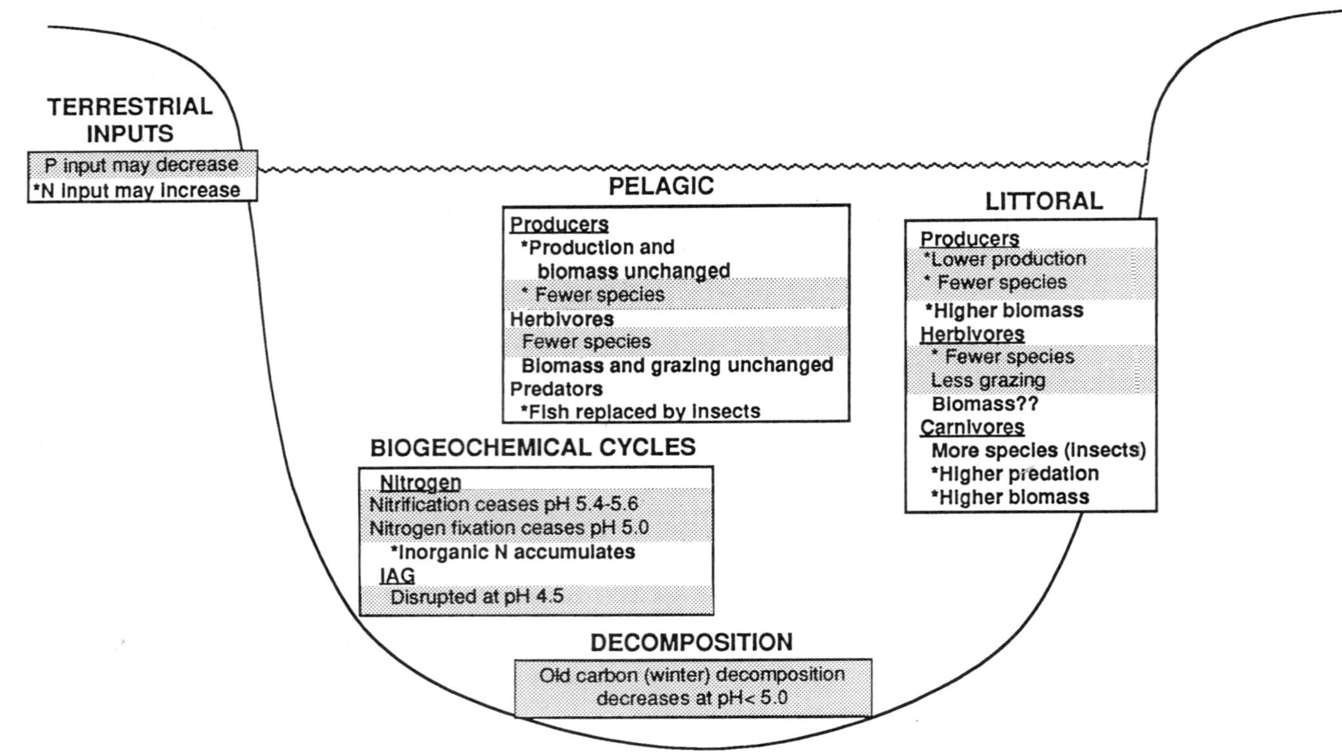


**Figure 11.1** A general model of how acidification causes biotic impoverishment and disrupts energy flow and biogeochemical cycling in lakes. Changes marked with * have been observed at several sites. Shaded changes would be regarded as evidence for oligotrophication by most scientists.

Lakes studied in enough detail to assess overall changes to community structure, productivity and nutrient cycling are still very few in number, so that it is difficult to generalize about how widespread and serious the observed changes in function might be. Analyses of pigments in lake sediments provide a record of both quantitative changes in lake phytoplankton and in the relative abundance of different phytoplankton groups (Gorham et al. 1974). Such analyses on near-surface sediments of acidified lakes, coupled with analysis of "hard" fossil remains, could shed useful light on the degree to which phytoplankton abundance has changed as the result of acidification.

## ACKNOWLEDGEMENTS

Work was supported by a Canadian National Science and Engineering Research Council operating grant during the period of writing. M.A. Turner provided an initial review of the manuscript and results of his periphyton studies. D. Brakke, J. Ford, and C.A. Kelly also provided helpful suggestions on the manuscript.

## REFERENCES

Almer, B., W.T. Dickson, C. Ekstrom and E. Hornstrom. 1978. Sulphur pollution and the aquatic ecosystem. In: Sulphur in the Environment: Part 2, ed. J. Nriagu, pp. 271–311. New York: Wiley.

Almer, B., W. Dickson, C. Ekstrom, E. Hornstrom and U. Miller. 1974. Effects of acidification on Swedish lakes. *Ambio* **3**:30–36.

Baker, J.P., D.P. Bernard, S.W. Christensen, M.J. Sale, J. Freda, K. Heltcher, D. Marmorek, L. Rowe, P. Scanlon, G. Suter, W. Warren-Hicks, and P. Welbourn. 1990. Biological effects of changes in surface water acid-base chemistry. NAPAP Report 13. In: National Acid Precipitation Assessment Program, Acidic Deposition: State of Science and Technology, vol. 2, ed. P.M. Irving. Washington, D.C.: NAPAP.

Broberg, O. 1987. Nutrient responses to the liming of Lake Gårdsjön. *Hydrobiologia* **150**:11–24.

Carpenter, S.R., ed. 1990. Complex Interactions in Lake Communities. New York: Springer.

Charles, D.F., et al. 1990. Paleoecological investigation of recent lake acidification in the Adirondack Mountains, New York. *J. Paleolimnol.* **3**:195–241.

Detenbeck, N.E., and P.L. Brezonik. 1991. Phosphorus sorption by sediments in a soft-water seepage lake. Parts 1 and 2. *Environ. Sci. Technol.* **25**:395–403 and **25**:403–409.

Dillon, P.J., N.D. Yan, and H.H. Harvey. 1984. Acidic deposition: Effects on aquatic ecosystems. *CRC Crit. Rev. Environ. Contam.* **13**:167–194.

Dillon, P.J., N.D. Yan, W.A. Scheider, and N. Conroy. 1979. Acidic lakes in Ontario Canada: Characterization, extent and responses to base and nutrient additions. *Arch. Hydrobiol.* **13**:317–336.

Ford, J. 1989. The effects of chemical stress on aquatic species composition and community structure. In: Ecotoxicology: Problems and Approaches, ed. S.A. Levin, M.A. Harwell, J.R. Kelly, and K.D. Kimball, pp. 99–144. New York: Springer.

Gorham, E., J.W.G. Lund, J.E. Sanger, and W.E. Dean, Jr. 1974. Some relationships between algal standing crop, water chemistry, and sediment chemistry in the English Lakes. *Limnol. Oceanogr.* **19**:601–617.

Grahn, O. 1985. Macrophyte biomass and production in Lake Gårdsjön: An acidified clearwater lake in SW Sweden. *Ecol. Bull. Stockholm* **37**:203–212.

Grahn, O., H. Hultberg, and L. Landner. 1974. Oligotrophication: A self-accelerating process in lakes subjected to excessive supply of acid substances. *Ambio* **3**:93–94.

Heitto, L. 1990. Macrophytes in Finnish forest lakes and possible effects of airborne acidification. In: Acidification in Finland, ed. Kauppi et al., pp. 963–972. Berlin: Springer.

Hendrey, G.R., K. Baalsrud, T.S. Traaen, M. Laake, and G. Raddum. 1976. Acidic deposition: Some hydrobiological changes. *Ambio* **15**:224–227.

Hornung, M. 1993. ENCORE: European network of catchments organised for research on ecosystems. In: Experimental Manipulations of Biota and Biogeochemical Cycling on Ecosystems, ed. L. Rasmussen, T. Brydges, and P. Mathy, pp. 37–45. Copenhagen: Commission of the European Communities.

Howell, T., M.A. Turner, R. France, and P.M. Stokes. 1990. Ecological features of acidification-induced growth of metaphyton. *Can. J. Fish. Aquat. Sci.* **47**:1085–1092.

Hultberg, H., and I. Anderson. 1982. Liming of acidified lakes: Induced long-term changes. *Water, Air, Soil Pollut.* **18**:311–331.

Jansson, M., G. Persson, and O. Broberg. 1986. Phosphorus in acidified lakes: The example of Lake Gårdsjön, Sweden. *Hydrobiologia* **139**:81–96.

Kelly, C.A., J.W.M. Rudd, A. Furutani, and D.W. Schindler. 1984. Effects of lake acidification on rate of organic matter decomposition in sediments. *Limnol. Oceanogr.* **29**:687–694.

Kelly, C.A., J.W.M. Rudd, and D.W. Schindler. 1990. Lake acidification by nitric acid: Future considerations. *Water, Air, Soil Pollut.* **50**:49–61.

Kippo-Edlund, P., and A. Heitto. 1990. Phytoplankton and acidification in small forest lakes. In: Acidification in Finland, ed. Kauppi et al., pp. 973–983. Berlin: Springer.

Lazarek, S. 1985. Epiphytic algal production in the acidified Lake Gårdsjön, SW Sweden. *Ecol. Bull. Stockholm* **37**:213–218.

Levine, S.N., M.P. Stainton, and D.W. Schindler. 1986. A radiotracer study of phosphorus cycling in a eutrophic Canadian shield lake, Lake 227, Northwestern Ontario. *Can. J. Fish. Aquat. Sci.* **43:**366–378.

Lydén, A., and O. Grahn. 1985. Phytoplankton species composition, biomass and production in Lake Gårdsjön: An acidified clearwater lake in SW Sweden. *Ecol. Bull. Stockholm* **37**:195–202.

Melzer, A., and E. Rothmeyer. 1983. Die Auswirkungen der Versauerung der beiden Arberseen in Bayerischen Wold auf die Makrophytenvegetation. *Ber. Bayer. Bot. Gesellsch.* **54:**9–18.

Mills, K.H., and D.W. Schindler. 1986. Biological indicators of lake acidification. *Water, Air, Soil Pollut.* **30**:779–789.

Minns, C.K., J.E. Moore, D.W. Schindler, P.G.C. Campbell, P.J. Dillon, J.K. Underwood, and D.M. Whelpdale. 1992. Expected reduction in damage to Canadian lakes under legislated decreases in sulfur dioxide emissions. Report 92–1, Committee on Acid Deposition, Royal Society of Canada, Ottawa. ISSN 1188–911X37, 37 pp.

Minns, C.K., J.E. Moore, D.W. Schindler, and M.L. Jones. 1990. Assessing the potential extent of danger to inland lakes in eastern Canada due to acidic deposition. III. Predicted impacts on species richness in seven groups of aquatic biota. *Can. J. Fish. Aquat. Sci.* **44(1)**:3–5.

Müller, P. 1980. Effects of artificial acidification on the growth of periphyton. *Can. J. Fish Aquat. Sci.* **37**:355–363.

Nilsson, J., and P. Grennfelt, eds. 1988. Critical Loads for Sulphur and Nitrogen. Miljorapport, p. 15. Copenhagen: Nordic Council of Ministers.

Ogburn, R. W., III. 1984. Phosphorus dynamics in an acidic softwater Florida Lake. Ph.D. Thesis. Gainesville, FL: Univ. of Florida.

Økland, J., and K.A. Økland. pH level and food organisms for fish: Studies of 1000 lakes in Norway. In: Ecological Impact of Acid Precipitation, ed. D. Drabløs and E. Tollan, pp. 326–327. Proc. of an Intl. Conf., Sandefjord, Norway, March 11–14, 1980. Oslo: SNSF Project.

Renberg, I., T. Hellberg, and M. Nilsson. 1985. Effects of acidification on diatom communities as revealed by analysis of lake sediments. *Ecol. Bull. Stockholm* **37**:219–223.

Rudd, J.W.M., C.A. Kelly, V.H. St. Louis, R.H. Hesslein, A. Furutani, and M. Holoka. 1986. Microbial consumption of nitric and sulfuric acids in acidified north temperate lakes. *Limnol. Oceanogr.* **31**:1267–1280.

Rudd, J.W.M., C.A. Kelly, D.W. Schindler, and M.A. Turner. 1988. Disruption of the nitrogen cycle in acidified lakes. *Science* **240**:1515–1517.

Rudd, J.W.M., C.A. Kelly, D.W. Schindler, and M.A. Turner. 1990. A comparison of the acidification efficiencies of nitric and sulfuric acids by two whole-lake addition experiments. *Limnol. Oceanogr.* **35**:663–679.

Schindler, D.W. 1971. Light, temperature, and oxygen regimes of selected lakes in the Experimental Lakes Area, northwestern Ontario. *J. Fish. Res. Bd. Canada* **28**:157–169.

Schindler, D.W. 1977. Evolution of phosphorus limitation in lakes. *Science* **195**:260–262.

Schindler, D.W. 1978. Factors regulating phytoplankton production and standing crop in the world's fresh waters. *Limnol. Oceanogr.* **23(3)**:478–486.

Schindler, D.W. 1980. Experimental acidification of a whole lake: A test of the oligotrophication hypothesis. In: Ecological Impact of Acid Precipitation, ed. D. Drabløs and E. Tollan, pp. 370–374. Proc. of an Intl. Conf., Sandefjord, Norway, March 11–14, 1980. Oslo: SNSF Project.

Schindler, D.W. 1988. Effects of acid rain on freshwater ecosystems. *Science* **239**:149–157.

Schindler, D.W. 1990. Experimental perturbations of whole lakes as tests of hypotheses concerning ecosystem structure and function. Proc. of 1987 Crafoord Symp. *Oikos* **57**:25–41.

Schindler, D.W., K.G. Beaty, E.J. Fee, D.R. Cruikshank, E.D. DeBruyn, D.L. Findlay, G.A. Linsey, J.A. Shearer, M.P.Stainton, and M.A. Turner. 1990. Effects of climatic warming on lakes of the central boreal forest. *Science* **250**:967–970.

Schindler, D.W., T.M. Frost, K.H. Mills, P.S.A. Chang, I.J. Davis, F.L. Findlay, D.F. Malley, J.A. Shearer, M.A. Turner, P.J. Garrison, C.J. Watras, K. Webster, J.M. Gunn, P.L. Brezonik, and W.A. Swenson. 1991. Freshwater acidification, reversibility and recovery: Comparisons of experimental and atmospherically acidified lakes. In: Acidic Deposition: Its Nature and Impacts, ed. F.T. Last and R. Watling, vol. 97B, pp.193–226. Proc. of the Royal Society of Edinburgh.

Schindler, D.W., R.E. Hecky, and K.H. Mills. 1993. Two decades of whole-lake eutrophication and acidification experiment. In: Experimental Manipualtions of Biota and Biogeochemical Cycling in Ecosystems, ed. L. Rasmussen, T. Brydges, and P. Mathy, pp. 294–304. Copenhagen: Commission of the European Communities.

Schindler, D.W., R.H. Hesslein, and G. Kipphut. 1977. Interactions between sediments and overlying waters in an experimentally eutrophied Precambrian Shield lake. In: Interactions between Sediments and Fresh Water, ed. H.L. Golterman, pp. 235–243. Proc. Intl. Symp. in Amsterdam, Sept. 6–10, 1976. The Hague: PUDOC.

Schindler, D.W., S.E. Kasian, and R.H. Hesslein. 1989. Losses of biota from American aquatic communities due to acid rain. *Environ. Monit. Assess.* **12**:269–285.

Schindler, D.W., K.H. Mills, D.F. Malley, D.L. Findlay, J.A. Shearer, I.J. Davies, M.A. Turner, G.A. Linsey, and D.R. Cruikshank. 1985. Long-term ecosystem stress: The effects of years of experimental acidification on a small lake. *Science* **228**:1395–1401.

Shearer, J.A., E.J. Fee, E.R. DeBruyn, and D.R. DeClercq. 1987. Phytoplankton primary production and light attenuation responses to the experimental acidification of a small Canadian Shield lake. *Can. J. Fish. Aquat. Sci.* **44**:83–90.

Steinberg, C., H. Hartmann, K. Arzet, and D. Krause-Dellin. 1988. Paleoindication of acidification in Kleiner Arbersee (F.R. Germany, Bavarian Forest) by chydorids, chrysophytes, and diatoms. *J. Paleolimnol.* **1**:149–157.

Stenson, J.A.E., and M.O.G. Eriksson. 1989. Ecological mechanisms important for the biotic changes in acidified lakes in Scandinavia. *Arch. Environ. Contam. Toxicol.* **18**:201–206.

Stenson, J.A.E., and H.G. Oscarson. 1985. Crustacean zooplankton in the acidified Lake Gårdsjön system. *Ecol. Bull. Stockholm* **37**:224–231.

Stokes, P.M. 1986. Ecological effects of acidification on primary products in aquatic systems. *Water, Air, Soil Pollut.* **30**:421–438.

Turner, M.A., E.T. Howell, M. Summerby, R.H. Hesslein, D.L. Findlay, and M.B. Jackson. 1991. Changes in epilithon and epiphyton associated with experimental acidification of a lake to pH 5. *Limnol. Oceanogr.* **36**:1390–1405.

Turner, M.A., M.B. Jackson, D.L. Findlay, R.W. Graham, E.R. De Bruyn, and E.M. Vandermeer. 1987. Early responses of periphyton to experimental lake acidification. *Can. J. Fish. Aquat. Sci.* **44(1)**:135–149.

Wetzel, R. G. 1983. Limnology. 2nd edition. Toronto: W.B. Saunders, Co.

Wile, I., G.E. Miller, G.E. Hitchin, and N.D. Yan. 1985. Species composition and biomass of the macrophyte vegetation of one acidified and two acid-sensitive lakes in Ontario. *Can. Field Nat.* **99**:308–312.

# 12

# Trace Organic Contaminants in Anthropogenically Acidified Surface Waters

J. FORD[1] and T.C. YOUNG[2]

[1]Department of Fisheries and Wildlife, Oregon State University, c/o U.S. EPA Environmental Research Laboratory, 200 SW 35th Street, Corvallis, OR 97333, U.S.A.

[2]Department of Civil and Environmental Engineering, Clarkson University, Potsdam, NY 13676, U.S.A.

## ABSTRACT

Trace organic contaminants are delivered to acidified and acidifying aquatic ecosystems by both long-range atmospheric transport and direct inputs from surface runoff. The interaction between trace organic contaminants and surface water acidification is complex and depends both on the properties of the contaminant and on the nature and quantity of organic carbon in the system. Thus, to understand the interaction between surface water acidification and the bioavailability and potential biological effects of trace organic contaminants, we must have some understanding of the parameters affecting the interaction of these with the dissolved and colloidal organic carbon in the system, as well as of how the process of acidification might affect these interactions.

In this chapter, we provide a conceptual context within which the potential impacts of acidification on trace organic bioavailability and uptake can begin to be discussed. Three properties that are important determinants of contaminant behavior in aqueous systems are the n-octanol-water partition coefficient ($K_{ow}$), aqueous solubility, and vapor pressure. We position 19 trace organic contaminants in a common 3-space defined by these properties and discuss how acidification might be expected to affect contaminants in each of four quadrants in this space. We theorize that when surface water acidification is associated with changes in concentrations of carbon pools, biological uptake of and exposure to certain classes of organic contaminants will be particularly affected. One specific prediction is that decreases in the size of the pool of dissolved carbon would be expected to increase biological exposure to contaminants with high $K_{ow}$, low solubility, and low volatility.

*Acidification of Freshwater Ecosystems: Implications for the Future*
Edited by C.E.W. Steinberg and R.F. Wright 

## INTRODUCTION

Although the co-occurrence of surface water acidification and trace organic contaminants can be significant, there is almost no literature that addresses the potential or actual interaction of these two types of surface water inputs and the consequences of these interactions for the biological integrity of surface waters. The purpose of this chapter is to initiate discussion and stimulate research on this potentially important set of cumulative impacts.

To the extent that this topic is addressed at all in the literature, the focus has primarily been on the potential overlap between surface water acidification and trace organic contaminants produced in association with the combustion of fossil fuels, particularly polycyclic aromatic hydrocarbons (PAHs). In some geographic regions, however, surface water acidification can also overlap significant atmospheric and surface water inputs of other types of industrial and/or agricultural chemicals, including persistent chlorinated hydrocarbons.

Atmospheric deposition of trace organic contaminants has been estimated for only a few areas. In Table 12.1 we summarize some of the currently available North American information in order to provide a frame of reference for assessing the significance of these inputs for acidifying systems. The sites in this table cover southern Canada, the Great Lakes region, and the Adirondack and Finger Lake regions of New York, all of which receive some level of acidic deposition and most of which include at least some areas of acid-sensitive landscapes.

Many trace organic contaminants are remarkably persistent in the environment, and it has been suggested that some of what were formerly considered to be permanent or dead environmental storage compartments may actually become long-term sources, via revolatization, of contaminants (e.g., Mackay et al. 1986). Several recent studies confirm the existence of this process (e.g., Achman et al. 1993). This situation suggests that modern exposure to trace organic contaminants will not necessarily end when the chemical is no longer applied or produced, which complicates calculations of both current ecological exposures to particular contaminants or of cumulative ecological exposures over time.

Acidification is likely to affect the partitioning, cycling, and toxicity of trace organic contaminants by influencing the quantity, nature, and association with pools of particulate (POC) and dissolved (colloidal) organic carbon (DOC). However, the effects of acidification on contaminant availability will also depend, in large part, on the characteristics of the contaminant. In this chapter we explore the likely behaviors of a suite of organic contaminants with changes in DOC, based on three properties that are recognized to influence the behavior of trace organic contaminants in the environment: the n-octanol/water partition coefficient, aqueous solubility, and vapor pressure. While this array of properties is not exhaustive, it should establish a starting point for subsequent discussions.

**Table 12.1** Deposition estimates for selected trace organic contaminants. Data in parentheses are recent, current best estimates for the Great Lakes, as estimated by a working group on atmospheric deposition of toxic chemicals to the Great Lakes (see text for details).

| Contaminant | Estimated Deposition ($\mu g\ m^{-2}\ yr^{-1}$) | Time Period | Area |
|---|---|---|---|
| Lindane | 0.4–7.3[a] | 1983–1986 | Southern Canada |
| Dieldrin | 0.01–0.51[a] | 1983–1986 | Southern Canada |
| HCB | 1.15[b] (0.05[g]) | 1983–1986 | Southern Lake Michigan |
| | 0.18[b] (0.05[g]) | 1983–1986 | Lake Huron, Northern Lake Michigan |
| | 0.04–0.10[a] | 1983–1986 | Southern Canada |
| p,p′DDE | 0.007–0.084[a] | 1983–1986 | Southern Canada |
| PCB 18 | 1.1[c] | 1983–1984 | Siskiwit Lake, Isle |
| PCB 101 | 0.75[c] | 1983–1984 | Royale, Lake Superior |
| ΣPCBs | 1.0–4.9[a] | 1983–1986 | Southern Canada |
| | 108–141[d] (1.91[g]) | 1983–1986 | Lake Superior |
| Phenanthrene | 0.01[e] (4.13[g]) | recent sediments | Lake Superior |
| | 0.03–0.20[e] | recent sediments | remote sites |
| | 1.8–4.6[e] | recent sediments | urban sites |
| Fluoranthene | 0.10[e] (5.5[g]) | recent sediments | Lake Superior |
| | 0.04[e] (5.5[g]) | recent sediments | Isle Royale, Lake Superior |
| | 0.04–0.50[e] | recent sediments | remote sites |
| | 4.0[f] | 1980–1981 | Cayuga Lake, New York |
| B(a)P | 0.02–0.2[e] | recent sediments | remote sites |
| | 0.04[e] (1.4[g]) | recent sediments | Lake Superior |
| | 1.7–14.0[e] | recent sediments | urban sites |
| | 2.1[f] | 1980–1981 | Cayuga Lake, New York |

[a]Strachan (1990), [b]Murphy et al. (1981), [c]Swackhamer et al. (1988), [d]Eisenreich et al. (1981), [e]Gschwend and Hites (1981), [f]Heit et al. (1988), [g]Eisenreich and Strachan (1992)

## Atmospheric Deposition of Trace Organic Contaminants

Stratigraphic studies of sediments from acidified lakes provide direct evidence for the co-occurrence of surface water acidification and elevated sedimentary concentrations of by-products of fossil fuel combustion (e.g., PAHs and historical patterns of sulfur emission; Hites et al. 1977; Tan and Heit 1981; Furlong et al. 1987; Heit et al. 1988; Steinberg et al. 1989). However, the issues of surface water acidification and atmospheric deposition of, for example, agricultural chemicals have rarely been linked in the scientific literature. This is partly because major agricultural areas tend to be rich in neutralizing capacity and therefore geographically distinct from acid-sensitive regions, and partly because agricultural chemicals are more often associated with environmental issues related to nonpoint source pollution (e.g., runoff) than those related to atmospheric deposition.

It is well known that many agricultural chemicals undergo regional or long-range atmospheric transport (e.g., Kurtz 1990), as do industrial compounds such as polychlorinated biphenyls (PCBs) and dibenzo-p-dioxins (PCDDs) and dibenzo-furans (PCDFs). Certainly, regions experiencing surface water acidification can also be expected to be simultaneously experiencing deposition of many kinds of anthropogenically produced trace organic chemicals, especially in moderately populated regions that sustain a mixture of urban, industrial, and agricultural activities.

Table 12.1 gives the ranges of deposition estimates for selected trace organic chemicals based on frequently cited literature. Since 1986 there has been considerable scientific activity to refine caculations for one region of particular concern, namely, the Great Lakes region that forms part of the international boundary between the United States and Canada. The most current deposition estimates made by a working group on atmospheric deposition of toxic chemicals to the Great Lakes (Eisenreich and Strachan 1992) suggests that many of the earlier direct estimates of atmospheric deposition were too high by at least one order of magnitude (and frequently more), but that even so "the atmosphere is a dominant contributor of many chemicals to the upper lakes...and at least an important contributor to the lower lakes" (Eisenreich and Strachan 1992). In Table 12.1, these 1992 best estimates of deposition are shown in parentheses.

Because atmospheric deposition does not respect geological boundaries, trace organic contaminants will be deposited over both acid-sensitive and acid-insensitive landscapes. One of the challenges for future research is to determine whether acidic precursors interact in any ecologically meaningful way with other atmospherically deposited contaminants in acid-insensitive landscapes. For the purpose of this paper, however, we consider only potential interactions in acidifying aquatic environments.

## CONTAMINANT CHARACTERISTICS AFFECTING MOBILITY AND PARTITIONING IN AQUEOUS ENVIRONMENTS

Determinants of partitioning are complex and specific to individual chemical species. Under steady-state conditions, phase partitioning of a chemical that distributes between water and suspended particulate matter is given by the Langmuir or, for hydrophobic organic solute partitioning, the Freundlich adsorption isotherm (Freundlich 1926):

$$S = K_F\, C^{(1/n)}, \tag{12.1}$$

where $S$ is the weight of the chemical bound to the solid phase per unit weight (or surface area) of the solid phase, $C$ is the aqueous phase concentration of contaminant, and $K_F$ and $1/n$ are conditional constants. In equilibrium systems with low total contaminant concentrations, the Freundlich equation simplifies to a linear adsorption isotherm where the slope of the resulting equation is the partition coefficient ($\pi$ or $K_p$). Because the size of many contaminants is about the same as that of humic acids, some authors prefer to think in terms of association, rather than partition, coefficients (Carter and Suffet 1982).

Under nonequilibrium conditions, the absolute rates of both surface sorption and desorption reactions become important. Experimental observations suggest that the surface adsorption rate is frequently much faster than the desorption rate (Karickoff and Morris 1985), although the latter may be of greater interest from the standpoint of potential biotic exposure. The extent of desorption hysteresis due to differential rates of adsorption and desorption is an important and little-studied aspect of trace organic partitioning in natural waters, particularly as these affect uptake and toxicity of contaminants.

The properties of trace organic contaminants vary widely. Figure 12.1 shows the distribution of a range of selected common contaminants for three fundamental physicochemical characteristics that affect contaminant mobility, and therefore pathways of biotic exposure. Table 12.2 provides the numerical values upon which Figure 12.1 is based. Aqueous solubility ($S_{aq}$) is a measure of the tendency of a given constituent to enter the dissolved phase in aqueous systems. Vapor pressure ($P_{vap}$) is a measure of the tendency of the constituent to volatilize. Subcooled liquid $P_{vap}$ was used in all cases where these data were available. Together, $S_{aq}$ and $P_{vap}$ determine the tendency of the constituent to move between the gaseous and aqueous phases; these characteristics jointly determine Henry's Law constant, which is widely used in models for air-water transfer. Finally, the n-octanol-water partition coefficient ($K_{ow}$) is a measure of the affinity of the constituent for hydrophobic compounds. $K_{ow}$, then, can be understood as a proxy measure of the tendency of a constituent to partition or bind to certain kinds of hydrophobic colloidal organic carbon (i.e., as a proxy measure of $K_p$) or to the lipophilic surfaces of biota (Swackhamer and Skoglund 1991). For the types of compounds considered in this chapter, $K_p$ correlates well with $K_{ow}$ (e.g., Banerjee et al. 1980; Chiou et al. 1977; Mackay et al. 1980), although as we discuss below, the actual value of $K_p$ may be as much as two orders of magnitude lower than $K_{ow}$. Chiou et al. (1977) have suggested that the use of $S_{aq}$ may be more practical than $K_{ow}$, as it is much easier to estimate and bears a strong quantitative relationship to $K_{ow}$, over several orders of magnitude.

It is important to emphasize that the numerical values upon which Figure 12.1 are based (i.e., the entries in Table 12.2) are simply the currently available best estimates. In addition, some of the compounds (e.g., chlordanes) are mixtures of several different congeners, each with somewhat different properties. In the case of toxaphene, the actual number of possible congeners in the environment is unknown, but is almost surely in the hundreds. We have used our best judgement in choosing values to report.

**Table 12.2** Constants used to construct Figure 12.1 (superscripts key numerical estimates of the given property to literature sources as indicated here; estimates with superscripts indicating multiple sources are geometric means of the most probable values identified in those sources).

| Compound | $P_{vap}$ kPa | Log[$S$] (mol m$^{-3}$) | Log [$K_{ow}$] | Abbreviation on Figure 12.1 |
|---|---|---|---|---|
| Trichloroethylene | 0.9[a,e] | 0.9[a,e] | 2.3[e] | TCE |
| Benzene | 1.1[a,e] | 1.4[a,e] | 2.3[e] | BNZ |
| Toluene | 0.6[a,e] | 0.8[a,e] | 2.7[e] | TOL |
| Aldrin | −5.7[a,b,e] | −3.9[a,b,e] | 2.8[b,e] | ALD |
| Chlordane | −5.9[b,e] | −3.5[b,e] | 2.9[b,e] | CLR |
| DDT | −7.0[a,b,e] | −5.0[a,b,e] | 5.5[b,e] | DDT |
| Dieldrin | −6.9[a,b,e] | −3.5[a,b,e] | 3.7[b] | DIE |
| Endrin | −7.6[b,e] | −3.2[b,e] | 4.4[b,e] | END |
| Lindane | −5.6[a,b,e] | −1.7[a,b,e] | 3.8[b,e] | LIN |
| Mirex | −7.0[b] | −6.9[b] | 6.9[b] | MRX |
| Toxaphene | −6.1[f] | −2.9[f] | 4.8[f] | TOX |
| Hexachlorobenzene | −5.9[a,b] | −3.7[a,b] | 6.8[b] | HCB |
| Fluoranthene | −4.8[a,c,e] | −2.7[a,e] | 5.5[e] | FrA |
| Phenanthrene | −3.4[a,c,e] | −1.4[a,e] | 4.5[e] | PhA |
| Benzo[a]pyrene | −8.5[c,e] | −4.8[e] | 6.0[e] | BaP |
| PCB 66 | −5.3[d] | −3.5[d] | 6.3[d,h] | PCB–66 |
| PCB 101 | −5.5[g] | −6.5[g] | 6.4[g,h] | PCB–101 |
| PCB 118 | −6.0[d] | −4.0[d] | 7.0[d,h] | PCB–118 |
| PCB 153 | −6.2[g] | −5.6[g] | 6.9[g,h] | PCB–153 |

[a]Mackay and Shiu (1981); [b]Suntio et al. (1988); [c]Bidleman and Foreman (1987); [d]Eisenreich (1987); [e]Mills et al. (1985); [f]T. Bidleman, Chemistry Dept., University of South Carolina, pers. comm. 1992; [g]Shiu and Mackay (1986); [h]Hawker and Connell (1988).

It must be noted that many values are controversial, sometimes by as much as two orders of magnitude. Indeed, the determination of physicochemical constants for many of these compounds is an area of active research.

For all these reasons, the topography of Figure 12.1 is not well-defined and should be expected to change as estimates of physical properties are refined and a wider range of chemicals is included. Despite these uncertainties, we feel that the general approach is a useful beginning for discussions of predicted behaviors of xenobiotics under specified changes in receptor systems (in this case, surface water acidification).

Structural attributes of contaminants will also affect contaminant fate and transport. For example, the planarity of PCB molecules has been demonstrated to affect uptake (i.e., transport across the cell membrane), with the more planar, less bulky molecules taken up more easily (Swackhamer and Skoglund 1991).

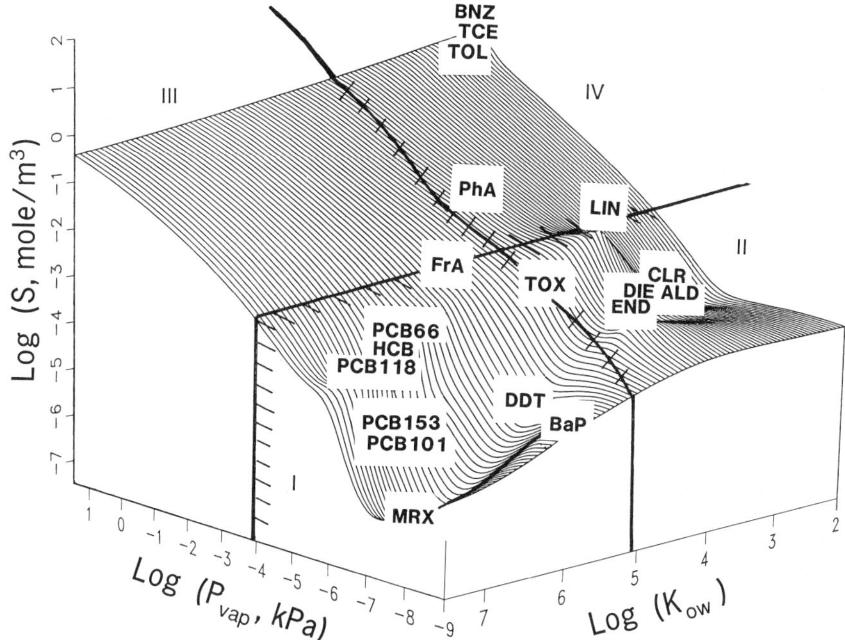


**Figure 12.1** Response surface illustrating the interdependence of aqueous solubility, vapor pressure, and octanol-water partition coefficient for oganic compounds listed in Table 12.2; labels show approximate coordinates of respective compounds on the response surface. We divide the response surface into four quadrants for convenience in discussion.

## Observed Influence of DOC on Partitioning of Trace Organic Contaminants

Properties of receptor aquatic ecosystems will be as important as physicochemical characteristics of contaminants in determining natural patterns of contaminant cycling and the influence of surface water acidification on those patterns. Figure 12.2 represents one way of describing generalized contaminant cycling in an aquatic ecosystem. Here we focus on the role of particulate and dissolved carbon in the system. Although there is actually no limit on the number of compartments into which contaminants can partition, a three-compartment model (freely dissolved compound, compound adsorbed onto POC [including biotic particles] and compound associated with DOC) has in the past been usefully applied in studies of contaminant partitioning (Pankow and McKenzie 1991; Hassett and Anderson 1979; Carter and Suffett 1982; Means and Wijayaratne 1982). $K_{ow}$ (or $S_{aq}$) indicates the tendency for a given compound to move out of the dissolved phase (concentration in the dissolved phase = $C_d$) and associate with either POC ($C_p$ = concentration associated with POC) or DOC ($C_c$ = concentration associated with colloidal DOC). Radiotracer studies (Eadie et al. 1990b) have

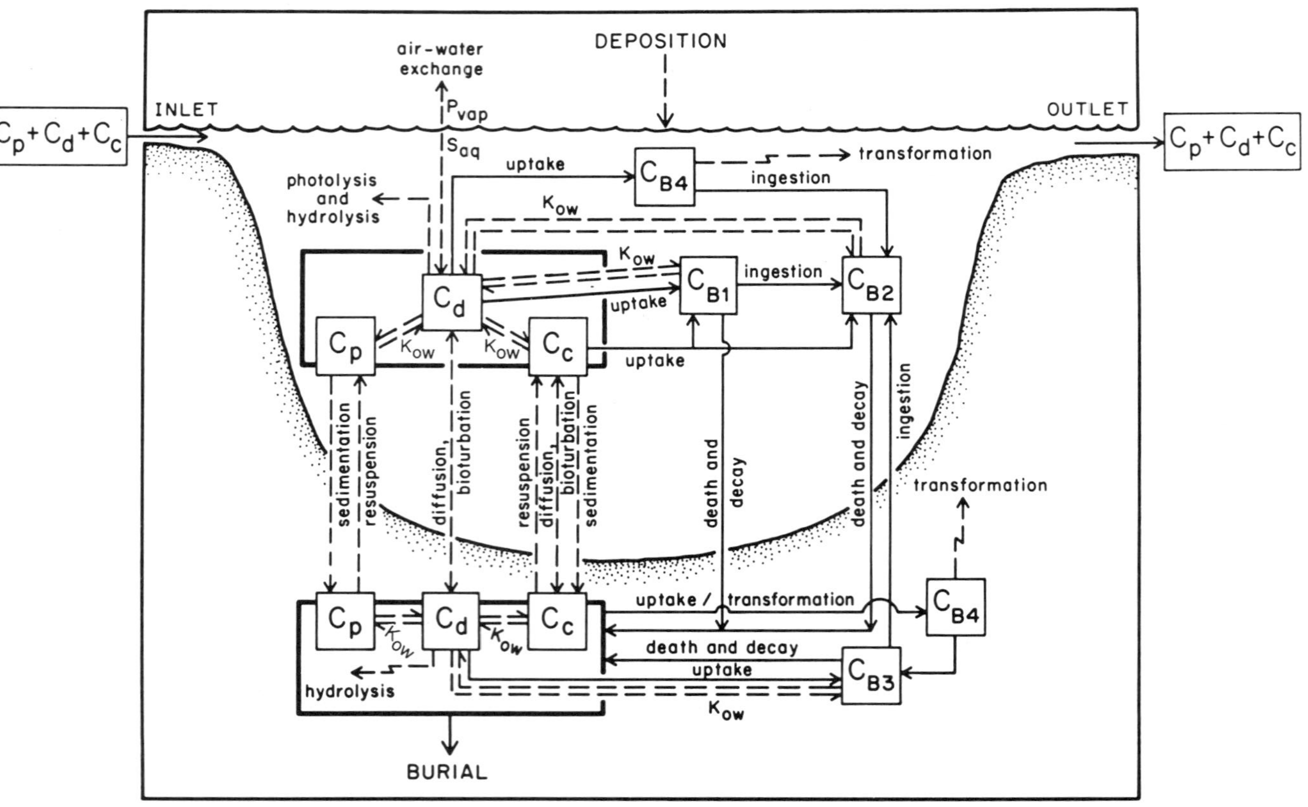


**Figure 12.2** Schematic representation of pools and pathways for trace organic contaminants. Boxes represent contaminant concentration in standing pools. $C_p$ = contaminant associated with particulate organic carbon; $C_d$ = freely dissolved, dissociated contaminant; $C_c$ = contaminant associated with dissolved organic carbon; $C_{B1-4}$ = contaminant associated with biotic pools (phytoplankton, planktivores, benthic organisms, and microbial populations, respectively)

confirmed the presence of these phase associations in natural aquatic systems and have shown that under at least certain kinds of conditions contaminant partitioning to POC is much more sensitive to contaminant solubility than is partitioning to DOC.

Both the identity and magnitude of the organic carbon fraction in sorbent phases are important determinants of contaminant behavior in natural waters (Karickhoff 1984; Weber et al. 1983; Schwarzenback and Westall 1981). For example, McCarthy et al. (1989) have demonstrated that the PAH benzo[a]pyrene (B[a]P) has a higher $K_p$ for acidic than for neutral hydrophobic DOC, whereas the reverse appears to be true for selected PCBs (Kukkonen et al. 1991). Acidic hydrophobic DOC is proportionately more abundant in the deeply stained natural waters commonly associated with peaty drainages relative to hydrophilic or neutral hydrophobic DOC. On the other hand, DOC in clearer lakes with lower DOC concentrations not only may have less total binding capacity but may also be more efficient at binding different types of contaminants, since a smaller proportion of the total DOC would be expected to be acidic hydrophobic DOC (e.g., Kukkonen et al. 1991). Thus, Eadie et al. (1990a, b) find that $C_d$ and $C_p$ (Figure 12.2) are the dominant compartments for a range of selected contaminants of low $P_{vap}$ and high $K_{ow}$, with a relatively small proportion of the total mass bound to DOC (i.e., in $C_c$). In this series of experiments, however, only low DOC water from lakes Michigan, Superior, and Huron was used. For the Great Lakes this is reasonable, as DOC concentrations do not become significant (e.g., > 5 mg DOC $L^{-1}$) except in situations where productivity is relatively high and cultural influences are significant (e.g., in Green Bay and Saginaw Bay; Stevens 1988; Zarull and Edwards 1990). However, different results might be obtained using water from higher DOC lakes with a different proportion of total DOC as acidic hydrophobic DOC.

Given these observations, it is not surprising to find that the value of the association coefficient $K_{doc}$ has been found to be different (about one order of magnitude lower) for naturally occuring DOC than it is for commercially prepared (Aldrich) DOC (HA) for several trace organic contaminants (e.g., B[a]P and p,p′-DDT [Landrum et al. 1984]; PCBs 52 and 153 [Evans 1988]). Humic acids (HA) appear to have a higher sorptive capacity than naturally occurring DOC, and thus systematically overestimates the fraction of contaminant bound to naturally occurring epilimnetic DOC (e.g., Evans 1991). Further, for commercial DOC, $K_{doc}$ can be up to an order of magnitude lower than $K_{ow}$. That is:

$$K_{doc(natural)} < K_{doc(HA)} < K_{ow}. \tag{12.2}$$

This is why published partition coefficients ($K_p$) for B[a]P, DDT, and PCBs 52 and 153 in natural waters are usually relatively low (3.48–5.04) when compared to $K_{ow}$ for the same compounds (Evans 1988). Despite the fact that quantitative predictions concerning bioavailability can be made using Aldrich HA, but not DOC from natural waters (Landrum et al. 1987), Equation 12.2 also suggests that the use of HA to provide quantitative assessments of actual contaminant bioavailability in natural aquatic systems is unsatisfactory.

The influence of organic carbon on the distribution of hydrophobic organic compounds has also been studied in sedimentary environments (e.g., Capel and Eisenreich 1990). However, experimental results describing partitioning in types of systems as different as sediments and open water are not necessarily comparable. In our discussion below, we consider primarily open-water environments.

## THE INFLUENCE OF ORGANIC CARBON ON CONTAMINANT AVAILABILITY, UPTAKE, AND TOXICITY

### Particulate Organic Carbon

Biological uptake can occur either passively or actively. In productive waters, much of the POC may be in the form of living and dead phytoplankton, both of which play important and apparently fairly equal roles in sorbing lipophilic contaminants (e.g., Swackhamer and Skoglund 1991), thereby making these contaminants accessible to higher trophic levels. Thus, $K_{ow}$ appears to be one important chemical determinant of passive uptake, especially when contaminants covering a wide range of $K_{ow}$s are being compared. However, Baker et al. (1991) observed no correlation between $K_{ow}$ and partition coefficient for PCB congeners in Lake Superior phytoplankton. Furthermore, normalization to lipid content does not always eliminate differences among species in bioconcentration factors (BCFs) for a given contaminant (Lederman and Rhee 1982). One of the problems is that "lipid" is only operationally defined. Swackhamer and Skoglund (1991) have shown that for the productive Green Bay, Lake Michigan there is a good correlation between BCF and $K_{ow}$ during the winter, but that this relationship deteriorates during the spring and summer high growth period, presumably because the system never comes to equilibrium. They conclude that the influence of $K_{ow}$ on contaminant partitioning to phytoplankton POC will be modified by both phytoplankton growth rates and seasonal changes in the lipid content or composition of the phytoplankton.

The affinity of the more hydrophobic organic contaminants for plankton, in general, is high enough to suggest that there may be a direct relationship between contaminant concentration in plankton and plankton biomass itself. In fact, field studies have indicated that when biomass is high, contaminant concentration in that biomass tends to be low, and when biomass is low, contaminant concentration tends to be high (Taylor et al. 1991; Larsson et al. 1992). Taylor et al. (1991) refer to this as the "biomass dilution effect"; not surprisingly, they find that the effect is strongest for the more hydrophobic compounds. Again, the influence of seasonal changes in phytoplankton growth rates may be expected to be important.

**Dissolved Organic Carbon/Dissolved Humic Material**

Several recent papers (e.g., Evans 1991; Kukkonen et al. 1991; Landrum et al. 1985; McCarthy et al. 1985) have included brief reviews of the fairly consistent finding that DOC appears to reduce the bioavailability and uptake of organic contaminants, particularly contaminants of relatively high $K_{ow}$. The reason appears to be related to the large molecular size of the complex between dissolved humic material (DHM) and contaminant, which prevents or reduces uptake (McCarthy et al. 1985). In contrast, the presence of DOC does not appear to affect the uptake of the lower $K_{ow}$ PAHs, such as phenanthrene, significantly (McCarthy and Jiminez 1985; McCarthy et al. 1985; Landrum et al. 1987).

Evans and Landrum (1989) have studied the toxicokinetics of B[a]P, 2,4,5,2′,4′,5′–hexachlorobiphenyl, and DDE, all of which probably reside in Region 1 of Figure 12.1. They find that although uptake clearances were similar for all three compunds for both the amphiphod *Ponteporeia hoyi* and the mysid *Mysis relicta*, overall BCFs were quite different for the three contaminants, especially for *M. relicta*. The reason for this difference was found to be related primarily to elimination rates for the three compounds, with B[a]P being the easiest to eliminate and PCB being the most difficult. In this case, BCFs, which are usually taken as a general measure of the tendency to bioconcentrate, were not strongly related to $K_{ow}$, although the range of $K_{ow}$ concerned was not large (5.7 to 6.7).

Contaminant bioavailability appears to vary with changes in contaminant concentration (e.g., Landrum et al. 1987, 1991; Servos et al. 1992), suggesting that DOC-bound contaminant functions as a long-term bank or reservoir rather than as a dead storage compartment, slowly releasing contaminant as the freely dissolved form is removed from solution by uptake or degradation (Landrum et al. 1987). In this way, it is possible that the presence of DOC in a system being externally dosed with trace organics may alter the nature of the biological exposure from acute, short-term exposures to lower-level chronic exposures.

There appears to be some indication that the controls over contaminant uptake may shift from direct equilibrium partitioning in the early phases of contaminant introduction to detrital food chain transfer as time goes on (Servos et al. 1992). For pulsed inputs, such as those probably associated with agricultural chemicals, the influence of DOC on contaminant partitioning and availability may therefore be greatest in the initial phases of contaminant introduction. Thus, seasonal DOC cycles need to be taken into account, especially as they relate to seasonality of trace organic inputs.

DHM has also been demonstrated to reduce the toxicity of certain PAHs directly (e.g., anthracene), by selectively attenuating active wavelengths of solar ultraviolet radiation that are important in photoinduction of toxicity (Oris et al. 1990). However, in other cases, DHM has also been demonstrated either to have no apparent impact on toxicity or to actually increase toxicity. For example,

Steinberg et al. (1993) have demonstrated this to be the case for a series of substituted anilines and phenols.

## POTENTIAL EFFECTS OF SURFACE WATER ACIDIFICATION ON UPTAKE AND TOXICITY OF TRACE ORGANIC CONTAMINANTS

### Direct Effects

Few papers examine the direct effects of pH on the uptake, bioconcentration, or toxicity of organic contaminants. However, both pH and temperature can certainly be expected to affect the rates at which chemicals are taken up and eliminated. Short-term laboratory studies of pH effects on uptake and toxicity of lindane by the cosmopolitan invertebrate *Chironomus riparius* indicate that lindane is significantly more toxic at pH 6 than at pH 4 or 8 (Fisher 1985). In these experiments, both the rates of breakdown of the parent compound and rates of uptake of lindane were important: less lindane is converted to metabolite at pH 6 than at pH 8, and lindane is taken up less rapidly at pH 4 than at pH 6 and 8.

### Effects Mediated by Dissolved Organic Carbon

Acidification can affect both the concentration and the temporal dynamics of POC and DOC. However, before we discuss the effects of decreased DOC on trace organic partitioning and uptake, a few words are in order with respect to the fundamental scenario of changes in DOC and/or transparency with surface water acidification.

The hypothesis that surface water acidification necessarily results in a direct decrease in DOC has been controversial. On one hand, there is both circumstantial regional evidence (Driscoll et al. 1989) and direct paleolimnological evidence (Davis et al. 1985; Steinberg 1991) that recent lake acidification may be accompanied by losses of organic matter. On the other hand, experimental studies (e.g., Hedin et al. 1990) often fail to find direct relationships between surface water acidification and decreases in organic carbon or DOC. However, acidification experiments based on microcosms, mesocosms, stream segments, or whole lakes might not be expected to result in decreases in DOC as watershed factors are excluded. One likely mechanism for decreasing surface water DOC or humicity with acidification involves flocculation by labile aluminum. Watershed acidification processes can mobilize aluminum from catchment soils, and labile aluminum should tend to complex with available DOC, removing it from aqueous solution. Steinberg and Kühnel (1987) have examined the potential validity of this mechanism in the laboratory using aqueous peat extracts. These authors confirm that aluminum, but not zinc, produces decreases in DOC and specific absorption at 254 nm. Furthermore, acidification experiments on whole watersheds, such as those at Bear Brook, Maine (U.S.A.), indicate that acidification of upland ecosystems by mineral acids can be associated with elevated exports of labile

Al (J.S. Kahl, pers. comm.). This finding from modern ecosystem studies is complemented by evidence from paleoecological studies of natural long-term (Holocene) ecosystem acidification, which have indicated that increasing watershed exports of labile aluminum are accompanied by biological evidence (diatoms, chrysophytes) for decreasing lakewater humicity (Ford 1990; Gibson 1986). The importance of this relationship for anthropogenically acidified ecosystems in any specific watershed will, of course, also depend on how watershed exports themselves are affected by acidic deposition.

The models presented in Figures 12.1 and 12.2 can be used to explore scenarios involving reductions in DOC. First, Figure 12.2 suggests that for any given concentration of contaminant, the proportion of contaminant in the dissolved compartment ($C_d$) must increase at least temporarily as the concentration $C_c$ of contaminant associated with colloidal DOC decreases. This, in turn, may be followed by increases in concentrations of particulate or plankton-bound contaminant or by loss of parent compound by volatilization, photolysis, or hydrolysis. Seasonality should be expected to play a role, with losses due to volatilization and photolysis most important during periods of high insolation.

As concentrations of DOC decrease, movement back into the dissolved phase may increase the availability of contaminant for uptake. This process would be expected to affect compounds in Region 1 of Figure 12.1 more than those in Regions 2 and 4, i.e., should be more important for high $K_{ow}$ compounds of relatively low solubility and vapor pressure (e.g., DDT, PCBs) than for less lipophilic compounds, such as lindane and chlordane (Chin et al. 1991). Within the group of compounds in Region 1, contaminants in the trough (e.g., Mirex) would be the most likely to immediately repartition onto POC, thus becoming available for biological transfer via detrital food webs. On the other hand, "hilltop" compounds of higher vapor pressure (e.g., HCB and PCB 66) (or higher Henry's Law constant) may be more likely to leave the system via volatilization. "Hilltop" compounds of lower vapor pressure (e.g., B[a]P) may spend more time in the system in dissolved phase, thereby increasing the overall concentration of contaminant available for direct biological exposure.

In addition to affecting actual concentrations of DOC, acidification can also affect the extent of its dissociation. For example, Petersen (1990) has found that whereas pH alone does not appear to influence the concentration of DOC exported from a catchment, it does influence dissociation, with less dissociation at lower pH levels (Petersen 1990). Less dissociation leads to increased lipophilicity, and consequently a higher affinity for high $K_{ow}$ contaminants. At the same time, however, aqueous solubility must decrease, which would tend overall to decrease contaminant transport. This effect on DOC dissociation appears to become important only at $pH < 5.0$ (Petersen 1990).

### Acidification Effects on Phytoplankton Species Composition and Implications for Contaminant Adsorption and Food Chain Transport

Changes in phytoplankton species composition with acidification will change the nature and quantity of POC available for contaminant adsorption. Changes in the size of POC pools ($C_p$ and $C_{B1}$ in Figure 12.2) with acidification would be expected to affect the mass of contaminant available to enter foodwebs by ingestion (e.g., by $C_{B2}$ in Figure 12.2). We have briefly discussed the issue of biomass dilution; however, available evidence indicates that no single simple generalization regarding plankton biomass changes with acidification is appropriate (Ford 1989; Schindler 1988 and this volume). Changes in the *nature* of the biomass—for example, in species composition—may affect contaminant concentrations by changing the surface area/volume ratio of the dominant plankton and/or their mean lipophilicity. Finally, in temporally dynamic systems with seasonal cycles of plankton biomass, changes in the timing of biomass distribution due to acidification also affect contaminant partitioning, especially for contaminants such as agricultural pesticides, whose inputs are expected to be strongly seasonal.

## CONCLUSIONS

Trace organic contaminants are delivered to acidified and acidifying aquatic ecosystems by both long-range atmospheric transport and, in certain geographic regions, direct inputs from surface runoff. It is important to begin discussions of how these different types of anthropogenic influences on the biological integrity of surface waters may interact. In this chapter, we have provided a conceptual context within which the potential impacts of acidification on trace organic bioavailability and uptake can be discussed, based primarily on selected characteristics of trace organic contaminants. The interaction between trace organic contaminants and surface water acidification depends, at a minimum, on both the properties of the particular contaminant of interest and the nature and quantity of organic carbon in the system. We suggest that the uptake of compounds with high $K_{ow}$, low solubility, and low volatility (or low Henry's law constant) may be particularly affected by anthropogenic acidification processes, primarily due to mediation by DOC/Al interactions. Clearly, however, we must know more about the changes in DOC/Al cycling associated with ecosystem acidification, if we are to make progress on understanding this issue.

## ACKNOWLEDGEMENTS

We thank C. Steinberg for providing the stimulus for preparing this contribution. T. Bidleman graciously permitted us to use his unpublished data on toxaphene and alerted us to several important pieces of current work concerning both physical and chemical constants, and deposition estimates. Reviews by H. Segner, C. Steinberg, D. Landers, D. Muir, L. Baker, and D. Coffey, and conversations with W. Taylor helped clarify several points and identify places where clarifications or expansions were needed. We are grateful for the important contributions to manuscript preparation made by J. Mello, S. Christie, C. Rued-Engel, and J. Lupp, and for preparation of Figure 12.2 by

L. Haygarth. This document was prepared at the EPA Environmental Research Laboratory in Corvallis, Oregon, and supported in part through cooperative agreement CR818187 with Oregon State University. It has been subjected to the Agency's peer and administrative review and approved for publication. Mention of trade names or commercial products does not constitute endorsement or recommendation for use.

## REFERENCES

Achman, D.R., K.C. Hornbuckle, and S.J. Eisenreich. 1993. Volatilization of polychlorinated biphenyls from Green Bay, Lake Michigan. *Environ. Sci. Technol.* **27(1)**:75–87.

Baker, J.E., S.J. Eisenreich, and D.L. Swackhamer. 1991. Field-measured associations between polychlorinated biphenyls and suspended solids in natural waters: An evaluation of the partitioning paradigm. In: Organic Substances and Sediments in Water, ed. R.A. Baker, vol. 2, pp. 79–89. Chelsea, MI: Lewis Publ.

Banerjee, S., S.H. Yalkowsky, and S.C. Valvani. 1980. Water solubility and octanol/water partition coefficients of organics. Limitations of the solubility-partition coefficient correlation. *Environ. Sci. Technol.* **14(10)**:1227–1229.

Bidleman, T.F., and W.T. Foreman. 1987. Vapor-particle partitioning of semivolatile organic compounds. In: Sources and Fates of Aquatic Pollutants, ed. R.A. Hites and S.J. Eisenreich, pp. 26–56. Washington, D.C.: Am. Chemical Society.

Capel, P.D., and S.J. Eisenreich. 1990. Relationship between chlorinated hydrocarbons and organic carbon in sediment and porewater. *J. Great Lakes Res.* **16**:245–257.

Carter, C.W., and I.H. Suffet. 1982. Binding of DDT to dissolved humic materials. *Environ. Sci. Technol.* **16**:735–740.

Chin, Y.-P., W.J. Weber, and C.T. Shiou. 1991. A thermodynamic partition model for binding of nonpolar organic compounds by organic colloids and implications for their sorption to soils and sediment. In: Organic Substances and Sediments in Water, ed. R.A. Baker, vol.1, pp. 251–273. Chelsea, MI: Lewis Publ.

Chiou, T.T., L.J. Peters, and V.H. Freed. 1977. Partition coefficient and bioaccumulation of selected organic chemicals. *Environ. Sci. Technol.* **11(5)**:475–478.

Davis, R.B, D.S. Anderson, and F. Berge. 1985. Palaeolimnological evidence that lake acidification is accompanied by loss of organic matter. *Nature* **316**:436–438.

Driscoll, C.T., R.D. Fuller, and W.D. Schecher. 1989. The role of organic acids in the acidification of surface waters in the eastern U.S. *Water, Air, Soil Pollut.* **43**:21–40.

Eadie, B.J., N.R. Morehead, J. Val Klump, and P.F. Landrum. 1990a. Distribution of hydrophobic organic compounds between dissolved and particulate organic matter in Green Bay waters. *J. Great Lakes Res.* **18(1)**:91–97.

Eadie, B.J., N.R. Morehead, and P.F. Landrum. 1990b. Three-phase partitioning of hydrophobic organic compounds in Great Lakes waters. *Chemosphere* **20**:161–178.

Eisenreich, S.J. 1987. The chemical limnology of nonpolar organic contaminants: Polychlorinated biphenyls in Lake Superior. In: Sources and Fates of Aquatic Pollutants, ed. R.A. Hites and S.J. Eisenriech, pp. 393–469. Washington, D.C.: Am. Chemical Society.

Eisenreich, S.J., G.J. Hollod, and T.C. Johnson. 1981. Atmospheric concentrations and deposition of polychlorinated biphenyls to Lake Superior. In: Atmospheric Pollutants in Natural Waters, ed. S.J. Eisenreich, pp. 425–444. Ann Arbor: Ann Arbor Science Publ., 512 pp.

Eisenreich, S.J., and W.M.J. Strachan. 1992. Estimating atmospheric deposition of toxic substances to the Great Lakes: An update. Proceedings of a workshop held at the Canada Centre for Inland Waters, Burlington, Ontario. Jan. 31–Feb. 2, 1992. Sponsored by The Great Lakes Protection Fund and Environment Canada, 59 pp.

Evans, H.E. 1988. The binding of three PCB congeners to dissolved organic carbon in freshwaters. *Chemosphere* **17**:2325–2338.

Evans, H.E. 1991. The influence of water column dissolved organic carbon on the uptake of 2,2′,4,4′,5,5′–hexachlorobiphenyl (PCB 153) by *Daphnia magna.* In: Organic Substances and Sediments in Water, ed. R.A. Baker, vol. 3, pp. 95–109. Chelsea, MI: Lewis Publ.

Evans, M.S., and P.F. Landrum. 1989. Toxicokinetics of DDE, benzo(a)pyrene, and 2,4,5,2′,4′,5′–hexachlorobiphenyl in *Pontoporeia hoyi* and *Mysis relicta. J. Great Lakes Res.* **15**:589–600.

Fisher, S.W. 1985. Effects of pH on the toxicity and uptake of [$^{14}$C]–lindane in the midge, *Chironomus riparius. Ecotoxicol Environ. Safety* **10**:202–208.

Ford, J. 1989. The effects of chemical stress on aquatic species composition and community structure. In: Ecotoxicology: Problems and Approaches, ed. S.A. Levin, M.A. Harwell, J.R. Kelly, and K.D. Kimball, pp. 99–144. New York: Springer.

Ford, M.S.(J.). 1990. A 10,000–year history of natural ecosystem acidification. *Ecol Monogr.* **60**:57–89.

Freundlich, H. 1926. Colloid and Capillary Chemistry. London: Methuen Ltd.

Furlong, E.T., L.R. Cessar, and R.A. Hites. 1987. Accumulation of polycyclic aromatic hydrocarbons in acid-sensitive lakes. *Geochim. Cosmochim. Acta* **51**:2965–2975.

Gibson, K.N. 1986. Mallomonadacean microfossils indicate recent acidification of Cone Pond, N.H. B.Sc. Thesis. Kingston, Ontario: Queens Univ., 47 pp.

Gschwend, P.M., and R.A. Hites. 1981. Fluxes of polycylic aromatic hydrocarbons to marine and lacustrine sediments in the northeastern United States. *Geochim. Cosmochim. Acta* **45**:2359–2367.

Hassett, J.P., and M.A. Anderson. 1979. Association of hydrophobic organic compounds with dissolved organic matter in aquatic systems. *Environ. Sci. Technol.* **13(12)**:1526–1529.

Hawker, D.W., and D.W. Connell. 1988. Octanol-water partition coefficients of polychlorinated biphenyl congeners. *Environ. Sci. Technol.* **22(4)**:382–387.

Hedin, L.O., G.E. Likens, K.M. Postek, and C.T. Driscoll. 1990. A field experiment to test whether organic acids buffer acid deposition. *Nature* **345**:798–800.

Heit, M., Y.L. Tan, and K.M. Miller. 1988. The origin and deposition history of polycyclic aromatic hydrocarbons in the Finger Lakes region of New York. *Water, Air, Soil Pollut.* **37**:85–110.

Hites, R.A., R.E. LaFlamme, and J.W. Farrington. 1977. Sedimentary polycyclic aromatic hydrocarbons: The historical record. *Science* **198**:829–831.

Karickoff, S.W. 1984. Organic pollutant sorption in aquatic systems. *J. Hydraulic Engin. ASCE* **110(6)**:707–735.

Karickhoff, S.W., and K.R. Morris. 1985. Sorption dynamics of hydrophobic pollutants in sediment suspensions. *Environ. Toxicol. Chem.* **4**:469–479.

Kukkonen, J., J.F. McCarthy, and A. Oikari. 1991. In: Organic Substances and Sediments in Water, ed. R.A. Baker, vol. 3, pp. 111–127. Chelsea, MI: Lewis Publ.

Kurtz, D.A. 1990. Long Range Transport of Pesticides. Chelsea, MI: Lewis Publ.

Landrum, P.F., B.J. Eadie, and W.R. Faust. 1991. Toxicokinetics and toxicity of a mixture of sediment-associated polycyclic aromatic hydrocarbons to the amphipod *Diporeia* sp. *Environ. Toxicol. Chem.* **10**:35–46.

Landrum, P.F., S.R. Nihgart, B.J. Eadie, and W.S. Gardner. 1984. Reverse-phase separation method of determining pollutant binding to Aldrich humic acid and dissolved organic carbon of natural waters. *Environ. Sci. Technol.* **18(3)**:380–387.

Landrum, P.F., S.R. Nihart, B.J. Eadie, and L.R. Herche. 1987. Reduction in bioavailability of organic contaminants to the amphipod *Pontoporeia hoyi* by dissolved organic matter of sediment interstitial waters. *Environ. Toxicol. Chem.* **6**:11–20.

Landrum, P.F., M.D. Reinhold, S.R. Nihart, and B.J. Eadie. 1985. Predicting the bioavailability of organic xenobiotics to *Pontoporeia hoyi* in the presence of humic and fulvic materials and natural dissolved organic matter. *Environ. Toxicol. Chem.* **4**:459–467.

Larsson, P., L. Collvin, L. Okla, and G. Meyer. 1992. Lake productivity and water chemistry as governors of the uptake of persistent pollutants in fish. *Environ. Sci. Tech.* **26**:346–352.

Lederman, T.C., and G.-Y. Rhee. 1982. Bioconcentration of hexachlorobiphenyl in Great Lake planktonic algae. *Can. J. Fish. Aquat. Sci.* **39**:380–387.

Mackay, D., A. Bobra, and W.Y. Shiu. 1980. Relationships between aqueous solubility and octanol/water partition coefficients. *Chemosphere* **9**:701–711.

Mackay, D., S. Paterson, and W.H. Schroeder. 1986. Model describing the rates of transfer processes of organic chemicals between atmosphere and water. *Environ. Sci. Technol.* **20(8)**:810–816.

Mackay, D., and W.Y. Shiu. 1981. A critical review of Henry's law constants for chemicals of environmental interest. *J. Phys. Chem. Ref. Data* **10(4)**:1175–1199.

McCarthy, J.F., and B.D. Jiminez. 1985. Interactions between polycyclic aromatic hydrocarbons and dissolved humic material: Binding and dissociation. *Environ. Sci. Technol.* **19(11)**:1072–1076.

McCarthy, J.F., B.D. Jimenez, and T. Barbee. 1985. Effect of dissolved humic material on accumulation of polycyclic aromatic hydrocarbons: Structure-activity relationships. *Aquat. Toxicol.* **7**:15–24.

McCarthy, J.F., L.E. Roberson, and L.W. Burrus. 1989. Association of benzo[a]pyrene with dissolved organic matter: Prediction of $K_{dom}$ from structural and chemical properties of the organic matter. *Chemosphere* **19**:1911–1920.

Means, J.C., and R. Wijayaratne. 1982. Role of natural colloids in the transport of hydrophobic pollutants. *Science* **215**:968–970.

Mills, W.B., D.B. Porcella, M.J. Ungs, S.A. Gherini, K.V. Summers, L. Mok, G.L. Rupp, G.L. Bowie, and D.A. Haith. 1985. Water Quality Assessment: A Screening Procedure for Toxic and Conventional Pollutants in Surface and Ground Water. EPA/600/6–85/002a,b,c. Cincinnati: U.S. Environmental Protection Agency, CERI/Technology Transfer.

Murphy, T.J., A. Schinsky, G. Paolucci, and C.P. Rzeszutko. 1981. Inputs of polychlorinated biphenyls from the atmosphere to Lakes Huron and Michigan. In: Atmospheric Pollutants in Natural Waters, ed. S.J. Eisenreich, pp. 445–458. Ann Arbor: Ann Arbor Science Publ., 512 pp.

Oris, J.T., A.T. Hall, and J.D. Tylka. 1990. Humic acids reduce the photo-induced toxicity of anthracene to fish and *Daphnia*. *Environ. Toxicol. Chem.* **9**:575–583.

Pankow, J.F., and S.W. McKenzie. 1991. Parameterizing the equilibrium distribution of chemicals between the dissolved, solid particulate matter, and colloidal matter compartments in aqueous systems. *Environ. Sci. Technol.* **25**:2046–2053.

Petersen, R.C. 1990. Effects of ecosystem changes (e.g., acid status) on formation and biotransformation of organic acids. In: Organic Acids in Aquatic Ecosystems, ed. E.M. Perdue and E.T. Gjessing, pp. 151–166. Dahlem Workshop Report LS 48. Chichester: Wiley.

Schindler, D.W. 1988. Effects of acid rain on freshwater ecosystems. *Science* **239**:149–157.

Schwarzenback, R.P., and J. Westall. 1981. Transport of nonpolar organic compounds from surface water to groundwater: Laboratory sorption studies. *Environ. Sci. Technol.* **15**:1360–1367.

Servos, M.R., D.C.G. Muir, and G.R. Barrie Webster. 1992. Bioavailability of polychlorinated dibenzo-p-dioxins in lake enclosures. *Can. J. Fish. Aquat. Sci.* **49**:735–742.

Shiu, W.Y., and D. Mackay. 1986. A critical review of aqueous solubilities, vapor pressures, Henry's law constants, and octanol-water partition coefficients of the polychlorinated biphenyls. *J. Phys. Chem. Ref. Data* **5(2)**:911–929.

Steinberg, C. 1991. Fate of organic matter during natural and anthropogenic lake acidification. *Water Res.* **25:**1453–1458.

Steinberg, C., W.Kalbfus, M. Maier, and K. Traer. 1989. Evidence of deposition of atmospheric pollutants in a remote high alpine lake in Austria. *Wasser-Abwasser-Forsch.* **22**:245–248.

Steinberg, C., and W. Kühnel. 1987. Influence of cation acids on dissolved humic stubstances under acidified conditions. *Water Res.* **21**:95–98.

Steinberg, C.E.W., A. Sturm, J. Kelbel, S.K. Lee, N. Hertkorn, D. Freitag, and A.A. Kettrup. 1992. Changes of acute toxicity of organic chemicals to *Daphnia magna* in the presence of dissolved humic material (DHM). *Acta Hydrochim. Hydrobiol.*, **20:**326–332.

Stevens, R.J.J. 1988. A Review of Lake Ontario Water Quality with Emphasis on the 1981–1982 Intensive Years. A report to the surveillance subcommittee of the Great Lakes Water Quality Board. Windsor, Ontario: International Joint Commission, Great Lakes Regional Office.

Strachan, W.M.J. 1990. Atmospheric deposition of selected organochlorine compounds in Canada. In: Long Range Transport of Pesticides, ed. D.A. Kurtz, pp. 233–240. Chelsea, MI: Lewis Publ.

Suntio, L.R., W.Y. Shiu, D. Mackay, J.N. Seiber, and D. Glotfelty. 1988. Critical review of Henry's Law constants for pesticides. *Rev. Environ. Contam. Toxicol.* **103**:1–59.

Swackhamer, D.L., B.D. McVeety, and R.A. Hites. 1988. Deposition and evaporation of polychlorobiphenyl congeners to and from Siskiwit Lake, Isle Royale, Lake Superior. *Environ. Sci. Technol.* **22**:664–672.

Swackhamer, D.L., and R.S. Skoglund. 1991. The role of phytoplankton in the partitioning of hydrophobic organic contaminants in water. In: Organic Substances and Sediments in Water, ed. R.A. Baker, vol. 2, pp. 91–105. Chelsea, MI: Lewis Publ.

Tan, Y.L., and M. Heit. 1981. Biogenic and abiogenic polynuclear aromatic hydrocarbons in sediments from two remote Adirondack lakes. *Geochim. Cosmochim. Acta* **45**:2267–2279.

Taylor, W.D., J.H. Carey, D.R.S. Lean, and D.J. McQueen. 1991. Organochlorine concentrations in the plankton of lakes in southern Ontario and their relationship to plankton biomass. *Can. J. Fish. Aquat. Sci.* **48**:1960–1966.

Weber, W.J., T.C. Voice, M. Pirbazaari, G.E. Hunt, and D.M. Ulanoff. 1983. Sorption of hydrophobic compounds by sediments, solid, and suspended solids. II. Sorbent evaluation studies. *Water Res.* **17(10)**:1143–1452.

Zarull, M.A., and C.J. Edwards. 1990. A Review of Lake Superior Water Quality with Emphasis on the 1983 Intensive Survey. A report to the surveillance subcommittee of the Great Lakes Water Quality Board. Windsor, Ontario: Intl. Joint Commission, Great Lakes Regional Office.

Standing, left to right:
Steve Norton, Dixon Landers, John Gunn, Suzanne Bayley, Michael Zahn

Seated, left to right:
Anke Lükewille, Jesse Ford, Josef Veselý, Christian Steinberg

# 13

# Group Report: Interactions among Acidification, Phosphorus, Contaminants, and Biota in Freshwater Ecosystems

D.H. LANDERS, Rapporteur

S.E. BAYLEY, J. FORD, J.M. GUNN,
A. LÜKEWILLE, S.A. NORTON,
C.E.W. STEINBERG, J. VESELÝ, M.T. ZAHN

## INTRODUCTION

Scientific investigations over the last two decades have examined the processes and effects of acidic deposition on surface waters and their catchments. These studies have been performed on various scales: laboratory benchtop, mesocosm/plot studies, whole system manipulations, and regional and national surveys. The processes and mechanisms of acidification of fresh waters have been studied intensively, and investigations have examined closely associated phenomena that influence the availability, transport, and biogeochemistry of acids and related substances. We have learned a great deal about the linkages among watershed processes, hydrology, microbial activities, and water chemistry. These factors interact to create the physical and chemical environment in which biota are exposed to an almost infinite array of conditions. Some of these conditions are so adverse that specific organisms can no longer survive; other conditions may result in sublethal effects that have less visible but important results. The purpose of this group report is to relate what we have learned about acidification and the fate, transport, and effects of other contaminants that may co-occur with acidic ecosystems.

Contaminants of interest to this report include trace metals, organic compounds, and phosphorus (P), all of which may be delivered to ecosystems via atmospheric transport and deposition processes similar to those that transport sulfur and

*Acidification of Freshwater Ecosystems: Implications for the Future*
Edited by C.E.W. Steinberg and R.F. Wright 

nitrogen associated with anthropogenic acidification. As with acidic precipitation, contaminants may be distributed over large regional landscapes at some distance from their sources. In other cases, local sources may yield spatially restricted occurrence of the contaminants.

There has been relatively little research on the interactions of contaminants with acidification in freshwater ecosystems, particularly with regard to organic contaminants. We have gained some basic knowledge about important components of this issue, however, from ancillary measurements taken as part of acidification research focused primarily on other topics. From our understanding of acidification processes, we have begun to develop some basic principles that can be applied to studies of how acidic conditions and related phenomena interact with a broad range of airborne contaminants. Investigations of the ultimate effects of these interactions on biota and the ecosystems in which they occur should be a major objective of future acidification-related research.

## WHAT IS THE ROLE OF HYDROLOGY IN ACIDIFICATION OF SURFACE WATERS?

Simply put, hydrology includes the tracking and quantification of water flow through the ecosystem. A thorough knowledge of the hydrology of an acidifying system, accompanied by chemical analysis of the water parcels and an understanding of the physicochemical processes involved, would enable us to link quantitatively the atmospheric inputs of water and contaminants with stream or lake outputs. Thus, our understanding could be expressed as a model explaining the observed acidification, contaminant mobilization, and transport. Complete knowledge of flow paths for all water parcels, water parcel residence times, and chemical, physical and biotic processes, coupled with accurate deposition estimates, would enable us to predict surface water quality. Most systems of interest, however, would probably not be studied at this level of intensity. In fact, because most ecosystems are quite heterogeneous, only in the simplest hydrogeologic setting is this predictability even, in principle, possible. For example, some models predicting the nutrient status of lakes have been shown to be quite successful (Schindler, this volume).

Hydrographs for surface drainage systems have traditionally been analyzed using concepts of "old" and "new" water, "slow" and "quick" flow, or "deep" and "shallow" flow routing. Techniques of flow separation originally equated "old" with "slow" and "deep." Conservative analytes contained in precipitation and soil/groundwater mixing models have been proposed to explain hydrographs. One of the first techniques used was $\partial\ ^{18}O/\ ^{16}O$ (Fritz et al. 1976). To this list have been added $^{222}Rn$ (Genereux and Hemond 1990), dissolved organic carbon (DOC; Schiff et al. 1990), silica (Heath et al. 1992), bromide (Kennedy et al. 1984), and others. When multiple hydrologic tracers are used, they commonly yield different inferences.

The empirical modeling of hydrologic response to changes in water input has been quite successful (e.g., Christophersen and Wright 1981); response can be shown to be a function of such watershed characteristics as physical (hydrologic) properties of soils, soil thickness, slope, and antecedent moisture condition. Generalization to unstudied but similar systems has also been quite successful (see section below on EXTENDING UNDERSTANDING TO DIFFERENT SPATIAL AND TEMPORAL SCALES). Wetland systems and vegetation, characteristics common to most ecosystems of interest, are complicating factors that make the prediction less reliable.

Chemical responses to variations in the input of water and acidic (e.g., $H_2SO_4$) or acidifying (e.g., $NH_4^+$) substances are typically monitored within streams or lakes. The transformation of atmospheric deposition into acidic surface waters with accompanying trace metals and other contaminants is dominated by processes within the soils (with the exception of seepage lakes). Scientists attempt to understand the processes of acidification and contaminant transport using information such as: (a) the spatial distribution of soil chemical characteristics and (b) soils solution chemistry—either from tension lysimeters ("micropore" water) or zero tension lysimeters ("macropore" water). These data have been used to construct complex mixing models that yield observed stream or lake water. In virtually all such attempts, successful modeling of the behavior of one group of ions has been accompanied by a failure to model other ions accurately (Christophersen and Wright 1981; Hooper et al. 1988). Furthermore, results of "old" vs. "new" water mixing, based on tracer evaluation, generally appear to be incompatible with mixing of lysimeter chemistries. Whereas we can predict watershed discharge well, we appear not to be able to predict water chemistry based on soil solution chemistry, particularly in the unsaturated zone. This inability indicates that we have either not gathered the appropriate information from within the system, in space or time, or that we have not appropriately interpreted and synthesized this information. Although there is almost a universal direct relationship between increasing discharge from a watershed and lower pH (e.g., Reuss et al. 1987), the sources of the associated increased concentrations of various metals (both atmospheric- and watershed-originated) are not well known.

Acidifying substances in precipitation reach groundwater as a result of infiltration into the soil matrix. As a result, regions with acidified surface waters may also experience aspects of groundwater acidification. Seasonal groundwater variations in the main acidifying substances and the magnitude of acidification are higher in surface waters than in groundwaters. In Bavaria (F.R. Germany), for example, the pH of groundwaters in granitic areas drops only occassionally below 4 and is most often between 4 and 5, while acidic surface waters in these areas normally have pH values of under 4 in spring and over 5 in summer. These acidity levels may lead to aluminum concentrations over 2 mg $L^{-1}$ (Zahn et al. 1992). Acidified groundwaters may not pass drinking water quality requirements and may also corrode water supply facilities if not processed properly. As a part of the freshwater ecosystem, groundwater participates in acidification dynamics of rivers and lakes through direct discharge. The difficulty in describing groundwater acidification in space and time is the large heterogeneity of soils in terms of chemical, physicochemical, and hydraulic properties. Subsurface flow path distribution is complex and affects residence times of the infiltration water in the various soil horizons. It seems that residence times of the aqueous phase often have an underestimated impact of buffering processes during the infiltration of groundwater recharge.

### Research Needs

1. Reevaluate the use of tracers to develop a better understanding of hydrologic routing and flow separation across different watersheds. Tracers understood to be conservative in some circumstances (e.g., Cl) may not be conservative under others, particularly in bogs (Bayley, pers. comm.). It may be possible to develop "new tracers," i.e., new techniques, for determining the source terms for water leaving a watershed. Routing may remain problematic.
2. Comparative evaluation of hydrological response of systems with distinctly different watershed characteristics (drainage vs. seepage, thick soils vs. thin soils, etc.) may enhance our understanding of the hydrologic routing responsible for imparting certain chemical characteristics to surface waters.
3. On a site-specific basis, a better understanding of hydrogeology is necessary to evaluate discharge from fractured bedrock aquifers and the recharge of bedrock aquifers with acidic surface waters.
4. We have not succeeded in locating those places in the terrestrial ecosystem that most greatly influence surface water chemistry before its emergence. Research should attempt to backtrack in the ecosystem to locate places "where the actions is," starting at the stream/lake border.

## SOURCES, MOBILIZATION, AND TRANSPORT OF TRACE METALS IN ACIDIFIED FRESH WATERS

Three main sources of trace metals are found in acidified surface waters:

1. atmospheric deposition and direct movement through the ecosystem,
2. chemical weathering of bedrock and soil parent material,
3. metals mobilized from pools of previously accumulated sinks (held in organic material, in exchangeable sites in soils, and in sediments), with metals originating from either source (1) or (2).

Mobile Pb and apparently Hg originate predominantly from the atmosphere; Zn comes from the atmosphere and accumulated pools; Cu commonly stems from both sources. Aluminum and Fe are derived predominantly from chemical weathering and reside as secondary hydroxides in the soil profile. These secondary minerals co-precipitate and absorb trace metals, including Mn, As, Cu, Zn, and Pb.

Controls on the concentrations of trace metals in surface waters include:

- Concentrations in precipitation.
- The pH of soil solutions and interstitial water in sediments, with decreasing pH associated with increasing solution concentrations due to desorption.
- The nature of DOC concentration: DOC increases the transport of metals but generally decreases the bioavailability of the metal.
- Bioaccumulation may significantly reduce the concentration of dissolved metal while at the same time increasing the total lake burden.

Stream survey data from many areas of the Northern Hemisphere demonstrate that flowing waters have been impacted by acidic deposition and show the general relationships seen in Czechoslovakia (Veselý, this volume). Trace metal concentrations typically increase with decreasing pH below 5 to 6, depending on the metal (Havas and Hutchinson 1982). What is unique about the Czech data is the pronounced and clear decline in concentrations of Be, Mn, Cd, and Zn in streams at very low pH (< 4.0). Assuming that source terms (1) and (2) are fairly constant, we hypothesize that source (3) is diminishing significantly with time.

Furthermore, we hypothesize that the decline in concentration of trace metals at about pH 4 (see Figure 8.1 in Veselý, this volume) is a result of continued and strong leaching of exchangeable metal pools with finite size; consequently, the pool becomes depleted. Such may be the case for Be, Mn, and Cd. For these elements, the removal rate by acidified water is greater than source (1). Ultimately, the concentration of these elements in streams would be lowered to a value controlled by source (1) plus (2).

For Pb, the rate of removal from soils into acidified waters is substantially less than the amounts in precipitation, resulting in Pb accumulation in ecosystems (Friedland 1984). The pH-concentration relationship for Pb should probably not change with time for a lake.

The concentration of several elements (e.g., Cd, Cu, Hg, and Pb) in acidic streams ($4.5 < pH < 5.3$) may be controlled more by concentration of DOC (Cu, Hg, and probably Pb) and bioaccumulation (Cu, Hg, and probably Cd) than by pH. Bioaccumulation would play a larger role in lakes. For example, Veselý (this volume) has found that filtered water from acidic lakes contained much lower amounts of Cd and Cu than unfiltered water, unlike the elements Be and Mn, which were virtually completely in the dissolved state.

## Research Needs

1. Test the hypothesis that stripping of mobilizable metal pools from watershed soils can be evaluated by examining strongly acidified catchments, particularly where elevational data on stream chemistry and soils are available or can be gathered (Matschulat et al. 1992).

2. Because trace metal concentrations seen in Veselý's data (this volume) are potentially biologically significant, we recommend that additional research focus on the controls of DOC and bioaccumulation on total trace metal lake burdens and concentrations.
3. Evaluate the different metal concentrations for lakes vs. streams, at the same pH, which suggests that either bioaccumulation or DOC effects are exerting more control in lakes than in streams.

## HOW DOES ACIDIFICATION INTERACT WITH TOXIC METALS AND ANTHROPOGENIC ORGANIC CONTAMINANTS TO AFFECT THE HEALTH OF ORGANISMS?

Acidifying systems contain a wide variety of potential toxicants, many of which are transported through atmospheric processes similar to those that transport acid precursors. Other toxicants are released or transferred from nontoxic to toxic forms by acid additions to watershed soils and sediments.

### Trace Metals

Numerous potentially toxic metals are associated with acidification (e.g., Al, Cd, Cu, Hg, Mn, Pb, Zn). However, most of those originally found in elevated quantities in low pH, low Ca waters appear not to reach lethal concentrations, with the exception of Al (Driscoll et al. 1980; Havas and Hutchinson 1982). Synergistic effects of metals at low pH have therefore focused largely on the toxicology of free forms of Al (i.e., those not bound to organic molecules). Recent information (Jagoe et al. 1993; Veselý et al. 1989) suggests that Be at low concentrations produced toxic effects similar to those of Al. Sublethal effects of metals also occur but are considerably less well understood.

The widespread Hg contamination problem has direct implications for both the health of natural ecosystems and the health of higher life forms (including humans). The inverse relationship between surface water pH and accumulation of methyl Hg in freshwater biota remains a controversial topic. Although in-lake methylation of Hg appears to be pH-dependent (Xun et al. 1987), the relevance of this process to biological effects, including bioaccumulation in acidic systems, is unclear. Mercury enters aquatic systems from both the original stored geological (e.g., soil) sources and directly from atmospheric transport. There is a strong spatial overlap between $SO_4$ and Hg deposition (Nater and Grigal 1992). In some areas, the atmospheric source of Hg is relatively large, appears to be increasing, and may represent the dominant source of contamination. In many acidic lakes, there is a strong relationship between the concentration of Hg in fish flesh and age and size structure. In these lakes, pH is the variable that accounts for most of the variation in Hg concentration in fish tissue. DOC, specific conductance, and Ca are secondary factors. In areas with less acidic surface waters, DOC is commonly a better predictor of body burdens of Hg than pH. There are also linkages between sulfur cycling and methylation, which demonstrate that watershed processes strongly influence the flux and transformation of this metal in acidic environments.

Alternate hypotheses for the accumulation of Hg in biota include (a) increased deposition or flux of Hg from watershed sources during acidification and (b) changes in the structure of the aquatic communities (size, longevity, abundance of herbivores and carnivores) that affect Hg partitioning. With the dramatic improvements in analytical techniques since about 1985, the usefulness of much of the historic data on Hg in air and water samples for examination of Hg transport rates and mechanisms has now been put into question. The problem deserves further research attention.

*Research Needs*

1. Additional information is needed on metal and organic chelators and their role in detoxifying Al solutions. Additional investigation into the forms of DOC and the identity and functions of inorganic ligands, including silicate, is warranted.
2. Mercury source quantification in acidifying aquatic systems is needed: (a) atmospheric/terrestrial/in-lake methylation and (b) gills/diet.
3. It is important to assemble information on metals that interact with contaminant uptake: Se/Hg and Se/organic contaminants.
4. Develop detailed information on the toxic effects of various metals on phytoplankton.
5. Determine the toxic effects of additional metals (i.e., Be, Cu, Cd) on fish and aquatic invertebrates in low pH, low Ca waters.
6. Identify sublethal indicators of stress: biochemical, metabolic, behavioral, and community.
7. Refine our understanding of the role of stored body burdens of contaminants on the health (physiology, bioenergetics, behaviors) of aquatic organisms.

## Acidification and Organic Contaminants

Similar to acidic precursors, organic contaminants such as polycyclic aromatic hydrocarbons (PAH), polychlorinated biphenyls (PCB), polychlorinated dibenzo-p-dioxins (PCDD), polychlorinated dibenzo-furans (PCDF), and pesticides may originate from remote point sources. Furthermore, nonpoint sources of organic contaminants may be important. For example, spray drift or surface runoff are important contributors to environmental burdens of persistent pesticides and herbicides. Once these organic contaminants enter the aquatic system, acidification may affect its environmental fate, the exposure of organisms to organic pollutants, and the subsequent toxic effects.

Humic materials, present in most freshwater ecosystems, possess the ability to photosensitize or photocatalyze abiotic degradation reactions of organic pollutants in aqueous systems (Perdue and Gjessing 1990; Frimmel and Christman 1988). By absorbing energy from UV-B (an increasing concern due to ozone depletion at high latitudes), humic materials provide so-called photoactive chemical species (such as singlet oxygen, organic and inorganic peroxides, OH radicals, solvated electrons, etc.) that attack organic pollutants. As a result, the degradation products have altered chemical as well as ecotoxicological properties. These photochemical reactions are strongly pH-dependent (Sukul et al. 1993; Minero et al. 1992).

Declining pH in acidifying systems can result in increasing protonation of ionic organic pollutants. In the case of phenolic pollutants, for instance, increases in protonation lead to corresponding rises in lipophilicity. If bioconcentration is a linear function of the organisms' lipid contents (Geyer et al. 1985), acidification will cause increases in bioavailability of such organic contaminants, resulting in direct effects on organisms.

Protonation of naturally occurring phenolic compounds, such as humic and fulvic acids, increases lipophilicity and, as a consequence, absorption of persistent lipophilic organic (and some metal) contaminants (e.g., PAH, PCB, organochlorines, and Hg). Super lipophilics, such as PCDDs/Fs and Mirex, may be absorbed to a higher degree. Because of the physicochemical properties of humic materials, especially solubility, associations between persistent organics and the protonated humics expected in acidified systems may be less soluble than associations with nonprotonated humics. For instance, in the sediments of acidified Großer Arbersee (Bavarian Forest, F.R. Germany), chironomid larvae are exposed to strongly elevated PAH concentrations (fluoranthene up to 4, benzo [k] fluoranthene up to 3.4, benzo [a] pyurene, up to 2.4, benzo [g,h,i] perylene up to 6, and ideno [1,2,3–cd] pyrene up to 3.4 mg/g dry weight; Steinberg et al. 1989). These figures are three to five times higher than in acidified Woods Lake, Adirondack Mountains (New York, U.S.A.). In the long term, an adverse effect on these benthic organisms might be anticipated.

Acidification may also affect biotic uptake and ecotoxicity. Acidification and associated changes in ionic composition of the aquatic medium put severe constraints on basic metabolic functions of the biota, possibly because of interference with ion balances (Havas and Likens 1985). Under acidic conditions, considerably more energy is consumed for maintenance of these basic functions than under nonacidic conditions (Rosseland, this volume). This means that less energy may be available for clearance activities (depuration as well as metabolism of organic pollutants) and/or the establishment of lipid reserves. Diminishing clearance rates in turn would lead to increasing bioconcentrations of organic pollutants. At the same time, decreasing lipid contents might be expected to result in corresponding increases in exposure to and possibly toxicity of organic micropollutants because they are not sequestered in the lipid content any longer (Geyer et al. 1993). Since the loading of organic contaminants is often a multiple contaminant issue, even low concentrations of single compounds may collectively display adverse effects on species, communities, or freshwater ecosystems. If acidification of fresh waters leads to "starvation" of fish due to alterations in the food chain, loss of body fat and release of organic contaminants sequestered in body fat could have synergistic impacts on organisms, thus affecting survival and other components of the ecosystem.

*Research Needs*

1. Determine the role of DOC in contaminant cycling in acidifying surface waters.
2. Determine which observed concentrations of trace metals and organic contaminants are sufficient to cause biological effects.
3. Determine how contaminants affect or are effected by organismal energetic considerations.

## ALTERATION OF THE PHOSPHORUS CYCLE BY ACIDIFICATION

The state of knowledge of P in acidified lakes is reviewed by Schindler (this volume). In brief, there is no evidence that the in-lake P cycle is disrupted by acidification. Decreased yields of P from acidified terrestrial catchments as a result of complexation by Al were observed at Lake Gårdsjön and other lakes in southwestern Sweden. However, analyses of the P cycle on a smaller scale have shown that internal processes can disrupt the lacustrine P cycle (Detenbeck and Brezonik 1991a, b; Dickson 1980). The subject deserves study in a wider variety of lakes.

Decreased P loading resulting from acidification has not been documented in terrestrial catchments outside southwestern Sweden, with the possible exception of the Czech Republic. In the Tatra Mountains (Czech Republic), J. Fott (pers. comm.) found that alpine lakes have recently become more oligotrophic; however, it remains to be determined whether this is P-linked. The number of studied sites where P yield has been studied is small, and the mechanisms of release have not been studied. Terrestrial catchments supply little of the P input to forest lakes, so that small changes of terrestrial P yield may not be detectable in lakes. It is possible, however, that acidification of terrestrial catchments can also affect the lacustrine P cycle. One hypothesis is that acidification may increase yields of iron from terrestrial catchments. This phenomena may combine with increased rates of Fe sedimentation, expected as the result of declining DOC, to remove P more efficiently from solution or prevent its return to the water column from sediments. Yet even in the most oligotrophic softwater lakes, the return of P from sediments to the hypolimnion is very small (Schindler 1977; Levine et al. 1986).

Long-term declines in P have been observed in acidified lakes in southwestern Sweden (Persson and Broberg 1985). Declining terrestrial inputs are believed to be the main reason for the decline. However, the bottom of Lake Gårdsjön is covered with a perennial algal mat that may intercept returning P from lake sediments. Such widespread perennial mats have not been observed in other countries, where littoral mats of Zygnematales (Chlorophyta) are seasonal in nature. Also, P in sediments appears to be efficiently intercepted at the sediment-water interface of oligotrophic boreal lakes, even where no algal mats are present (Levine et al. 1975).

It has also been suggested that changes in the ability of algae to utilize P may accompany acidification. For example, the general decline of pelagic diatoms at low pH is believed to be caused by the lack of acid phosphatases in this group (Charles, pers. comm.). The matter deserves further study.

One method of examining the linkage between disruption of P cycling, iron inputs, and productivity in a variety of systems would be to analyze dated (preferably varved) sediment cores for deposition of algal pigments, Fe, and P. Changes in Fe/P or pigment/P ratios would indicate whether a more thorough study of mechanisms is needed.

## ANALYZING THE ECOLOGICAL IMPACTS OF MULTIPLE XENOBIOTIC COMPOUNDS

In moving past the issue of single stressors (e.g., $H_2SO_4$, $HNO_3$) to a consideration of multiple stressors, we essentially seek to understand the issue of "cumulative impacts" on ecosystems. Three problems need to be addressed:

1. The effects of multiple toxicants are inherently difficult to address because individual dose-response relationships are not strictly additive or even multiplicative in an obvious way.
2. In any given situation, diagnosis (relative impact to a singe toxicant or set of toxicants) is difficult because the relative roles of individual contaminants may vary depending on the background chemical matrix in that specific situation.
3. Current analytical capabilities limit our ability to identify and quantify contaminants in complex mixtures.

During this workshop, the value of creating empirically derived cause-effect relationships has been repeatedly questioned, even in the absence of a mechanistic understanding of underlying processes. The consensus has been that simple empirical models that realistically describe the behavior of systems may often be more useful than heavily parameterized models that attempt to detail and quantify all known aspects of the problem. This concept can be fruitfully developed using a medical analogy to ecosystem health.

Essentially, assessing the dimensions of the problems raised by exposure of ecosystems to various chemical "soups" is a two-step process. First, we must develop general indicators of ecosystem health so that we can detect potentially worrisome changes in target ecosystems. Thus, we need integrative tools that function, for example, in the role of "basal body temperature" for the system as a whole. Second, we need to be able to diagnose the reason for the observed departure of the ecosystem from its baseline condition of health. This process of diagnosis probably requires the development of a number of screening tools that could be likened, for example, to various kinds of blood tests, urine tests, allergy tests, etc., used as primary diagnostic tools in modern medicine. In carrying this analogy further, heavily stressed ecosystems of central Europe may be likened to diagnosing a "patient" suffering from a serious, possibly fatal, disease. Any discussion about the validity of various models of the disease's progress should, at a minimum, be accompanied by some effort to diagnose effectively and treat the patient, lest the patient succumb before our studies are complete. The consequence of accepting this analogy is that we will almost surely lose some "patients." In order for their loss to not be in vain, we must be diligent in our autopsies and remain committed to improving our diagnostic and treatment capabilities.

Another, more practical, corollary of accepting this analogy as a model for dealing with multiple chemical stresses on ecosystems is that we must begin to view our most traditional tool, experimental toxicity testing, as simply one tool among many. Experimental work, in general, can at best help fill in the m × n matrix (Figure 13.1), which can be further expanded in laboratory studies.

| | | SYSTEM | | | | |
|---|---|---|---|---|---|---|
| | | 1 | 2 | 3 | . . . | n |
| STRESSOR | 1 | | | | | |
| | 2 | | | | | |
| | 3 | | | | | |
| | . . . | | | | | |
| | m | | | | | |

| | | **System 1** Species | | | | **System 2** Species | | | | **System n** Species | | | |
|---|---|---|---|---|---|---|---|---|---|---|---|---|---|
| | | 1 | 2 | . . . | x | 1 | 2 | . . . | y | 1 | 2 | . . . | n |
| STRESSOR | 1 | | | | | | | | | | | | |
| | 2 | | | | | | | | | | | | |
| | . . . | | | | | | | | | | | | |
| | m | | | | | | | | | | | | |

**Figure 13.1** m and n matrix.

Although the kinds of information contained in this matrix may continue to have useful applications, the approach truly represents a shorthand approach to the larger issue of diagnosing ecosystem health. Some of the test pairs might produce relevant information. However, it will be difficult, if not often virtually impossible, to know a priori which approach, in the future, will be the most effective in identifying important relationships between environmental stressors and biotic response.

Additional tools that have traditionally been effective in advancing our understanding of environmental issues include: (a) survey work across natural spatially distributed environmental gradients and human-influenced chemical gradients, (b) survey work over time (both real-time studies, including monitoring work, and retrospective historical studies using paleoecological techniques), and (c) modeling. With particular regard to paleoecological work, we recognize that stratigraphic studies are capable of giving both qualitative and quantitative information about the importance of single or multiple pollutants. However, successful application of paleolimnology as a quantitative tool requires the creation of appropriate quantitative regionally calibrated transfer functions that relate species assemblages to the parameters of interest for reconstruction purposes. Only in this way can observed patterns of change in species assemblages be reliably interpreted over time with respect to changes in environmental forcing functions.

## Research Needs

Air pollution in Europe is currently a very large-scale, multidimensional "experiment," in terms of multiple contaminant stresses. We suggest that environmental scientists take advantage of this unfortunate situation in the following ways:

1. Develop a catalog of existing survey, monitoring, and paleoecological studies in regions characterized by heavy inputs of multiple contaminants.
2. Develop a catalogue of existing watershed and site-specific modeling studies in regions characterized by heavy inputs of multiple contaminants, including the following programs: EC Encore (Hornung 1992), UN ECE Integrated Monitoring of Air Pollution Effects on Ecosystems (UN/ECE), and EC NITREX (Dise and Wright 1992).
3. Focus new work, including paleoecological studies, on these regions in such a way as to take good advantage of existing data.

A potentially successful approach in the study of the effects of single contaminants within a multiple stressed environment would be to focus gradient studies geographically near known point sources of specific pollutants within such an environment. For example, many regions of the Czech Republic seem to receive relatively high Cd burdens from local sources. Similarly, gradient studies near central European chlor-alkali plants would be informative regarding the relative role of Hg within an otherwise already stressed environment.

Finally, we recognize the value of dealing with multiple chemical stresses within the framework of "adaptive management," that is, applying and learning from management actions. This approach carries with it the belief that while multiple contaminant situations are difficult, complex, and generally poorly characterized, we must still do the best we can, given the current state of knowledge, and continue to build in the flexibility to change our approaches when new information warrants. Concomitant with this is the responsibility to ensure that sufficient, well-focused environmental research is being funded and carried out to provide future guidance to policymakers and resource managers.

## EXTENDING UNDERSTANDING TO DIFFERENT SPATIAL AND TEMPORAL SCALES

The natural variability inherent in the attributes of lakes, streams, and watersheds is a function of the landscapes and climates in which they occur. To make optimal use of information pertaining to a specific system, we must interpret this information within the context of some larger group of systems. When results from a specific laboratory, mesocosm, or watershed studies are compared to appropriate regional databases, the results aid in refining our understanding and establishing the relevance of specific systems to larger groupings. Various aggregations of systems can be viewed as subpopulations, which may have geographic specificity and/or have specific attributes in common (Kaufmann et al. 1992).

By relating results obtained from site-specific studies to a larger, possibly regional context (i.e., scaling-up), we may be able to establish the spatial bounds or other inherent characteristics that are correlated with a specific set of attributes. Scaling-up enables us to evaluate assumptions inherent in selecting research sites or systems and forces us to view information in a broader context. Moreover, by examining regional data, we may be able to observe relationships that are not possible to discern from a few site-specific studies. Scaling-up is an exploratory research tool that can be used to identify unexpected relationships and develop hypotheses. This process identifies gaps in our understanding, providing an approach to determine the type of systems that should be selected for future site-specific research and monitoring. Thus, by interpreting specific results in a larger context, we test our overall understanding.

An example of the importance of spatial context may prove useful. Due to the importance of hydrology on lake chemistry, studies on a seepage lake in the higher elevations of the Adirondack mountains of New York (U.S.A.) would not be particularly relevant to most of the lakes in this region because seepage lakes represent a relatively rare hydrologic lake type for this region (Brakke et al. 1988). Similarly, studies of large drainage lakes in northern Wisconsin (U.S.A.) would not provide information on the most prevalent and sensitive hydrologic lake types in that area, which are seepage lakes (Eilers et al. 1988). Without the regional data, this context would be lost on the researcher as well as the policymaker.

The last twenty years of acidic deposition research provides an opportunity to examine the question of how the issue of scale might be approached, and North American studies provide an example. In 1975, there were few regional-scale data available in the United States for the remote, small, low alkalinity streams and lakes that were eventually determined to be most at risk from acidic precipitation. Moreover, the number of lakes or their basic physical, chemical, and biological characteristics (e.g., size, depth, hydrologic type, ionic strength, pH, elevation, etc.) were unknown. Without this information, it was not possible to determine the number or surface area of lakes that were likely to be at risk for acidification or were already acidic.

Eventually, quantitative probability-based surveys were conducted by U.S. federal and state governments, which provided the ability to interpret the many excellent, more intensive research projects performed on individual or groups of lakes on a regional scale (Landers et al. 1988). These surveys also provided databases that assisted in developing new hypotheses about surface water acidification, thus providing direction for future site-specific research. These small-, moderate-, and large-scale studies were complementary, providing a thorough spatial understanding of the important processes affecting surface water acidification.

Based on this experience and the knowledge that in the future society faces other regional environmental threats (e.g., global warming, deposition of metals and organic contaminant, etc.), it is reasonable to attempt to develop more general baseline data on ecosystem status (Messer et al. 1991). The baseline data should be based on measurements of ecosystem "health" and should serve as early indicators of change that might herald the onset of detrimental effects coincident with a regional perturbation. To assist in reaching this objective, long-term (large-scale) experiments are a useful tool. Such investments in environmental science would provide an "insurance policy," not specific to a particular environmental threat but designed to detect change using robust ecological indicators. Resulting regional databases would also provide a context for the continued interpretation and understanding of more specific and intense research. In the future, remote sensing tools from satellite platforms may make aspects of this data collection job easier. However, these approaches must not supplant the continuing need for a strong knowledge base derived from intense site-specific studies on ecosystem processes and function. These studies will provide the all-important "ground truth" linkage to the functioning ecosystems required to validate and interpret remotely sensed data.

## REFERENCES

Brakke, D.F., D.H. Landers, and J.M. Eilers. 1988. Chemical and physical characteristics of lakes in the northeastern United States. *Environ. Sci. Tech.* **22**:155–163.

Christophersen, N., and R.F. Wright. 1981. Sulfate budget and a model for sulfate concentrations in streamwater at Birkenes, a small forested catchment in southenmost Norway. *Water Resour. Res.* **17**:377–389.

Detenbeck, N.E., and P.L. Brezonik. 1991a. Phosphorus sorption by sediments from a soft-water seepage lake. 1. An evaluation of kinetic and equilibrium models. *Environ. Sci. Tech.* **25**:395–403.

Detenbeck, N.E., and P.L. Brezonik. 1991b. Phosphorus sorption by sediments from a soft-water seepage lake. 2. Effects of pH and sediment composition. *Environ. Sci. Tech.* **25**:403–409.

Dickson, W.T. 1980. Properties of acidified water. In: Ecological Impact of Acid Precipitation, ed. D. Drabløs and A. Tollan, pp. 75–83. Oslo: SNSF Project.

Dise, N.B., and R.F. Wright. 1992. The NITREX project. Ecosystems Research Report 2. Brussels: Comm. of the European Community, 101 pp.

Driscoll, C.T., J.J. Baker, J.R. Bisogni, and C.L. Schofield. 1980. Effect of aluminum speciation on fish in dilute acidified water. *Nature* **284**:161.

Eilers, J.M., D.F. Brakke, and D.H. Landers. 1988. Chemical and physical characteristics of lakes in the upper Midwest, United States. *Environ. Sci. Tech.* **22**:164–172.

Friedland, A.J., A.H. Johnson, and T.T. Siccama. 1984. Trace metal content of the forest floor in the Green Mountains of Vermont: Spatial and temporal patterns. *Water, Air, Soil Pollut.* **21**:161–170.

Frimmel, I.H., and R.F. Christman, eds. 1988. Humic Substances and Their Role in the Environment. Dahlem Workshop Report LS 41. Chichester: Wiley.

Fritz, P., J.A. Cherry, K.U. Weyer, and M. Sklash. 1976. Storm runoff analysis using environmental isotope and hydrochemical data in groundwater hydrology, pp. 111–130. Vienna: Intl. Atomic Energy Agency.

Genereux, D.P., and H.F. Hemond. 1990. Naturally occurring radon-222 as a tracer for streamwater generation: Steady state methodology and field example. *Water Resour. Res.* **26**:3065–3076.

Geyer, H.J., I. Scheunert, R. Brüggemann, W. Schütz, M. Matthies, C.E.W. Steinberg, and A. Kettrup, A. 1993. The relevance of aquatic organisms' lipid content to the toxicity of lipophilic chemicals: Toxicity of lindane (y–HCH) to different fish species. *Ecotoxicol. Environ. Safety*, in press.

Geyer, H., I. Scheunert, and F. Korte. 1985. Relationship between the lipid content of fish and their bioconcentration potential of l,2,4-trichlorobenzene. *Chemosphere* **14**:545–555.

Havas, M., and T.C. Hutchinson. 1982. Aquatic invertebrates from the Smoking Hills. N.W.T.: Effects of pH and metals on mortality. *Can. J. Fish. Aquat. Sci.* **39**:890–903.

Havas, M., and G.E. Likens. 1985. Changes in $^{22}$Na influx and outflux in *Daphnia Magna* (Straus) as a function of elevated Al concentrations in soft water at low pH. *Proc. Natl. Acad. Sci. USA* **82**:7345–7349.

Heath, R.H., J.S. Kahl, S.A. Norton, and I.J. Fernandez. 1992. Episodic stream acidification caused by atmospheric deposition of sea salts at Acadia National Park, Maine, U.S.A. *Water Resour. Res.* **28**:1081–1088.

Hooper, R.P., A. Stone, N. Christophersen, E. de Grobois, and H.M. Seip. 1988. Assessing the Birkenes model of stream acidification using a multi-signal calibration methodology. *Water Resour. Res.* **24**:1308–1316.

Hornung, M. 1992. ENCORE: European Network of Catchments Organized for Research on Ecosystems. International symposium on experimental manipulations of biota and biogeochemical cycling in ecosystems: Approach, methodologies, findings. Copenhagen, 18–20 May, 1992, in press.

Jagoe, C.H., V.E. Matey, T.A. Haines, and V.T. Komov. 1993. Beryllium on fish in acid water is analogous to aluminum toxicity. *Aquat. Toxicol.* **24**:241–256.

Kaufmann, P.R., A.T. Herlihy, and L.A. Baker. 1992. Sources of acidity in lakes and streams of the United States. *Environ. Pollut.* **77**:115–122 .

Kennedy, V.C., A.P. Jackman, S.M. Zand, G.W. Zellweger, and R.J. Avanzino. 1984. Transport and concentration controls for chloride, strontium, potassium, and lead in Ovas Creek, a small cobble-bed stream in Santa Clara County, California, U.S.A. 1. Conceptual model. *J. Hydrol.* **75**:67–110.

Landers, D.H., W.S. Overton, R.A. Linthurst, and D.F. Brakke. 1988. Eastern Lake Survey Regional estimates of Lake Chemistry. *Environ. Sci. Tech.* **22**:128–135.

Levine, S. 1975. Orthophosphate concentration and flux within the epilimnia of two Canadian Shield lakes. *Verh. Internat. Verein Limnol.* **19**:624–629.

Levine, S.N., M.P. Stainton, and D.W. Schindler. 1986. A radiotracer study of phosphorus cycling in an eutrophic Canadian Shield Lake, L227, Northwestern Ontario. *Can. J. Fish. Aquat. Sci.* **43**:366–378.

Matschulat, J., H. Andreae, D. Lessman, V. Malessa, and U. Siewers. 1992. Catchment acidification from the top down. *Environ. Pollut.* **77**:143–150.

Messer, J.J., R.A Linthurst, and W.S. Overton. 1991. An EPA program for monitoring ecological status and trends. *Environ. Monit. Assess.* **17**:67–78.

Minero, C., E. Pramauro, E. Pelizetti, M. Dolei, and A. Marchesini. 1992. Photosensitized transformations of atrazine under simulated sunlight in aqueous humic acid solution. *Chemosphere* **24**:1597–1606.

Nater, E.A., and D.F. Grigal. 1992. Regional trends in mercury distribution across the Great Lake States, north central U.S.A. *Nature* **358**:139–141.

Perdue, E.M., and E.T. Gjessing. 1990. Oroganic Acids in Aquatic Ecosystems. Dahlem Workshop Report LS 48. Chichester: Wiley.

Persson, G., and O. Broberg. 1985. Nutrient concentrations in the acidified lake Gårdsjön: The role of transport and retention of phosphorus, nitrogen, and DOC in watershed and lake. *Ecol. Bull.* **37**:158–175.

Reuss, J.O., B.J. Cosby, and R.F. Wright. 1987. Chemical processes governing soil and water acidification. *Nature* **329**:27–32.

Schiff, S.L., R. Aravena, S.E. Trumbore, and P.J. Dillon. 1990. Dissolved organic carbon cycling in forested watersheds: A carbon isotope approach. *Water Resour. Res.* **26**:2949–2957.

Schindler, D.W. 1977. Evolution of phosphorus limitation in lakes. *Science* **195**:260–262.

Steinberg, C., Kalbfus, W., Maier, M. and Traer, K. 1989. Evidence of deposition of atmospheric pollutants in a remote high alpine lake in Austria. *Z. Wasser-Abwasser-Forsch.* **22**:245–248.

Sukul, P., Moza, P.N., Hustert, K. and Kettrup, A. 1993. Photochemistry of Metalaxyz. *J. Agric. Food Chem.*, in press.

UN/ECE Programme on Integrated Monitoring of Air Pollution Effects on Ecosystems. Manual for Integrated Monitoring. Programme Phase 1993-96. Environmental Report No. 5, Environment Data Centre, National Board of Waters and Environment. Helsinki, Finland, 114 pp.

Veselý, J., P. Benes, and K. Sevcik. 1989. Occurence and speciation of Beryllium in acidified freshwaters. *Water Res.* **23**:711–717.

Xun, L. N.E.R. Campbell, and J.W.M. Rudd. 1987. Measurements of specific rates of net methyl mercury production in the water column and surface sediments of acidified and circumneutral lakes. *Can. J. Fish. Aquat. Sci.* **44**:750–757.

Zahn, M.T., J. Bittersohl, and H. Sager. 1992. Atmospheric deposition as a factor of freshwater pollution (in German). In: XVI Konf. der Donauländer über hydrologische Vorhersagen und hydrologisch-wasserwirtschaftlicher Grundlagen, pp. 401–408. WMO. Koblenz: Natl.komitee der Bundesrepublik Deutschland für das IHP der UNESCO.

# 14

# Biological Processes that Affect Water Chemistry

C. KELLY
Department of Microbiology, University of Manitoba,
Winnipeg, Manitoba R3T 2N2, Canada

## ABSTRACT

Microbial processes occurring within lakes have significant effects on the concentrations of $SO_4^{2-}$, $NO_3^-$, $NH_4^+$, and $H^+$, with the degree of significance increasing with increasing water residence time. In watersheds, microbial and large vegetation activities have large effects on $NO_3^-$ and $NH_4^+$; however, $SO_4^{2-}$ appears to be affected more by adsorption processes than by biological ones. An exception to this is in wetlands, where $SO_4^{2-}$ reduction is important. In lakes, the nature of the microbial processes removing sulfuric acid ($H^+$ and $SO_4^+$) and nitric acid ($H^+$ and $NO_3^-$) is such that nitric acid is removed more efficiently than sulfuric acid, but that neither acid is 100% removed and thus both lead to acidification. The behavior of several in-lake processes changes as lakes become markedly acid, and some changes that occur below pH 5 may affect recovery.

## INTRODUCTION

Many factors combine to produce the particular chemical composition of the water in a lake: (a) the chemistry of the precipitation in the region, (b) geochemical and biological reactions that occur in the watershed, (c) geochemical and biological reactions that occur within the lake itself, and (d) hydrologic characteristics, especially water residence time, which is a major factor affecting the relative importance of (b) and (c). In studying the effects of acidic precipitation, reactions that decrease (neutralize) or increase acidity are of particular interest. Of these reactions, the geochemical ones tend to be more finite in nature, i.e., there is a certain amount of readily available cation exchange capacity in soils and sediments that is used up as acidity enters the system. In contrast, biological reactions (primarily microbial ones) are generally more renewable in nature, occurring over and over again. Because of

*Acidification of Freshwater Ecosystems: Implications for the Future*
Edited by C.E.W. Steinberg and R.F. Wright 

this, biological reactions may be especially important in aiding recovery of acidified lakes after deposition is reduced.

In this chapter, I will review the microbiological activities that particularly affect lake chemistry as it pertains to acidity, focusing on three major questions:

1. What factors control microbial rates of acid neutralization?
2. Under what circumstances are these rates significant compared to geochemical reactions?
3. How susceptible are these microorganisms and their activities to low pH conditions?

Much of the work on these activities has been done in North America, and the discussion reflects this. However, the relative importance of the various microbial activities is affected by the amount and composition of the deposition inputs, and many European sites have greater inputs of both nitrate and ammonia. However, this is also becoming a recognized problem in parts of North America.

## EFFECT OF MICROBIAL ACTIVITIES ON LAKE SULFATE ($SO_4^{2-}$) AND ASSOCIATED HYDROGEN ION ($H^+$) CONCENTRATIONS

The major microbial reactions involving $SO_4^{2-}$ are sulfate reduction and sulfate assimilation. Both of these reactions also consume $H^+$:

$$SO_4^{2-} + 2\,H^+ + 2\,CH_2O = 2CO_2 + 2\,H_2O + H_2S \tag{14.1}$$

$$SO_4^{2-} + 2\,H^+ + \text{metabolic precursor} = \text{methionine or cysteine.} \tag{14.2}$$

Reaction 14.1 is carried out solely by sulfate-reducing bacteria, which are obligate anaerobes and which use $SO_4^{2-}$ as the terminal electron acceptor in energy-producing reactions. Reaction 14.2 is carried out by most microorganisms, including algae. This, however, is a biosynthetic reaction, and, in general, much smaller amounts of substrate are used in biosynthesis than in energy-producing reactions. A key point in understanding the relative importance of these two reactions in systems receiving elevated $SO_4^{2-}$ inputs is that biosynthesis is controlled by factors other than $SO_4^{2-}$ concentration (productivity, growth rates; Jassby 1975; Cuhel et al. 1982), whereas $SO_4^{2-}$ reduction is usually limited only by $SO_4^{2-}$ availability, as long as there are anaerobic habitats (Cook and Schindler 1983; Kelly and Rudd 1984; Rudd et al. 1990).

The input of $SO_4^{2-}$ to lakes is determined primarily by the rate of $SO_4^{2-}$ deposition, with watershed characteristics (anion adsorption capacity of the soil and, possibly, $SO_4^{2-}$ reduction) playing secondary roles (Baker et al. 1990). Sulfate concentrations in wet precipitation vary by region. For example, annual average concentrations range in North America from 12 μeq $L^{-1}$ at remote Canadian stations to 100 μeq $L^{-1}$ around Lake

Erie (in 1980; Barrie and Hales 1984). In the northeastern U.S. (Likens 1981; Church et al. 1990) and the Canadian Shield (Schindler et al. 1976), weathering reactions add very little $SO_4^{2-}$ to water as it passes through the watershed. Also, very little $SO_4^{2-}$ is retained, on an average net annual basis, although uncertainty about the quantity of $SO_4^{2-}$ derived from dry deposition means that there may be some unquantified retention. At present, however, neither $SO_4^{2-}$ reduction nor anion adsorption are considered to be important mechanisms in these northeastern and Canadian Shield watersheds.

In contrast, there is a pattern of high net annual $SO_4^{2-}$ retention in watersheds located in the southern U.S. (near the Blue Ridge Mountains) and a pattern of variable retention in the mid-Appalachian regions, with areas of high deposition having low retention (Church et al. 1990). The greater retention in these regions has been attributed to higher anion adsorption capacity (not to $SO_4^{2-}$ reduction) in these older soils (Church et al. 1990).

While $SO_4^{2-}$ reduction does occur in soils, with sulfur accumulating largely in the organic pool (Swank et al. 1984), the quantitative significance of this reaction compared to anion adsorption is not established. In forests of glaciated areas, the dominant form of sulfur in all soil horizons is organic sulfur (Mitchell et al. 1992). Movement into and out of this large pool is affected by biological activities, and only a small percentage change involving uptake or release of $SO_4^{2-}$ could have a significant effect on the smaller $SO_4^{2-}$ pool (Mitchell et al. 1992). This is obviously a topic that could benefit from further study, especially since $SO_4^{2-}$ reduction in soils may be sensitive to pH (Connell and Patrick 1968). One type of watershed where $SO_4^{2-}$ reduction is known to be important is wetlands. Wetlands provide a good habitat for $SO_4^{2-}$ reduction; *Sphagnum* also takes up $SO_4^{2-}$, with a net annual $SO_4^{2-}$ retention ranging from 22–77% (Hemond 1980; Bayley et al. 1986).

In upland forests, the trees, rather than microorganisms, may be the most important biological species affecting the output of $SO_4^{2-}$ to streams and lakes because of the role they play in enhancing dry deposition, with conifers more efficient than broadleaf species (see Kreutzer, this volume, for a discussion of this and other effects of vegetation).

Although these factors influence the proportion, to some degree, of $SO_4^{2-}$ in precipitation that enters streams and lakes, in all of the North American regions where streams have been surveyed, higher $SO_4^{2-}$ in deposition is still correlated with higher $SO_4^{2-}$ in stream water. The median $SO_4^{2-}$ concentrations in streams of different regions ranges from 100–175 $\mu$eq $L^{-1}$ in the northeast to 10–50 $\mu$eq $L^{-1}$ in the central and southern regions of the U.S. (Kaufmann et al. 1991).

In lakes, the role of biological $SO_4^{2-}$ uptake in affecting $SO_4^{2-}$ and $H^+$ concentrations is fairly well understood. First, $SO_4^{2-}$ is consumed for biosynthesis (Equation 14.2), by both algae and bacteria in the water column and in the sediments (Jassby 1975; Cuhel et al.1982). Algal uptake can be significant in lakes with low $SO_4^{2-}$ inputs; however, when inputs increase, this uptake becomes a relatively insignificant portion of the total $SO_4^{2-}$ consumption (Rudd et al. 1990). Second, sulfate-reducing bacteria use $SO_4^{2-}$ as their terminal electron acceptor (instead of oxygen, as used by aerobic organisms).

Because sulfate reducers are obligate anaerobes, their aquatic habitat is in the sediments, below the depth of oxygen penetration, which may be anywhere from a few mm to 15 mm below the sediment surface in sediments overlain by oxygenated water (Sweerts 1990). In sediments in anaerobic hypolimnia, their habitat is just below the sediment surface (Cook 1981). Epilimnetic (shallow water) sediments are an especially important site of $SO_4^{2-}$ reduction because these sediments are in contact with surface water, which has relatively high $SO_4^{2-}$ concentrations year round (Kelly and Rudd 1984). In contrast, the hypolimnion often becomes depleted of $SO_4^{2-}$ during summer stratification, causing $SO_4^{2-}$ reduction to cease (Cook and Schindler 1983).

Sulfate reducers are distributed universally in aquatic sediments, and $SO_4^{2-}$ consumption in sediments has been observed in most lakes (e.g., Herlihy and Mills 1985; Rudd et al. 1986b). Most porewater profiles of $SO_4^{2-}$ show a decrease in $SO_4^{2-}$ concentration just below the surface, to a minimum concentration of 1–2 μmol $L^{-1}$ at 3–4 cm (Figure 14.1; Cook 1981; Rudd et al. 1986b; Carignan 1987). The only exceptions that have been seen are in Lake Hovvatn, Norway (Figure 14.1), where the sediments are also unusual in that they are just as acidic as the overlying water (pH = 4.5 in both the water and the sediment), and in McNearney Lake (Cook et al. 1990), which is highly oligotrophic and thus organic supply might limit $SO_4^{2-}$ reduction. Another factor that may lead to a decreased $SO_4^{2-}$ reduction, on a net annual basis, is the development of metaphytic algae above the sediments (see below).

There is a seasonal pattern to $\mathbf{SO_4^{2-}}$ reduction that affects the amount and chemical nature of net annual S accumulation in sediments resulting from this activity. Rates of $SO_4^{2-}$ reduction are higher in spring, summer, and fall than in winter (Rudd et al. 1990). Also, reduced S is stored in the sediments in more than one form: as ferrous sulfide (FeS), pyrite ($FeS_2$), and organic S (Landers et al. 1983; Nriagu and Soon 1985; Rudd et al. 1986a). In the winter, microbial oxygen consumption slows, allowing oxygen to penetrate more deeply into the sediments. This means that some of the reduced sulfur stored during the summer is exposed to oxidizing conditions and a significant portion of it is returned to the water column as $SO_4^{2-}$, along with $H^+$ (Rudd et al. 1986a; Giblin et al. 1990). Interestingly, the organic S forms appear to be the most resistant to reoxidation (Rudd et al. 1986a).

While seasonal differences occur, the net annual rate of $SO_4^{2-}$ removal in many lakes occurs as a first-order reaction, meaning that removal rate is directly related to $SO_4^{2-}$ concentration in the overlying water (Cook and Schindler 1983; Kelly and Rudd 1984). This is demonstrated by the relatively constant relationship between $SO_4^{2-}$ concentration and the sediment removal rate (Table 14.1), expressed mathematically as:

$$S_s = \text{m yr}^{-1} = \frac{\text{removal rate } (\mu\text{mol m}^{-2}\text{ yr}^{-1})}{\text{concentration } (\mu\text{mol m}^{-3}).} \tag{14.3}$$

This expression is referred to as a mass transfer coefficient, or a piston velocity.

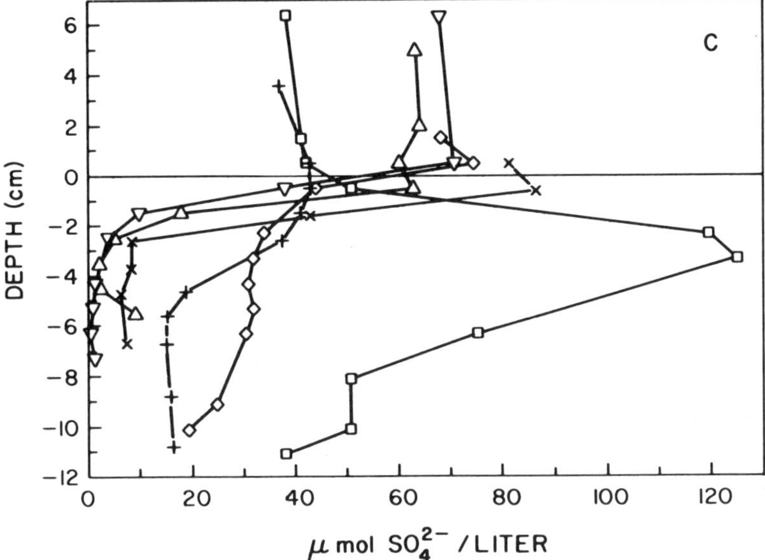


**Figure 14.1** Depth profiles of concentrations of sulfate in the porewater of epilimnetic sediments of Big Moose Lake (◇) and Woods Lake (△) in the Adirondacks region of New York State; Lake 302S (▽) at the Experimental Lakes Area, Canada; Lille Hovvatn (□) and Hovvatn (+), Norway; Chubb Lake in central Ontario, Canada (×). In all of the lakes, except the Hovvatn lakes, the pH of the sediments was greater than 5.0 at the surface, and 6.0–6.8 at –6 cm; in the Hovvatn lakes the sediment pH's were only 4.5–5.2 throughout. From Rudd et al. (1986b).

**Table 14.1** Removal coefficients ($S_S$) comparing $SO_4^{2-}$ removal rates with $SO_4^{2-}$ concentrations in surface water, estimated by different methods.

| Method | $S_S$ (m $yr^{-1}$) |
|---|---|
| Sediment $SO_4^{2-}$ profiles[a] (epilimnetic) | 0.17–0.36 |
| Whole-lake $SO_4^{2-}$ budgets[b] | 0.3–0.5 |
| Long-term S accumulation in sediments (hypolimnetic)[c] | 0.5–0.8 |

[a]Rudd et al. (1986a); [b]Baker et al. (1986), Kelly et al. (1987), Baker and Brezonik (1988); [c]Norton et al. (1988), Cook and Kelly (1992)

The value of $S_s$ has been determined in a variety of ways, summarized in Table 14.1. First, sediment $SO_4^{2-}$ removal rates derived from summertime porewater profiles were compared to $SO_4^{2-}$ concentrations in overlying waters in a variety of lakes, and assumptions about net annual removal were made from a lake where a more detailed year-round measurements were done (Kelly et al. 1987). Second, $SO_4^{2-}$ removal was calculated from mass balance budgets for a number of lakes (Baker et al. 1986; Kelly et al. 1987). Third, long-term sulfur accumulation rates in sediments of a number of Adirondack lakes (Norton et al. 1988) were compared to $SO_4^{2-}$ concentrations in the overlying water. The somewhat higher values from this third method may result from the fact that accumulation was determined at the center of each lake, where sediment focusing may result in higher than average accumulation rates. Altogether, these data approximate the relationship between concentration and net annual removal to be about 0.5 ± 0.3 m yr$^{-1}$ (Table 14.1) for a broad group of softwater lakes (mostly clearwater, drainage and seepage, pH > 5), making these lakes easy to model. (This model is not meant to apply to eutrophic lakes with large anaerobic hypolimnia; in these lakes, values for $S_S$ will be significantly higher.)

Even though $SO_4^{2-}$ reduction occurs in the sediments of a wide variety of lakes, this activity will have little effect on the water chemistry of lakes with very short water residence times and/or great mean depths. In contrast, lakes with long water residence times and/or shallow mean depths will have significantly lower $SO_4^{2-}$ concentrations, and higher alkalinities, than expected from $SO_4^{2-}$ and $H^+$ inputs alone. The mathematical expression of these factors is:

$$R_s = \frac{S_s}{\frac{z}{t_w} + S_s} \tag{14.4}$$

where $R_S$ is the proportion of incoming $SO_4^{2-}$ that is removed, $S_s$ is the mass transfer coefficient for $SO_4^{2-}$, $z$ is the mean depth, and $t_w$ is the water residence time (Baker et al. 1986; Kelly et al. 1987). This phenomenon was observed in a general way in the National Surface Water Survey in North America (Baker et al. 1990). In lakes affected by acid precipitation, $SO_4^{2-}$ concentrations ranged from 40–200 μeq L$^{-1}$, and were lower than expected from evapoconcentration alone, by 23–37% (median percentages for drainage and seepage lakes).

Acidified lakes that deviate from the general model, in which $S_s$ = 0.5 m yr$^{-1}$ in Equation 14.4, are probably most easily identified by porewater profiles that do not show decrease to near zero by 3–4 cm. This was the case in Lake Hovvatn, Norway, where $SO_4^{2-}$ even increased below the sediment-water interface and where the sediments were as acidic as the overlying water (Figure 14.1; Rudd et al. 1986b).

In lakes where $SO_4^{2-}$ decreases with depth in the sediments, the pH of the sediment is higher than the overlying water, due to a variety of anaerobic $H^+$-consuming reactions (Herlihy and Mills 1986; Rudd et al. 1986b). These elevated sediment pHs occur even in lakes with water column pHs as low as 2.5–3.5 (Herlihy and Mills 1986). Thus, low water column pH alone does not explain the Hovvatn situation.

At the Experimental Lakes Area, study of a lake as it was acidified from neutral conditions to pH 5.0 showed that $SO_4^{2-}$ reduction increased as $SO_4^{2-}$ concentration increased (Rudd et al. 1990). However, when the pH was reduced further to 4.5, there was a dramatic decrease in net annual $SO_4^{2-}$ reduction, and in the associated $H^+$

removal (C.A. Kelly, J.A. Amaral, J.W.M. Rudd, M.A. Turner, D.W. Schindler, and M.P. Stainton, in prep.). Above pH 5, the seasonal $SO_4^{2-}$ cycle always showed some return of $SO_4^{2-}$ under winter ice. Below pH 5, however, this winter return increased greatly. Detailed study suggests that this change was linked to the development of filamentous metaphytic algae, whose growth and decay at the sediment-water interface caused the major zone of $SO_4^{2-}$ reduction to occur closer than usual to the sediment-water interface (Amaral 1991). This would increase sulfur oxidation in the winter, because the proportion of S oxidized is greatest in the sediments closest to the sediment-water interface (Rudd et al. 1986a).

The widespread development of metaphytic algae in the more severely acidified lakes (e.g., Howell et al. 1990) means that their effect on underlying sediment microbial activities needs to be known. The effect of moss development (Grahn 1986; Melzer and Rothmeyer 1983) on sediment surfaces should be also be investigated. Sulfate reduction in the sediments is often the most important in-lake process for neutralizing acid inputs, and the above evidence indicates that this process is likely decreased in very acidic lakes. Lakes below pH 5 are common in many European regions: in southern Norway, 40% have a pH $< 5$ (Henriksen and Brakke 1988) while in western Germany, 55% of the softwater lakes were $< 5$ (Arzet 1987). In addition, effects on $SO_4^{2-}$ reduction at low pH have important implications for recovery of these more severely acidified systems.

## THE EFFECT OF MICROBIAL ACTIVITIES ON NITRATE ($NO_3^-$) AND AMMONIUM ($NH_4^+$) AND ASSOCIATED $H^+$

As precipitation water passes through watersheds and lakes, the effect of biological activities (of both plants and microorganisms) on the concentrations of $NO_3^-$ and $NH_4^+$ is large and complex. In addition, major differences in the types and magnitudes of N reactions occur in watersheds, as compared to lakes, because of general differences in nutrient limitation in terrestrial and aquatic ecosystems: terrestrial systems are generally considered to be N-limited while lakes are generally P-limited.

All of the biological N reactions, except N fixation, also involve $H^+$:

$$4NO_3^- + 4H^+ + 5CH_2O = 5CO_2 + 2N_2 + 7H_2O \qquad (14.3)$$

Denitrification

$$NH_4^+ + 2O_2 = NO_3^- + 2H^+ + H_2O \qquad (14.6)$$

Nitrification

$$NO_3^- + H^+ + \text{cell precursor} = \text{organic N} \qquad (14.7)$$

N assimilation

$$NO_3^- + H^+ + \text{cell precursor} = \text{organic N.} \qquad (14.8)$$

N assimilation

Equations 14.5 and 14.6 are bacterial reactions while 14.7 and 14.8 are biosynthetic reactions that can be carried out by plants, algae, and bacteria. In contrast to the lack of role that $SO_4^{2-}$ concentration has in affecting biosynthetic rates of uptake, concentrations of N species in some systems are low enough to limit uptake.

$NO_3^-$ and $NH_4^+$ in precipitation vary regionally (e.g., Barrie and Hales 1984), with concentrations in North America ranging from less than 5 μmol $L^{-1}$ $NO_3^-$ or $NH_4^+$ in remote locations to over 50 μmol $L^{-1}$ $NO_3^-$ and 40 μmol $L^{-1}$ $NH_4^+$ in areas heavily affected by anthropogenic emissions. The effective concentrations at high elevation areas can be considerably higher, due to acidic fog (for discussion, see Stoddard 1993). A subject of great interest has been whether increased N deposition has caused forested watersheds to change from systems that are "N-limited" and highly efficient at removing nitrogen compounds from precipitation, to systems that are "N-saturated," meaning that deposition has exceeded the biological requirements for nitrogen and removal efficiency is thus decreased. Any change in watershed removal efficiency will change the concentrations of N compounds reaching surface waters.

The results of the National Stream Survey indicate that in regions where N deposition is less than about 35 meq $m^{-2}$ $yr^{-1}$ (total $NO_3^-$ and $NH_4^+$), stream N concentrations are uniformly low (less than 10 μeq $L^{-1}$ $NO_3^-$ + $NH_4^+$; Kaufmann et al. 1991). However, in higher deposition areas, higher N deposition is related to higher N concentrations in streams (up to over 40 μeq $L^{-1}$; Kaufmann et al. 1991). Furthermore, there is evidence that stream N concentrations have been increasing with time, as N deposition has increased (Stoddard 1993). The relationships are not exactly linear with deposition, as would be expected from differences in growth stages (i.e., growing, mature, or declining forest stocks). In Europe, N deposition rates are as much as 180 meq $m^{-2}$ $yr^{-1}$, and drainage waters are up to 600 μeq $L^{-1}$ $NO_3^-$ and $NH_4^+$ (Driscoll et al. 1989). Data from the Evaluation of Nitrogen and Sulfur Fluxes (ENSF) sites indicate that drainage waters at many sites show elevated N concentrations where deposition exceeds 70 meq $m^{-2}$ $yr^{-1}$ (N. Dise, pers. comm.).

Where stream nitrogen is elevated, its form is almost entirely as $NO_3^-$ (Sale et al. 1988). This is expected from the greater mobility of this ion in soils, compared to $NH_4^+$. Also, $NH_4^+$ is nitrified to $NO_3^-$ at high rates in soils, even in acidic watersheds (Tietema and Verstraten 1991). Thus, the major consequence to streams and lakes of higher N ($NO_3^-$ + $NH_4^+$) deposition onto watersheds is higher $NO_3^-$ input.

Increased deposition is not the only factor that may lead to increased watershed output of $NO_3^-$. Any change in watershed removal of N will also be important. In Norwegian lakes, during a decade in which N deposition did not change much, nitrate concentrations in lakes doubled, suggesting that decreased watershed removal was the cause (Henriksen et al. 1988).

Nitrogen removal in watersheds may be controlled by more than vegetation growth. Denitrification has recently been shown to be the dominant removal route of N from at least one watershed with high N deposition (Tietema and Verstraten 1991). The variability of this activity in different watersheds under different conditions needs to

**Table 14.2** Removal coefficients for nitrate ($S_N$ = nitrate removal rate in $\mu$mol m$^{-2}$ yr$^{-1}$ per nitrate concentration in $\mu$mol m$^{-3}$) in lakes with low (first three rows) and high nitrate loadings (excerpted from Kelly et al. 1990).

| Lake | Years | $S_N$ (m yr$^{-1}$) |
|---|---|---|
| 302S (pre-acid) | 1981 | 440 |
| 302N (pre-acid) | 1981 | 220 |
| Crystal (Wisconsin) | 1984 | 42 |
| Harp (central Ontario) | 1984–1986 | 6.8 |
| Langtjern (Norway) | 1972–1978 | 6.8 |
| 302N (nitric acid added) | 1982–1985 | 5 |
| Dart's (New York) | 1982–1984 | 0.89 |

be studied further. For example, the proportion of wet soils, both in space and time, probably affects the amount of denitrification that occurs.

The biological utilization of $NO_3^-$ in lakes is also affected by nitrogen loading rate, but the effects apparently occur at much lower levels of increased loading than in watersheds. In contrast to watershed systems, biological productivity in lakes tends to be P-limited (e.g., Schindler 1977). Thus, when lakes receive increased N inputs in the absence of increased P inputs, the algal community cannot increase its biomass (Equations 14.7 and 14.8) because of P limitation (Rudd et al. 1990).

The NSS streams discussed above with high $NO_3^-$ concentrations did not have elevated P concentrations (Sale et al. 1988), so the increased $NO_3^-$ would not likely be utilized by algae for biosynthesis (Equation 14.7). Rather, this $NO_3^-$ is available for denitrification (Equation 14.5; Rudd et al. 1990). Reactions 14.7 and 14.5 have greatly different rates in lakes. The relationship of removal rate of $NO_3^-$ to concentration ($S_N$) can be expressed similarly to Equation 14.3 for $SO_4^{2-}$. Kelly et al. (1990) showed that in lakes receiving close to natural nitrogen inputs, $NO_3^-$ removal was rapid ($S_N > 25$ m yr$^{-1}$; Table 14.2) and complete, with summer $NO_3^-$ concentrations at or below the detection level (0.5 meq L$^{-1}$). In contrast, lakes receiving recently elevated N inputs removed $NO_3^-$ at much slower rates (< 7 m yr$^{-1}$), and $NO_3^-$ concentrations in summer were greater than 1 $\mu$eq L$^{-1}$. The difference in removal rates is due to the fact that algal uptake is rapid and occurs throughout the surface water, while denitrification rate is limited by the position of these bacteria in the sediments, similar to the situation of $SO_4^{2-}$ reducers (Rudd et al. 1990). Denitrifiers are located closer to the sediment-water interface than are $SO_4^{2-}$ reducers (Rudd et al. 1986b) and this fact, together with the greater free energy release of denitrification compared to $SO_4^{2-}$ reduction, results in higher relative rates of $NO_3^-$ removal compared to $SO_4^{2-}$ removal (i.e., $S_N > S_S$).

The difference in $NO_3^-$ and $SO_4^{2-}$ removal rates is important in determining the relative acidification efficiency of nitric and sulfuric acid inputs. Obviously, the efficiency of nitric acid inputs in acidifying lakes should be less, and this was shown

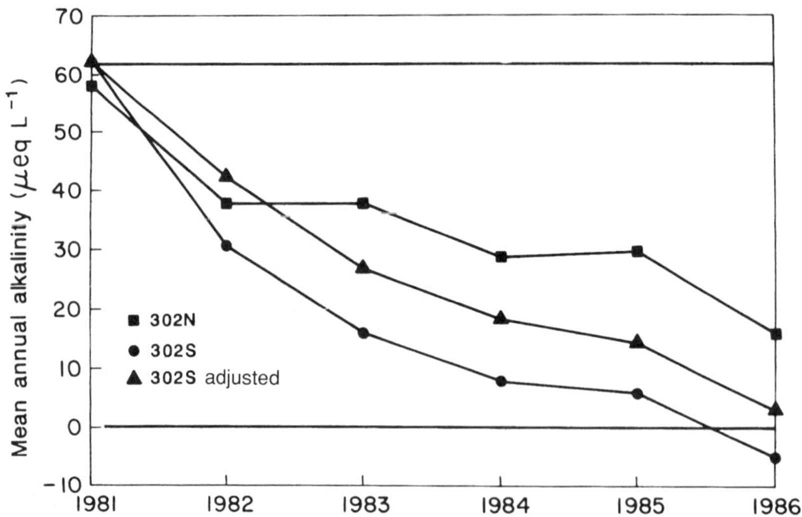


**Figure 14.2** Measured losses of alkalinity from the north basin of Lake 302 (acidified with nitric acid only) and the south basin (acidified with sulfuric acid only), and alkalinity losses for the south basin adjusted to account for the slightly higher rates of acid addition. The most meaningful comparison, therefore, is between the 302N line (■) and the 302S adjusted line (▲). From Rudd et al. (1990.)

experimentally in a double basin acidification experiment, where one basin received only nitric acid and the other basin received only sulfuric acid (Figure 14.2; Rudd et al. 1990). Nitric acid was about 70% as efficient as sulfuric acid, per equivalent added. It is important to note that *both* acids were able to acidify; biological reactions that remove $NO_3^-$ cannot be 100% efficient when inputs exceed algal requirements. In a number of acid lakes, $NO_3^-$ can contribute significantly to the total acid anions, with concentrations up to 35 μeq $L^{-1}$ in some Adirondack lakes (Driscoll et al. 1989), 14–28 μeq $L^{-1}$ in southern Norwegian lakes (Henriksen et al. 1988), and 70–90 μeq $L^{-1}$ in the Schwarzwald region of Germany (Zoettl et al. 1985).

An internally generated effect of microbial activity on N water chemistry in acidified lakes comes from the special sensitivity of aquatic nitrification (Equation 14.6) to pH below 5.4–5.6 (Rudd et al. 1988). There are only three known genera of aquatic nitrifiers, and they have been shown in the laboratory to require $NH_3$ rather than $NH_4^+$ (Suzuki et al. 1974). Thus, as the pH declines, the chemical form that they require becomes less and less available. Inhibition of nitrification can be observed by examining winter profiles of $NO_3^-$ and $NH_4^+$ (Figure 14.3; Rudd et al. 1988).

When nitrification is *not* inhibited, $NH_4^+$ does not accumulate in surface water, even though it is constantly being produced by decomposition in the sediments. This has important consequences for acid-base chemistry. In the spring, after ice-off, the dominant N species in the water will be $NO_3^-$, produced over winter by nitrification

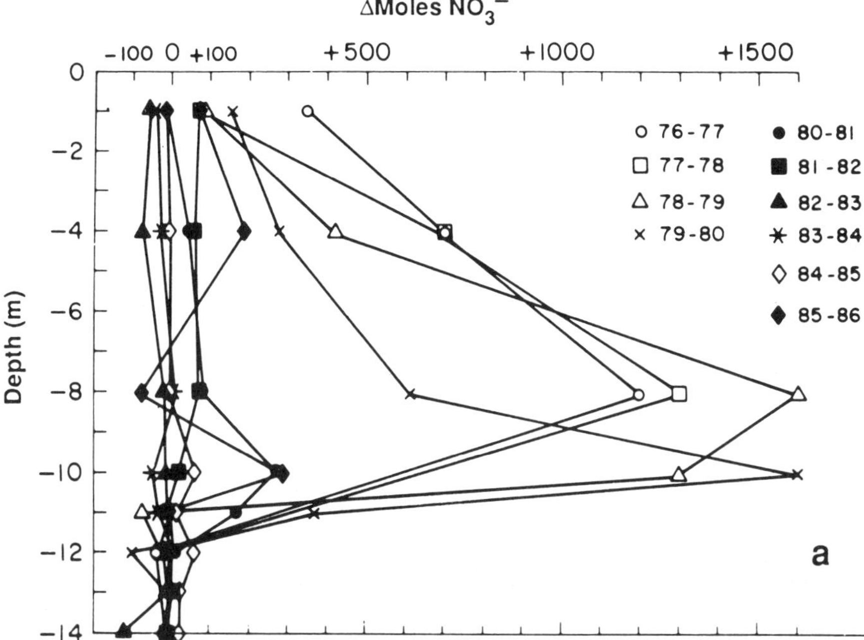


**Figure 14.3** Changes of nitrate mass with depth in Lake 223 at the Experimental Lakes Area under ice cover. The average pH values of the lake water during these winters were: 6.2 in 1976/77, 5.9 in 1977/78, 5.6 in 1978/79 and 1979/80, 5.4 in 1980/81 and 1981/82, 5.2 at 1982/83, 5.1 in 1983/84. In 1984/85 and 1985/86 the lake was in recovery phase and the winter pH was 5.5. Resumption of nitrification did not occur in 1984/85 and was only slight in 1985/86. From Rudd et al. (1988).

(Rudd et al. 1988) and also coming in with the spring stream flow. Any $NO_3^-$ taken up by algae will also result in uptake of $H^+$ (Equation 14.7) and tend to offset acid inputs at that time. Also, over the summer, excess $NO_3^-$ can be denitrified (Rudd et al. 1990) and thus removed from the lake to the atmosphere as $N_2$. This tends to keep the supply of N in approximate balance with the supply of P.

On the other hand, when nitrification *is* inhibited, $NH_4^+$ accumulates from decomposition over winter (Rudd et al. 1988). Consequently, in spring and summer, a significant amount of algal N uptake involves $NH_4^+$, which releases, instead of consuming, $H^+$ (Equation 14.7). Also, there is no "exit" route for $NH_4^+$, except the lake outflow, and $NH_4^+$ concentrations would be expected to increase. Elevated $NH_4^+$ has been observed in Lake 302S (M. Turner and J. Rudd, pers. comm.), which has been experimentally acidified to pH 4.5 with sulfuric acid. In this lake, $NH_4^+$ has increased greatly, apparently due to increased inhibition of nitrification as the water and, more recently, the sediments, have become more acidic. Chemical exchange of $H^+$ for $NH_4^+$ in the sediments was examined but not found (L. Camaro and J. Rudd, pers. comm.).

Thus, even though this lake was acidified only by sulfuric acid, effects on microbial reactions involving N species have caused $NH_4^+$ to become a dominant ion in the water chemistry, and uptake and release of $NH_4^+$ through algal growth and decomposition have become major determinants of seasonal changes in alkalinity and pH.

A major question is why aquatic nitrification is so much more susceptible to acid conditions than is terrestrial nitrification. One suggestion is that terrestrial nitrification is carried out primarily by heterotrophic nitrifiers, which remove $NH_4^+$ from organic material and nitrify it intracellularly; however, $^{15}N$ studies suggest that the $^{15}NH_4^+$ added is also nitrified under acidic conditions (Stams et al. 1991). A further question is why nitrification is not inhibited at low pH in Lake Orta, Italy, which receives high inputs of $NH_4^+$ in waste effluent and where nitrification occurs slowly at pHs as low as 3.5 (Gerletti and Provini 1978). Are there genetically different strains in these systems, or are the environmental conditions somehow different? There is some evidence that nitrifiers can grow under less favorable conditions if the population is already well developed (Gerletti and Provini 1978).

In contrast to nitrification, there is no evidence that denitrification is sensitive to low pH (Rudd et al. 1986b). As mentioned above, net annual $SO_4^{2-}$ reduction may be significantly reduced in severely acidified lakes. As this is a major acid-neutralizing activity (Cook et al. 1986), this phenomenon and its consequences should be thoroughly investigated in areas where there are high numbers of lakes below pH 5 (e.g., southern Scandinavia, central Europe). Decomposition of organic carbon supports both denitrification and $SO_4^{2-}$ reduction, and it is also affected at pH 5 and below (McKinley and Vestal 1982; Kelly et al. 1984). Thus, above pH 5, negative effects are observed on only one major microbial reaction that affects water chemistry (nitrification), while below pH 5 the effects multiply.

## ACKNOWLEDGEMENTS

I thank D. Landers, J. Rudd, M. Turner, and D. Schindler for helpful reviews.

## REFERENCES

Amaral, J.A. 1991. Sulfate reduction and organic sulfur formation in lake sediments. Ph. D. Thesis. Manitoba: Univ. of Manitoba.

Arzet, K. 1987. Diatomeen als pH-Indikatoren in subrezenten Sedimenten von Weichwasserseen. *Diss Abt. Limnol Innsbruck* **24**:1–266.

Baker, L.A., and P.L. Brezonik. 1988. Dynamic model of in-lake alkalinity generation. *Water Resour. Res.* **24**:65–74.

Baker, L.A., P.L. Brezonik, and C.D. Pollman. 1986. Model of internal alkalinity generation: Sulfate retention component. *Water, Air, Soil Pollut.* **31**:89–94.

Baker, L.A., P.R. Kaufmann, A.T. Herlihy, and J.M. Eilers. 1990. Current status of surface water acid-base chemistry. NAPAP Report 9. In: Acid Deposition: State of Science and Technology, vol. II. Washington, D.C.: Natl. Acid Precipitation Assessment Program.

Barrie, L., and J.M. Hales. 1984. The spatial distributions of precipitation acidity and major ion wet deposition in North America during 1980. *Tellus* **36B**:333–355.

Bayley, S.E., R.S. Behr, and C.A. Kelly. 1986. Retention and release of S from a freshwater wetland. *Water, Air, Soil Pollut.* **31**:101–114.

Carignan, R. 1987. Quantitative importance of alkalinity flux from the sediments of acid lakes. *Nature* **317**:158–160.

Church, M.R., P.W. Shaffer, and B.P. Rochelle. 1990. Retention of atmospherically deposited sulfur in watersheds of the eastern United States. Intl. Conf. Acidic Deposition. Edinburgh: Royal Society of Edinburgh.

Connell, W.E., and W.H. Patrick, Jr. 1968. Sulfate reduction in soil: Effects of redox potential and pH. *Science* **159**:86–87.

Cook, R.B. 1981. The biogeochemistry of sulfur in two small lakes. Ph.D. Dissertation, Columbia University, 246 pp.

Cook, R.B., and C.A. Kelly. 1992. Sulphur cycling and fluxes in temperate dimictic lakes. In: Sulphur Cycling on the Continents, ed. R.W. Howarth, J.W. B. Stewart, and M.V. Ivanov, pp. 145–188. SCOPE. Chichester: Wiley.

Cook, R.B., C.A. Kelly, D.W. Schindler, and M.A. Turner. 1986. Mechanisms of hydrogen ion neutralization in an experimentally acidified lake. *Limnol. Oceanogr.* **31**:134–148.

Cook, R.B., R.B. Kreis, Jr., J.C. Kingston, K.E. Camburn, S.A. Norton, M.J. Mitchell, B. Fry, and L.C.K. Shane. 1990. Paleolimnology of McNearney Lake: An acidic lake in northern Michigan. *J. Paleolimnol.* **3**:13–34.

Cook, R.B., and D.W. Schindler. 1983. The biogeochemistry of sulfur in an experimentally acidified lake. *Ecol. Bull. Stockholm* **35**:115–127.

Cuhel, R.L., C.D. Taylor, and H.W. Jannasch. 1982. Assimilatory sulfur metabolism in marine microorganisms: sulfur metabolism, protein synthesis, and growth of *Alteromonas luteo-violaceus* and *Pseudomonas halodurans* during perturbed batch growth. *Appl. Environ. Microbiol.* **43**:151–159.

David, M.B., M.J. Mitchell, and J.P. Nakas. 1982. Organic and inorganic sulfur constituents of a forest soil and their relationship to microbial activity. *Soil Sci. Soc. Am. J.* **46**:847–852.

Driscoll, C.T., D.A. Schaefer, and J.L. Malanchuk. 1989. Discussion of European and North American Data. In: The Role of Nitrogen in the Acidification of Soils and Surface Waters, ed. J.L. Malanchuk and J. Nilsson. Miljorapport 1989, vol 10. Denmark: Nordic Council of Ministers.

Gerletti, M., and A. Provini. 1978. Effect of nitrification in Orta Lake. *Prog. Water Tech.* **10**:839–851.

Giblin, A.E., G.E. Likens, D. White, and R.W. Howarth. 1990. Sulfur storage and alkalinity generation in New England lake sediments. *Limnol. Oceanogr.* **35**:852–869.

Grahn, O. 1986. Vegetation structure and primary production in acidified lakes in southwestern Sweden. *Experientia* **42**:465–70.

Hemond, H. 1980. Biogeochemistry of Thoreau's Bog, Concord, Massachusetts. *Ecol. Monogr.* **50**:507–526.

Henriksen, A., and D.F. Brakke. 1988. Increasing contributions of nitrogen to the acidity of surface waters in Norway. *Water, Air, Soil Pollut.* **42**:182–201.

Henriksen, A., L. Lien, T.S. Traaen, I.S. Sevaldrud, and D.F. Brakke. 1988. Lake acidification in Norway: Present and predicted chemical status. *Ambio* **17**:259–266.

Herlihy, A.T., and A.L. Mills. 1985. Sulfate reduction in freshwater sediments receiving acid mine drainage. *Appl. Environ. Microbiol.* **49**:179–186.

Herlihy, A.T., and A.L. Mills. 1986. The pH regime of sediments underlying acidified waters. *Biogeochem.* **2**:95–99.

Howell, E.T., M.A. Turner, R.L. France, M.B. Jackson, and P.M. Stokes. 1990. Comparison of Zygnematacean (Chlorophyta) algae in the metaphyton of two acidic lakes. *Can. J. Fish. Aquat. Sci.* **47**:1085–1092.

Jassby, A.D. 1975. Dark sulfate uptake and bacterial productivity in a subalpine lake. *Ecology* **56**:627–636.

Kaufmann, P.R., A.T. Herlihy, M.E. Mitch, and J.J. Messer. 1991. Stream chemistry in the Eastern United States. 1. Synoptic survey design, acid-base status, and regional patterns. *Water Resour. Res.* **27**:611–627.

Kelly, C.A., and J.W.M. Rudd. 1984. Epilimnetic sulfate reduction and its relationship to lake acidification. *Biogeochem.* **1**:63–77.

Kelly, C.A., J.W.M. Rudd, A. Furutani, and D.W. Schindler. 1984. Effects of lake acidification on rates of organic matter decomposition in sediments. *Limnol. Oceanogr.* **29**:687–694.

Kelly, C.A., J.W.M. Rudd, and D.W. Schindler. 1990. Acidification by nitric acid-future considerations. *Water, Air, Soil Pollut.* **50**:49–61.

Kelly, C.A., et al. 1987. Prediction of biological acid neutralization in acid-sensitive lakes. *Biogeochem.* **3**:129–140.

Landers, D.H., M.B. David, and M.J. Mitchell. 1983. Analysis of organic and inorganic sulfur constituents in sediments, soils, and water. *Intl. J. Environ. Anal. Chem.* **14**:245–256.

Likens, G.E., F.H. Bormann, and N.M. Johnson. 1981. Interactions between major biogeochemical cycles in terrestrial ecosystems. In: Some Perspectives of the Major Biogeochemical Cycles, ed. G.E. Likens, pp. 93–112. New York: Wiley.

McKinley,V.L., and J.R. Vestal. 1982. Effects of acid on plant litter decomposition in an arctic lake. *Appl. Environ. Microbiol.* **43**:1188–1195.

Melzer, A., and E. Rothmeyer. 1983. Die Auswirkung der Versauerung der beiden Arberseen im Bayerischen Wald auf die Makrophytenvegetation. *Ber. Bayer. Bot. Ges.* **54**:9–18.

Mitchell, M.J., M.B. David, and R.B. Harrison. 1992. Sulphur dynamics of forest ecosystems. In: Sulphur Dynamics of Forest Ecosystems, ed. R.W. Howarth, J.B. Stewart, and M.V. Ivanov, pp. 215–254. SCOPE. Chichester: Wiley.

Norton, S.A., M.J. Mitchell, J.S. Kahl, and G.R. Brewer. 1988. In-lake alkalinity generation by sulfate reduction: A paleolimnological assessment. *Water, Air, Soil Pollut.* **39**:33–45.

Nriagu, J.O., and Y.K. Soon. 1985. Distribution and isotopic composition of sulfur in lake sediments of northern Ontario. *Geochim. Cosmochim. Acta 49*:823–834.

Rudd, J.W.M., C.A. Kelly, and A. Furutani. 1986a. The role of sulfate reduction in long-term accumulation of organic and inorganic sulfur in lake sediments. *Limnol. Oceanogr.* **31**:1281–1291.

Rudd, J.W.M., C.A. Kelly, V. St. Louis, R.H. Hesslein, A. Furutani, and M.H. Holoka. 1986b. Microbial consumption of nitric and sulfuric acids in acidified north temperate lakes. *Limnol. Oceanogr.* **31**:1267–1280.

Rudd, J.W.M., C.A. Kelly, D.W. Schindler, and M.A. Turner. 1988. Disruption of the nitrogen cycle in acidified lakes. *Science* **240**:1515–1517.

Rudd, J.W.M., C.A. Kelly, D.W. Schindler, and M.A. Turner. 1990. A comparison of the acidification efficiencies of nitric and sulfuric acids by two whole-lake addition experiments. *Limnol. Oceanogr.* **35**:663–679.

Sale, M.J., P.R. Kaufmann, H.I. Jager, J.M. Coe, K.A. Cougan, A.J. Kinney, M.E. Mitch, and W.S. Overton. 1988. Chemical characteristics of streams in the mid-Atlantic and southeastern United States. Vol. II: Streams Samples, Descriptive Statistics, and Compendium of Physical and Chemical Data. EPA/600/3–88/021b. Washington, D.C.: U.S. EPA.

Schindler, D.W. 1977. Evolution of phosphorus limitation in lakes. *Science* **195**:260–262.

Schindler, D.W., R.W. Newbury, K.G. Beaty, and P. Campbell. 1976. Natural water and chemical budgets for a small Precambrian lake basin in central Canada. *J. Fish. Res. Board Can.* **33**:2526–2543.

Stams, A.J.M., H.W.G. Booltink, I.J. Lutke-Schipholt, B. Beemsterboer, J.R.W. Woittiez, and N. Van Breemen. 1991. A field study on the fate of $^{15}$N-ammonium to demonstrate nitrification of atmospheric ammonium in an acid forest soil. *Biogeochem.* **13**:241–255.

Stoddard, J.L. 1993 Long-term changes in watershed retention of nitrogen: Its causes and aquatic consequences. In: Environmental Chemistry of Lakes and Reservoirs. Advances in Chemistry Series No. 237. Washington, D.C.: American Chemical Society, in press.

Suzuki, I., U. Dular, and S.C. Kwok. 1974. Ammonia or ammonium ion as substrate for oxidation by *Nitrosomonas europaea* cells and extracts. *J. Bacteriol.* **120**:556–558.

Swank, W.T., J.W. Fitzgerald, and J.T. Ash. 1984. Microbial transformation of sulfate in forest soils. *Science* **223**:182–184.

Sweerts, J.P.R.A. 1990. Oxygen consumption processes, mineralization and nitrogen cycling at the sediment-water interface of north temperate lakes. Ph.D. Thesis. Amsterdam: Univ. of Amsterdam.

Tietema, A. and J.M. Verstraten. 1991. Nitrogen cycling in an acid forest ecosystem in the Netherlands under increased atmospheric nitrogen input. *Biogeochem.* **15**:21–46.

Zoettl, H.W., K.-H. Feger, and G. Brahmer. 1985. Chemismus von Schwarzwaldgewässern während der Schneeschmelze 1984. *Naturwiss.* **72**:268–270.

# 15

# Extrapolating from Toxicological Findings to Regional Estimations of Acidification Damage

J.M. GUNN[1] and N. BELZILE[2]

[1]Ontario Ministry of Natural Resources, Cooperative Freshwater Ecology Unit, Laurentian University, Biology Department, Sudbury, Ontario P3E 2C6, Canada

[2]Chemistry Department, Laurentian University, Sudbury, Ontario P3E 2C6, Canada

## ABSTRACT

The decade of the 1980s was mainly a period of testing and refining existing models of the interactive effect of acidification and toxic metals. Most work focused on the interaction between $H^+$ and inorganic Al, and on the ameliorating effects of base cations (particularly $Ca^{2+}$) and dissolved organic compounds. Results of these studies have not been widely used to estimate resource damages from acid deposition. Simple pH (or acid-neutralizing capacity [ANC]) thresholds models, such as the pH/species richness or pH/species occurrence models are still the basis for regional estimations of the damaging effects of acidification.

## INTRODUCTION

During the early decades of acidification research (1960s and 1970s) there was considerable evidence that $H^+$ alone could not account for the observed biological damage that occurred as lakes and streams acidified. Laboratory toxicity tests, designed to determine pH thresholds under controlled conditions, generally underestimated effects observed in the field (e.g., Sadler 1983).

Mortality of aquatic biota in natural systems appeared to be correlated not only to pH, but to a suite of chemical variables. Acidified waters were characteristically clear, low-calcium waters with elevated concentrations of several potentially toxic metals (Al, Cd, Cu, Pb, Hg, Mn, and Zn; Campbell et al. 1985). Most metals were in free ion

*Acidification of Freshwater Ecosystems: Implications for the Future*
Edited by C.E.W. Steinberg and R.F. Wright 

forms rather than bound to organic molecules (Campbell and Stokes 1985). Interactions between $H^+$, toxic metals, and base cations (particularly $Ca^{2+}$) were therefore hypothesized as having a significant role in the resource damages that were occurring.

The year 1980 was a benchmark year in the study of interactive effects of acidification and toxic metals. The International Conference at Sandefjord, Norway, in March, 1980 (Drabløs and Tollan 1980), provided a synthesis of biological effects of acidification that was far more complete than at the first international conference in Columbus, Ohio, in 1975. For example, in the 1980 proceedings, blood ion regulatory failure was recognized as the principal toxic mechanism for aquatic biota in low pH environments, and the high sensitivity of early life stages to acid stress was a common phenomenon for several species of biota (Muniz and Leivestad 1980). However, a "new" discovery was also highlighted at Sandefjord: aluminum has a toxic effect on freshwater biota in low pH waters at realistic concentration levels (Driscoll et al. 1980). The aluminum finding triggered a great deal of work on biological effects of metals in acid stressed systems.

In preparing this paper, we reviewed the recent literature on interactive effects of acidification and metals on freshwater biota. We chose to focus mainly on the research findings from the 1980s (i.e., work that followed from Sandefjord). The 1980s were a period when large national programs, such as NAPAP (National Acid Precipitation Assessment Program) in the U.S.A. and LRTAP (Long Range Transport of Atmospheric Pollutants) in Canada, were launched to determine the need for pollutant abatement programs (Schindler 1992).

For the purposes of this workshop, we discuss the importance of the findings on interactive effects of toxic metals and acidification from a management perspective.[1] By this, we mean that we recognize that most of the national acid rain programs had very applied research goals. For example, managers needed information on the rate and extent of biological damage or recovery in order to select the most cost-effective pollutant control program (Rubins et al. 1992). Therefore, one important criterion that should be applied when examining the significance of acidification-related research findings is whether they contributed to achieving these goals.

We use a question-and-answer format to focus our discussion and draw heavily on the fishery literature for examples. This emphasis on fisheries also reflects the management perspective that we have adopted. It is well recognized that the loss of sportfish populations was one of the principal factors that initiated much of the concern and subsequent research on acidification.

---

1 Editor's note: The basic difference between science and management is a crucial factor in planning mitagation procedures and in policymaking. See also Rosseland and Staurnes, this volume.

## WHAT SIGNIFICANT "DISCOVERIES" CONCERNING $H^+$/TOXIC METAL INTERACTIONS WERE MADE IN THE 1980s?

The 1980s seem to have been mainly a period of testing and refining existing models. When we compared the state of knowledge about acidification and metal interactions, as summarized in review articles from before (Schofield 1976; Muniz and Leivestad 1980; Haines 1981; Harvey et al. 1981; Spry et al. 1981) and after (Baker et al. 1990; RMCC 1990; Spry and Wiener 1991; Wren and Stephenson 1991) the research activities of the 1980s, we had to conclude that very few substantive "discoveries" were made. The intensive research period generated few, if any, findings that represented radical departures from past ways of thinking about interactive effects with metals. This appears to be also true for several other aspects of the study of the physiology and toxicology of acid stress.

Several metals (e.g., Al, Mn, Zn, Be, Cu, Cd, Fe, Ni) were found to occur in elevated or measurable concentrations in acidic waters (reviewed by Veselý, this volume), but of these, only Al was clearly contributing to the loss of freshwater biota in acidifying systems (Baker et al. 1990). Therefore, toxicity research in the 1980s focused mainly on the interactive effects of $H^+$, $Ca^{2+}$, and inorganic Al (Figure 15.1).

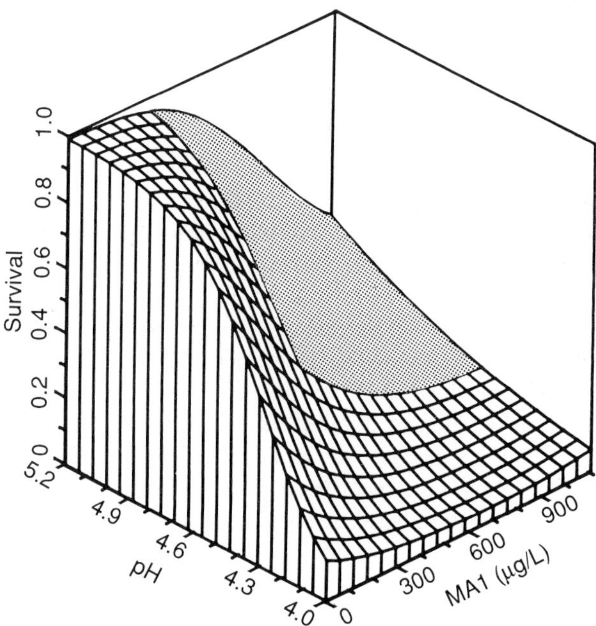


**Figure 15.1** A typical response surface diagram illustrating the interactive effects of pH and inorganic monomeric aluminum (MAI) on survival of brook trout fry in low calcium waters (Ca 2 mg $L^{-1}$). Figure from Mount and Marcus (1989).

Toxicity studies of the interactive effects of $H^+$, $Ca^{2+}$, and Al confirmed and improved our knowledge of the ameliorating effects of Ca and other base cations on $H^+$ stress; they also contributed to our understanding of how dissolved organic substances (humic material) reduce the toxicity of Al. However, Al chemistry in natural waters remains poorly understood, and the observed effects of Al on freshwater biota are, to some extent, variable (Rosseland et al. 1990; Rosseland and Staurnes, this volume). For example, Al in low pH solutions can have severe detrimental effects, no effects, or even marked beneficial effects on test organisms, depending on exposure conditions, life stage, and species. It has also proven difficult to identify the relative toxicity of various forms of inorganic aluminum (Campbell and Stokes 1985); therefore, most recent toxicity models (Baker et al. 1990) have related mortality or growth effects simply to the sum of the various inorganic species of Al (i.e., inorganic [labile] monomeric Al).

Prior to the 1980s, toxicity testing was applied to a relatively small number of freshwater biota, usually the economically important fish species. In the 1980s, such testing expanded (albeit in a piecemeal, rather than systematic, way) to include a large and varied array of species: from microbes and periphyton to amphibians and shoreline birds. However, a great deal of attention was also given to a number of "key" species. Using a battery of test procedures—from whole-lake manipulations, to *in situ* bioassays, to highly controlled laboratory tests—strong evidence was obtained on toxic conditions for a number of species of freshwater biota. For species such as lake trout (*Salvelinus namaycush*), brook trout (*S. fontinalis*), fathead minnow (*Pimephalis promelas*), atlantic salmon (*Salmo salar*), *Hyalella azteca*, and *Daphnia galeata mendotae* much replication was conducted and refined toxicity models now exist.

In addition, at the physiological level, attention primarily focused on the effects of pH, inorganic Al, and $Ca^{2+}$ (McDonald et al. 1989). This physiological work advanced our knowledge of basic physiological processes, particularly at the organ and tissue level. However, it has not been used to predict population effects or to address resource damage issues, because of the traditional problems of determining the ecological significance of physiological and biochemical indicators.

In terms of bioaccumulation, we have seen improvements in sampling and metal speciation techniques that have made it possible to construct empirical models based on geochemical considerations and the free metal concept. They better predict contaminant burdens of some metals, such as Cd (Tessier et al. 1992). However, with accumulation of mercury—the metal receiving the most attention with regard to acidification—the association between bioaccumulation and lakewater pH still remains questionable (Richman et al. 1988). Mercury deposition is now clearly an air pollution problem, with considerable overlap between the areas of high acid and mercury deposition (Nater and Grigal 1992). Rates of methylation (Veselý, this volume) and bioaccumulation (Ponce and Bloom 1991) appear to be pH-dependent, but it remains difficult to explain the high variability in body burdens of Hg in freshwater biota (Lathrop et al. 1991).

## DO THE DISCREPANCIES BETWEEN LABORATORY AND FIELD pH THRESHOLDS STILL EXIST? HAS THE WORK ON INTERACTIVE EFFECTS WITH TOXIC METALS HELPED US EXPLAIN THESE APPARENT DISCREPANCIES?

### North America

The original premise that pH alone was not sufficient for predicting biological effects, and that we needed information on interactive effects with toxic metals, may have been incorrect, particularly in a North American context, where base cation concentrations are relatively high. Under North American conditions, $H^+$ rather than inorganic Al appears to be the dominant toxicant (Hutchinson et al. 1990).

With improvements in laboratory test procedures (e.g., appropriate life stages, water hardness, acclimation procedures) the discrepancies between laboratory and field pH thresholds have been largely eliminated for many North American species. The good agreement in Figure 15.2 between lab LC50s (generated for pH only) and field observation of the pH threshold for extinction of ten common fish species and one zooplankton species (Hutchinson et al. 1990), demonstrates that we can now use laboratory findings to predict population effects for these and perhaps many other species. The high predictability of the occurrence of the ubiquitous crustacean zooplankton, *Daphnia galeata mendota*, using simple pH thresholds derived from the lab, is a good example of these improvements (Hutchinson et al. 1990; Keller et al. 1990). Such findings illustrate the methodological refinements that have occurred during the 1980s; however, more importantly, for our discussion of interactive effects with metals, they present a strong case against the importance of the toxic effects of Al or other metals in acidifying systems of North America.

One should note that in the Hutchinson et al. (1990) study there were three species of fish (lake trout; brook trout; fathead minnow) for which large discrepancies (lab/field) between pH tolerance limits still existed. The authors suggest that even in these cases, particularly for the large long-lived salmonids, problems with the test protocols (use of short duration, early life stage testing) rather than water chemistry, may explain the discrepancies. However, these discrepancies remain interesting research challenges. Recent findings on the potential toxicity of low-level concentrations of Be (Vesely et al. 1989) and Cu (Sayer et al. 1991; Welsh et al. 1992) in dilute, low-dissolved organic carbon waters, may be key to solving some of these problems.

From our management perspective, empirical relationships also offer solutions to some of these discrepancies. The lake trout is a good example where, even in the absence of agreement between lab and field toxicity results, managers still do not appear to need data on toxic metals to predict resource damages. There is now very strong empirical evidence, from extensive surveys, stocking experiments, and whole-lake manipulations through acid additions directly to the water column (i.e., no watershed leaching of Al), that field pH measurements alone can be used to predict the health of lake trout populations accurately (Beggs et al. 1985; Baker et al. 1990).

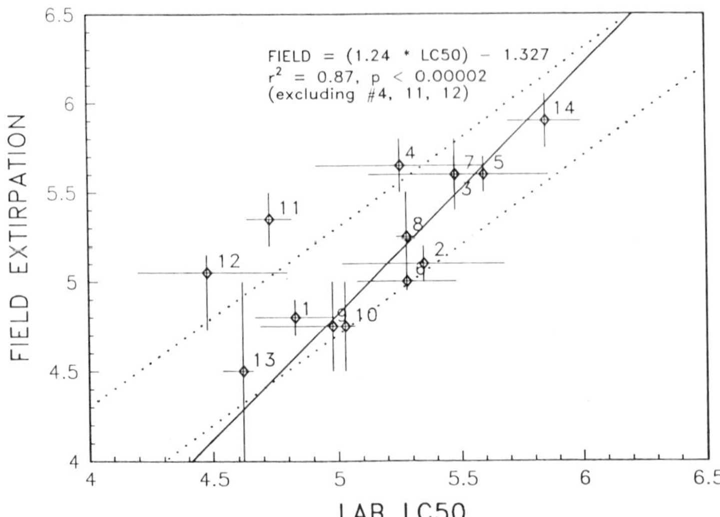


**Figure 15.2** Comparison of LC50 for pH determined in the laboratory (i.e., without addition of metals) and pH of extirpation of fish and *Daphnia galeata mendotae* determined from the occurrence of species in field surveys. The regression line for the 11 matching species are shown. Dotted lines are drawn ± 0.3 pH units about the 1:1 line (not illustrated). The numbers correspond to the following species: 1–golden shiner, 2–creek chub, 3–bluntnose dace, 4–fathead minnow, 5–bluntnose minnow, 6–redbelly dace, 7–common shiner, 8–walleye, 9–smallmouth bass, 10–white sucker, 11–lake trout, 12–brook trout, 13–lake whitefish, 14–*Daphnia galeata mendotae*. Figure from Hutchinson et al. (1990).

A pH threshold of approximately 5.5 separates lakes that can and cannot support lake trout.

## Scandinavia

In Scandinavia, the situation appears to be quite different. Many of the acidic systems in southern Norway and other parts of Scandinavia have very high concentrations of inorganic Al and very low concentrations of $Ca^{2+}$ (< 1 mg $L^{-1}$). In these systems inorganic Al appears to be the dominant toxicant (Rosseland and Staurnes, this volume); however, the low calcium concentrations can also be considered an important stress factor for freshwater biota. In fact, the low calcium concentrations in some waters may greatly delay chemical and biological recovery even after substantial reductions in acid deposition are achieved. Using the MAGIC model projections of lakewater quality in southern Norway, under various acid reduction scenarios, Skeffington and Brown (1992) have predicted that potential benefits to aquatic biota, from the initial increases in lakewater pH, will be largely offset by declining $Ca^{2+}$ concentrations (through reduced leaching rates). Recent results from the Thousand Lakes Survey in Norway tend to support this prediction (B. Rosseland, pers. comm.).

## HAVE TOXICITY MODELS BEEN SUCCESSFUL IN INCORPORATING THE COMBINED EFFECTS OF LOW pH AND ELEVATED METALS TO PREDICT RESOURCES DAMAGES OR RISKS?

Most of the published aquatic resource damage models, developed through the research programs of the 1980s, do not make use of interactive effects of $H^+$ and toxic metals. Instead, they mainly use the now well-known negative relationships between species richness and acidity to estimate species losses on a regional scale (e.g., Minns et al. 1990). The toxicity components of these models are usually very simple threshold relationships, with measured pH, or its correlate ANC, used to predict presence or absence of individual species or groups of organisms. In most of these species richness models, measures of biological damage are assigned to lakes when pH drops below approximately 6.0, the threshold below which sensitive species of fish and invertebrates decline and/or acid-tolerant nuisance algae begin to proliferate.

More complex toxicity models, which make use of the interactive effects of $H^+$, Ca, and Al and combine laboratory and field data, have been recently developed as part of the NAPAP regional assessments of effects of acidification on fish communities (Baker et al. 1990). These models appear to improve our predictive ability of the occurrence of some species (e.g., brook trout, a species for which laboratory pH thresholds still differ from field observations). However, it is difficult to apply these models to make resource damage estimates, even for the areas of the U.S.A., where they were developed, for the following reasons:

- lack of background information on the distribution of fish or other aquatic resources for regional extrapolations of toxicity models;
- lack of measured inorganic monomeric Al data from regional surveys—this variable was not included in most of the large regional surveys in the U.S.A. and Canada;
- remaining uncertainty about the role of episodic acidification events.

Another difficulty with incorporating toxicity models into predictions of regional-scale resource damages (or recovery) is the lack of direct linkage to the current group of hydrogeochemical models (e.g., MAGIC). The principal outputs of these models are estimates of average ANC, not toxicity parameters, such as pH or inorganic Al. Therefore, again it appears that empirical relationships (i.e., observed relationships between simple chemical variables [e.g., critical limits of ANC] and the health of the aquatic community) may have greater predictive value than the output of toxicity-based models (Cosby et. al., this volume).

Additional research is needed, however, to examine ways of linking these models in order to deal with problems such as site-specific or episodic acidification.

## WHAT HAVE WE LEARNED AND WHERE DO WE GO FROM HERE?

The published literature on interaction of low pH and toxic metals on freshwater biota is now very extensive (see partial list of references in Wang 1987 and Baker et al. 1990). Given the lack of real breakthroughs and discoveries during the last decade, one could question whether this was a particularly productive research field. It appears that the research community locked on to the $H^+$/inorganic Al/Ca/early life-stage paradigm rather early on (N. Hutchinson, pers. comm.) and that no alternate hypothesis was seriously considered (possible exception of episodic effects). Toxicological research in the 1980s seemed to have been caught in a treadmill of testing species after species in a similar pH, Ca, Al matrices in response to a need, or perceived need, for additional proof of the damaging effects of acidification.

Why did this happen? On one hand you can blame much of the redundancy on the short-sighted nature of research funding or the isolation of the research community. On the other, in many instances, management and policymakers were not consistent and clear in stating their research needs (Rubins et al. 1992; Schindler 1992).

It is interesting to note that the summary documents of national acid rain programs in North America (Baker et al. 1990; RMCC 1990) do not contain a call for more toxicological research. Instead, there appears to be a recognition that mechanistic toxicology models have limited applications for predicting the health of natural populations, and that we need more direct measures of the state of the environment.

Empirical models, such as the simple threshold models, and the results of biological and chemical surveys appear to be the types of tools or data sets that resource managers need and/or are willing to use to address industrial pollution problems. Therefore it is important that we invest in inventory programs, collecting extensive data on the distribution of biota and the chemical and physical characteristics of their habitats over wide geographic areas (e.g., random stratified surveys of the type conducted by Kretser et al. 1989). Similarly, long-term chemical and biological monitoring programs are needed to identify the effects of anthropogenic stresses, such as acidification, and to be able to distinguish them from natural fluctuations in aquatic communities.

## CONCLUSION

Prior to 1980, biologists had a reasonably good understanding of the toxicity of acid-stress. The intensive research effort of the 1980s did not greatly change our understanding of the problem. However, there were substantial improvements and refinements to existing models, as well as several significant spin-off benefits. Trace metal chemistry in natural waters and the physiology of gill respiration are two basic science areas that greatly benefited from the study of the interactive effects of acidification and trace metals on freshwater biota.

## ACKNOWLEDGEMENT

We greatly appreciate comments and ideas from R. Bradley, J. Ford, N. Hutchinson, W. Keller, G. McDonald, B. Rosseland, A. Tessier and many of the Dahlem workshop participants. However, the opinions expressed in this manuscript should be considered ours alone. N. Hutchinson kindly permitted us to use Figure 15.2 from an unpublished manuscript. This is contribution 92–09 of the Ontario Ministry of Natural Resources, Fisheries Research, Maple, Ontario, Canada.

## REFERENCES

Baker, J.P., D.P. Bernard, S.W. Christensen, and M.J. Sale. 1990. Biological effects of changes in surface water acid-base chemistry. State of Science/Technol. Report Nr. 13, Natl. Acid Precip. Assess Prog., Report for the U.S. Environ. Protection Agency.

Beggs, G.L., J.M. Gunn, and C.H. Olver. 1985. The sensitivity of Ontario lake trout (*Salvelinus namaycush*) and lake trout lakes to acidification. Ont. Fish. Tech. Report Ser. No. 17, 24 pp.

Campbell, P.G.C., and P.M. Stokes. 1985. Acidification and toxicity of metals to aquatic biota. *Can. J. Fish. Aquat. Sci.* **42**:2304–2049.

Campbell, P.G.C., P.M. Stokes, and J.N. Galloway. 1985. Acidic deposition: Effects on geochemical cycling and biological availability of trace metals. Subgroup on Metals of the Tri-Academy Committee on Acidic Deposition. Washington, D.C.: Natl. Academy Press.

Drabløs, D., and A. Tollan. 1980. Ecological Impact of Acid Precipitation., Proc. Intl. Conf., Sandefjord, Norway, March 11–14, 1980. Oslo: SNSF Project.

Driscoll, C.T., J.P. Baker, J.J. Bisogni, and C.L. Schofield. 1980. Aluminum speciation in dilute acidified waters and its effects on fish. *Nature* **284**:161–164.

Haines, T.A. 1981. Acidic precipitation and its consequences for aquatic ecosytems: A review. *Trans. Amer. Fish. Soc.* **110**:669–707.

Harvey, H.H., R.C. Pierce, P.J. Dillon, J.R. Kramer, and D.M. Whelpdale. 1981. Acidification in the Canadian environment. Natl. Research Council of Canada Report. No. 18475.

Hutchinson, N., N. Yan., D. Spry, W. Keller, and K. Holtze. 1990. Responses of *Daphnia galeata mendotae* and native fish species demonstrate relevance of laboratory bioassays in predicting population responses to acidification. Proc. Intl. Symp. Aquatic Ecosys. Health. Univ. of Waterloo, Waterloo, Canada, July 1990.

Keller, W., N.D. Yan., K.E. Holtze, and J.R. Pitblado. 1990. Inferred effects of lake acidification on *Daphnia galeata mendotae*. *Environ. Sci. Technol.* **24**:1259–1261.

Kretser, W., J. Gallagher, and J. Nicoletter. 1989. Adirondack Lakes Study 1984–1987: An evaluation of fish communities and water chemistry. Ray Brook, New York: Adirondack Lakes Survey Corporation.

Lathrop, R.C., P.W. Rasmussen, and R.R. Knaner. 1991. Mercury concentrations in walleye from Wisconsin (U.S.A.) lakes. *Water, Air, Soil Pollut.* **56**:295–307.

McDonald, D.G., J.P. Reader, and T.R.K. Dalziel. 1989. The combined effects of pH and trace metals on fish ionoregulation. In: Acid Toxicity and Aquatic Animals, ed. R. Morris, E.W. Taylor, and D.J.A. Brown, pp. 221–242. Cambridge: Cambridge Univ. Press.

Minns, C.K., J.E. Moore, D.W. Schindler, and M.L. Jones. 1990. Assessing the potential extent of damage to inland lakes in eastern Canada due to acidic deposition. III. Predicting impacts on species richness in seven groups of aquatic biota. *Can. J. Fish. Aquat. Sci.* **47**:821–830.

Mount, D.R., and M. D. Marcus, eds. 1989. Physiologic, toxicologic and population responses of brook trout to acidification. Report for Elec. Power Res. Inst. (EPRI), Report. Nr. 2346, Laramie, WY: Univ. Wyoming.

Muniz, I.P., and H. Leivestad. 1980. Acidification—Effects on freshwater fish. In: Ecological Impact of Acid Precipitation, ed D. Drabløs and A. Tollan, pp. 84–92. Oslo: SNSF Project.

Nater, E.A., and D. F. Grigal. 1992. Regional trends in mercury distribution across the great lakes states, north central U.S.A. *Nature* **385**:139–141.

Ponce, R.A., and N.S. Bloom. 1991. Effect of pH on the bioaccumulation of low-level, dissolved methylmercury by rainbow trout (*Oncorhynchus mykiss*). *Water, Air, Soil Pollut.* **56**:631–640.

Richman, L.A., C.D. Wren, and P.M. Stokes. 1988. Facts and fallacies concerning mercury uptake by fish in acid stressed lakes. *Water, Air, Soil Pollut.* **37**:465–473.

RMCC. 1990. The 1990 Canadian long-range transport of air pollutants (LRTAP) and acid deposition report: Aquatic Effects. Federal/Provincial Research and Monitoring Coordination Committee (RMCC) Report Nr. 4. Burlington, Ont.: Can. Dept. of Fish. and Oceans, 151 pp.

Rosseland, B.O., T.D. Eldhuset, and M. Staurnes. 1990. Environmental effects of aluminum. *Environ. Geochem. Health.* **12**:17–27.

Rubins, E.S., L.B. Lave, and M.G. Morgan. 1992. Keeping climate research relevant. *Iss. Sci. Tech.* **Winter 91–92**:47–55.

Sadler, K. 1983. A model relating the results of low pH bioassay experiments to the fishery status of Norwegian lakes. *Freshwater Biol.* **13**:453–463.

Sayer, M.D.J., J.P. Reader, and R. Morris. 1991. Embryonic and larval development of brown trout, *Salmo trutta* L: Exposure to aluminum, copper, lead or zinc in soft, acid water. *J. Fish. Biol.* **38**:431–455.

Schindler, D.W. 1992. A view of NAPAP from north of the border. *Ecol. Appl.* **2(2)**:124–130.

Schofield, C.L. 1976. Acid precipitation: Effects on fish. *Ambio* **5**:228–230.

Skeffington, R.A., and D.J.A. Brown. 1992. Timescales for recovery from acidification: Implications of current knowledge for aquatic organisms. *Environ. Pollut.* **77**:227–234.

Spry, D.J., and J.G. Wiener. 1991. Metal bioavailability and toxicity to fish in low-alkalinity lakes: A critical review. *Environ. Pollut.* **71**:243–304.

Spry, D.J., C.M. Wood, and P.V. Hodson. 1981. The effects of environmental acid on freshwater fish with particular reference to the softwater lakes in Ontario and the modifying effects of heavy metals: A literature review. Can. Tech. Rep. Fish. Aquat. Sci. No. 999.

Tessier, A., Y. Couillard, P.G.C. Campbell, and J.C. Auclair. 1992. Modelling Cd partitioning in oxic lake sediments and Cd burdens in the freshwater. *Limnol. Oceanogr.*, in press.

Vesely, J., P. Benes, and K. Sevcik. 1989. Occurrence and speciation of beryllium in acidified freshwater. *Water Res.* **23**:711–717.

Wang, W. 1987. Factors affecting metal toxicity to (and accumulation by) aquatic organisms: Overview. *Environ. Intl.* **13**:437–457.

Welsh, P.G., J.F. Skidmore, D.J. Spry, D.G. Dixon, P.V. Hodson, N.J. Hutchinson, and B.E. Hickie. 1992. Effect of pH and dissolved organic carbon on the toxicity of copper to larval fathead minnow (*Pimephalis promelas*) in natural surface waters of low alkalinity. *Can. J. Fish. Aquat. Sci.* **49**, in press.

Wren, C.D., and G.L. Stephenson. 1991. The effect of acidification on the accumulation and toxicity of metals to freshwater invertebrates. *Environ. Pollut.* **71**:205–241.

# 16

# Physiological Mechanisms for Toxic Effects and Resistance to Acidic Water: An Ecophysiological and Ecotoxicological Approach

[1]B.O. ROSSELAND and M. STAURNES[2]
[1]Norwegian Institute for Water Research, P.O. Box 69 Korsvoll, 0808 Oslo 8, Norway
[2]SINTEF Applied Chemistry, Center of Aquaculture, 7034 Trondheim, Norway

## ABSTRACT

Decline in fish populations is explained by what happens with critical life processes (e.g., reproduction, migrations) at critical life stages (e.g., embryo, smolt, spawners) during critical environmental situations (e.g., occurrence of episodes, "mixing zones"). The explanation of fish disappearance has to be looked for by an effect on the fish itself, i.e., effects caused by biochemical/physiological failure(s). The identification and understanding of these failures is of fundamental importance for the prediction of future changes for the optimalization of mitigation techniques. Many laboratory studies have been performed at irrelevant water chemical conditions. What really causes death to a free swimming fish is probably the sum of a variety of disturbances; hydrogen ions, aluminum, and calcium are still considered to be most important. Toxicants affecting the calcium homeostasis are of special concern. The primary target organs for toxicants are superficial sensory organs and the gills. Effects on olfactory and taste organs are likely to cause behavioral effects related to avoidance, escape, migration, and appetite/food search reactions. Precipitated aluminum complexes can irritate the gill and cause inflammation, edema, swelling, and sometimes irradiation of the secondary lamella, initiating a series of resistance mechanisms. Potential sites of aluminum interaction at the lamellar epithelium can explain changing membrane fluidity and permeability characteristics, as well as intracellular uptake of aluminum and secondary effects on enzyme systems, intracellular calcium, and energy metabolism. The net result of the effects on these processes is, in most cases, osmoregulation failure. As organisms themselves strongly influence their chemical

*Acidification of Freshwater Ecosystems: Implications for the Future*
Edited by C.E.W. Steinberg and R.F. Wright 

microenvironment near their body surface (i.e., along the gills), the basic chemistry of the external water will therefore not necessarily represent the true toxic components. An exact chemical threshold level for a certain physiological response might therefore be based on wrong assumptions. Although both chronic and episodic changes in water quality are important, the "mixing zone chemistry" might be the most important selection parameter for all life history stages. More attention should be paid to resistance mechanisms: avoidance or escape reactions, exclusion, removal, neutralization, excretion, and/or repair of damage caused by the toxicant. The interpretation of possible environmental implications has to be based on a proper understanding of the fundamental basic processes involved, preferentially discussed in conceptual frameworks, and with as much knowledge about the natural conditions as possible.

## INTRODUCTION

The environmental problem of aquatic acidification involves effects on all organizational levels of life. Most attention, however, has been on the decrease and loss of fish populations, especially salmonids. Although all trophic levels are disturbed, including the disappearance of important food organisms, fish populations are not eliminated primarily due to lack of food (reviewed by Rosseland 1986; see also Schindler, this volume). The major cause of fish disappearance must be related to the effects on the fish itself, i.e., effects caused by biochemical/physiological failure(s) (Figure 16.1). The identification and understanding of these failures is therefore of fundamental importance for the prediction of future changes of reductions/increases in pollution loads and to the optimalization of mitigation techniques concerning liming and restocking strategies.

For the ecophysiologist, the key question then is to identify these biochemical/physiological failures, thereby contributing to the explanation of the decline or disappearance of fish populations in acidic waters (Figure 16.1). Unfortunately, many laboratory studies, which intend to explain important life processes, are not relevant for natural conditions. These studies have most often been performed at irrelevant water chemical conditions, e.g., at concentrations of $H^+$, Al, and Ca that are too high.

When discussing the effects of toxicants, resistance mechanisms must also be considered. On an individual level, the effects of chemical toxicants can be resisted in a number of ways (Calow 1991): avoidance or escape reactions, exclusion (e.g., excretion of more mucous onto exposed surfaces of aquatic animals), removal (incoming toxicants might be actively pumped out), neutralization or complexation (i.e., by complexation with protective proteins, e.g., metallothioneins), excretion, and/or repair of damage caused by the toxicant.

A life-history study of a fish population in a lake undergoing acidification involves all physiological mechanisms in the area of toxic effects and resistance. All temporary and chronic changes in water chemistry impact individual fish and fish populations differently, depending on the species, the life-history stages represented, previous acid-acclimation history, year-class composition, population size, spawning strategy and spawning facilities, as well as competition between other fish species in the lake or stream. Only to a certain extent, however, do controlled laboratory studies represent what occurs in nature.

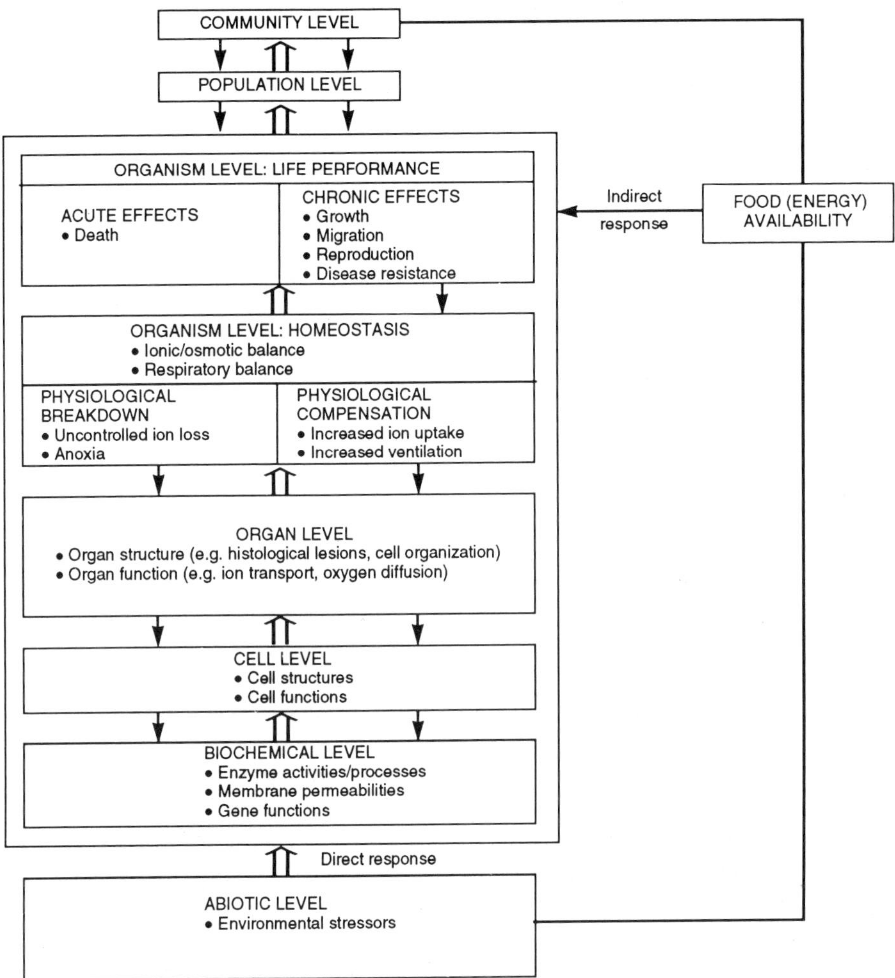


**Figure 16.1** A conceptual model for levels of integration in effect studies of environmental stressors. Some examples are given of the processes or effects involved.

The actual cause of death to a free-swimming fish is probably the result of various disturbances, of which only a few are known today. Laboratory, field laboratory, and population studies must therefore be strongly correlated and should span over a broad set of water chemistry challenges. To link the toxic substances involved to the ecological effects observed, cooperation must exist between chemists, geologists, ecophysiologists, ethologists, and ecologists (cf. Figure 16.1). The identification of Al as a primary toxicant in acidic waters is an example of this (see review by Rosseland et al. 1990).

## THE ABIOTIC ELEMENTS INVOLVED

Today, three elements are considered to be most important for the toxicity of acidic water to freshwater biota (reviewed by Wood and McDonald 1987; Rosseland et al. 1990): $H^+$ (pH), aluminum (Al), and calcium (Ca). The effects of $H^+$ and Al are dependent on animal species, the life-history stage of the animals, and their previous acclimation history. In the field, the effects of Al alone are difficult to isolate from a variety of potentially interrelated adverse factors. Especially during episodes of high water flow, and in lakes and streams where different water qualities mix, large variations in pH, Al species distribution, Ca (and other ions) and metals, and organic substances relevant for biological responses do occur (Henriksen et al. 1984). When the pH of an acidic body of water rises, e.g., when acidic Al-rich water mixes with limed or neutral waters, low molecular inorganic forms of Al will be transformed to high molecular weight forms and hence precipitate. In such mixing zones, rapid Al precipitation onto fish gills, osmoregulation failure, inhibition of enzyme activities, and gill lesions have been observed, the water in the mixing zone being more toxic than the original acidic water (Rosseland et al. 1992).

The situation is complicated even more by the fact that the organisms themselves strongly influence their chemical microenvironment near their body surface, including the gills (see Exley et al. 1991). The basic chemistry of the external water will therefore not necessarily represent the true toxic components relevant to the fish. Today, our knowledge regarding the exact chemical threshold level for a certain physiological response might, therefore, be based on the wrong assumptions.

$Ca^{2+}$ has a fundamental biological importance for "gill- and skin-breathing" animals. $Ca^{2+}$ is a key factor in the permeability of all membranes, including gill epithelium. In many acidified areas, the Ca concentration is so low that it is close to the concentration limit for even the softwater-tolerant salmonid species. In such areas, one must be aware of any substance having an adverse effect on the Ca metabolism.

## PRIMARY PHYSIOLOGICAL EFFECTS

A toxicant affects an organism immediately upon contact. In fish and other gill- and skin-breathing animals, the primary target organs for toxicants are the superficial sensory organs and gills. During early life-history stages, the skin carries out major regulatory processes between the animal and environment. When the gills and circulatory system develop, these functions are gradually taken over by the highly specialized gill organ, and the skin becomes nearly impermeable.

Possible effects of pollutants on digestive tract functions are of great importance for the nutritional status and accumulation of toxicants. Any disturbances in such fundamental processes can influence immune mechanisms as well as reduce growth. However, reports of direct effects on digestive tract functions are almost completely lacking, except for indications of a reduced evacuation rate in brown trout exposed to acidic water (Åtland and Barlaup 1991).

Moreover, the accumulation of toxicants (e.g., mercury) and the disturbance of internal organ functions are not primary causes of fish death in acidic waters. For that reason, they will not be discussed here.

**Effects on Sense Organs**

Superficial sense organs, such as olfactory and taste organs, are not protected by "external barriers" and are therefore vulnerable to pollutants. Toxicants may disrupt normal chemosensory functions by masking or counteracting biologically relevant chemical signals, or they may cause direct morphological and physiological damage to the receptors. Such effects on olfactory and taste organs are likely to cause behavioral disturbances, e.g., interference with avoidance and escape reactions related to chemical perception (which are primary resistance mechanisms), social interactions (e.g., reproduction), schooling behavior, predator avoidance, territoriality, as well as interfering with the ability to search for food, which could contribute to a reduced appetite. Another possible disturbance, with great environmental implications, would be the inhibition of the olfactory system in seaward-migrating smolts of the anadromous species. As olfaction is one of the main senses believed to be important for homing in Atlantic salmon, a reduced imprinting (caused by acidic or "mixing zone" water quality upon leaving their home river) may be hypothesized to increase straying upon return. Effects on chemoreception and chemical communication are sublethal per se; however, they may have important implications for long-term survival of fish populations.

In spite of the importance of the sensory system, relatively few studies on the effect of acidic waters on sensory organs exist, and they need to be encouraged. Low pH alone has been found to reduce the olfactory response to amino acids and increase the mucous layer in the olfactory organ (Klaprat et al. 1988). Adding Al to the water depresses olfactory response, even more, and causes histopathological changes, such as irradiation of the microvilli, swelling, and disformation of the olfactory epithelium (Klaprat et al. 1988). Even in cases where no changes in structure of the chemosensory tissue have been observed, complete elimination of feeding response has occurred at low pH (Lemly and Smith 1987). Behavioral effects can also be caused by homeostatic changes. Avoidance reactions to low pH waters have been observed when plasma cation concentrations have been moderately reduced by acidic water exposures (see Rosseland et al. 1990).

During episodic changes in water quality, related to snowmelt or heavy rain, fish often gather at the outlet of, or have migrated into, a less acidic brook or stream (Rosseland 1986). In a limed lake, during reacidification, brown trout migrated into an adjacent pond with better water quality in spite of higher fish density and thus increased competition for food (Barlaup et al. 1989). These studies indicate that the avoidance/escape reactions are important resistance mechanisms under acid conditions and are therefore important for survival and selection of more tolerant fish. A sensitive olfactory organ, but still insensitive to the negative effects of pH and Al, might thus be one of the most important factors for natural selection and resistance within species in nature. More research into the field of avoidance is needed.

## Effects on Gills and Homeostasis

In gill- and skin-breathing animals, gills are the primary organ for respiration, iono- and osmoregulation, acid-base regulation, and nitrogen excretion. Any environmental stressor influencing the function of this organ may therefore cause homeostatic disorders. Breakdown of some physiological functions may result in rapid death or may induce physiological compensatory responses to maintain homeostasis, at least for some time, thus greatly interfering with the overall life performance of the animal (Figure 16.1). Toxicity thresholds are species- and life-history stage-specific; however, the resultant physiological disturbances are generally similar across species. Prominent physiological disturbances for fish exposed to acidic waters include iono/osmoregulatory, acid-base regulatory, respiratory, and circulatory failure. Most of these effects can be directly attributed to the effects on gill functions or structure or indirectly via general stress responses (not discussed here; see Mazeaud and Mazeaud 1981).

Environmental irritants, including toxicants, will generally affect gill functions through direct interference with main biochemical/physiological processes and by causing structural changes (gill lesions). When contact is made with such an irritant, one of the first responses will be increased mucous secretion, to protect the epithelial cells and prevent entrance of the toxicant (e.g., Segner et al. 1988). In addition to hypersecretion and proliferation of mucous cells, prominent structural changes include: lifting the epithelial cells from the gill lamellae and the interlamellar zones of the gill filament; epithelial necrosis; epithelial hypertrophy; epithelial hyperplasia; epithelial rupture; lamellar fusion; bulbing of lamella; changes in chloride cells (number and size); and changes in gill vascularization (Mallat 1985; Figure 16.2). Most of these lesions can be documented in fish exposed to acidic Al-rich waters both in laboratory and field studies (Rosseland et al. 1990), where the main target of the $Al/H^+$ effects was chloride cells. However, except for mucous secretion and resulting respiratory effects, studies on the direct effects of such gill lesions in biochemical/physiological processes are few and warrant further research.

## Toxic Effects of $H^+$

The acute toxicity of $H^+$ on fish in acidic waters (reviewed, e.g., by McDonald 1983; Exley and Phillips 1988) is mostly explained by the reducing power of $H^+$ that causes the enhanced dissolution and loss of tissue salts in acidic solutions. In gills, this results in the loss of Ca from the important binding sites in the gill epithelium, thus reducing the ability of the gill to control ion permeability and thereby causing ion-regulatory disturbance and loss of ions (primarily $Na^+$ and $Cl^-$). An additional effect of $H^+$ is blood acidosis. Very high concentrations (not likely to be found in nature) cause interlamellar mucous clogging and result in hypoxia and severe gill lesions. However, acidity per se has only rarely been the preeminent factor in the decline of fish in acidic waters, and Al is now recognized as the principal toxicant.

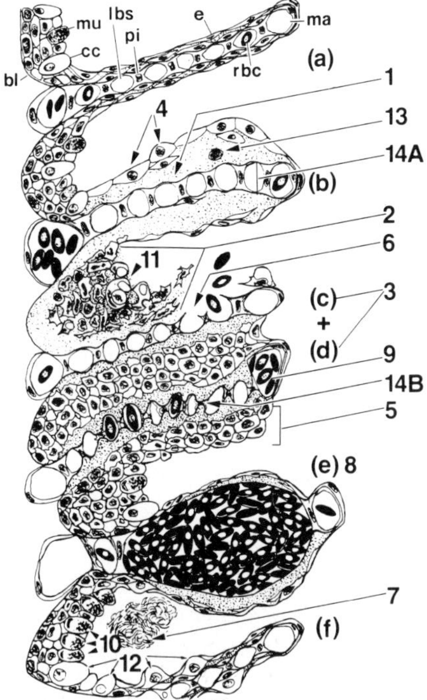


**Figure 16.2** Composite diagram of the common irritant-induced gill lesions. Six respiratory lamellae are shown (a–f), the top one of which is normal (*Oncorhynchus mykiss*). The lesions are numbered as followed: 1–epithelial lifting; 2–necrosis; 3–lamellar fusion (c and d); 4–hypertrophy; 5–hyperplasia; 6–epithelial rupture and bleeding into pharynx; 7–mucous secretion; 8–clavate lamella or lamellar aneurism (e); 9–vascular congestion; 10–mucous cell profileration; 11–chloride cells damage early; 12–chloride cell profileration; 13–leukocyte infiltration of epithelium; 14A–lamellar blood sinus dilates; 14B–lamellar blood sinus constricts. Abbrevations: bl, basal lamina; cc, chloride cell; e, typical lamellar epithelial cells; lbs, lamellar blood sinus; ma, marginal blood channel; mu, mucous cell; pi, pillar cell; rbc, erythrocyte. After Mallat (1985).

## Aluminum

### *Symptoms of Toxic Effects*

The toxic effects of Al on fish physiology are numerous (reviewed by Wood and McDonald 1987; Exley and Phillips 1988; Rosseland et al. 1990). The main consequences, all of which contribute to death of fish when exposure concentrations are acute, appear to be:

1. Respiratory disturbances due to interlamellar mucous clogging and Al precipitation and reduced membrane fluidity;

2. Iono/osmoregulatory disturbances due to decreased uptake and increased loss of ions (e.g., $Na^+$, $Cl^-$, and $Ca^{2+}$) caused by Al binding to gill surface, intracellular Al accumulation and inhibition of uptake mechanisms, increased membrane permeability, and damage of epithelium cells;
3. Circulatory disturbances characterized by very high levels of hematocrit due to reduced blood plasma volume, erythrocyte swelling, and release from the spleen. This, in addition to increased blood protein concentration, increases the blood viscosity.

These effects result from the combined effects of Pts. 1 and 2 and from general stress-induced responses (Mazeaud and Mazeaud 1981).

Therefore, at our present state of knowledge, the main factor involved in $H^+$ toxicity is thought to be ionoregulatory failure; in Al/$H^+$ toxicity, respiratory and circulatory distress also play a role.

### *Mode of Toxic Action*

Although the symptoms of Al toxicity have, to some extent, been characterized, the mechanisms are poorly understood. Al is found on the epithelial surface and inside the epithelium cells (e.g., see Rosseland et al. 1990). Several potential sites of Al interaction are suggested in the toxic mechanism model recently suggested by Exley et al. (1991; Figure 16.3).

To explain how Al affects the gill, we need to take into account the basic chemical properties of Al, the basic properties of the gill epithelium and its surface microenvironment, as well as the fact that Al interferes with basic biochemical and physiological functions. Some general, prominent intracellular effects of Al are the interference with high energy metabolism, enzyme activities, Ca metabolism, neurotransmission, cell nuclei functions, microtubuli functions, and lipid membrane fluidity (reviewed by Ganrot 1986).

Furthermore, to explain Al precipitation and binding, the important biochemical properties of the gill epithelium include its net negative charge, the glycoproteins and

---

**Figure 16.3** Schematic representation of the potential sites of aluminum interaction at the lamellar epithelium of the gill: mbl–mucous boundary layer; ml–mucous layer; gbl–gill boundary layer; ij–intercellular junction; ap–ATPase pump; cc–chloride cell; ac–accessory cell; ec–epithelial cell. Numbers denote interaction sites: (1) apical surface of chloride cell; (2) basally located active transport system; (3) narrow apical junction connecting chloride and accessory cells; (4) wide apical junction connecting accessory and epithelial cells; (5) polyanionic mucous layer including the enzyme carbonic anhydrase; (6) apical surface of epithelial cell; (7) intracellular effects on intercellular junctions; (8) intracellular accumulation of aluminum; (9) apically located active transport system; (10) apical membrane channels; (11) extracellular effect on intercellular junctions. After Exley et al. (1991).

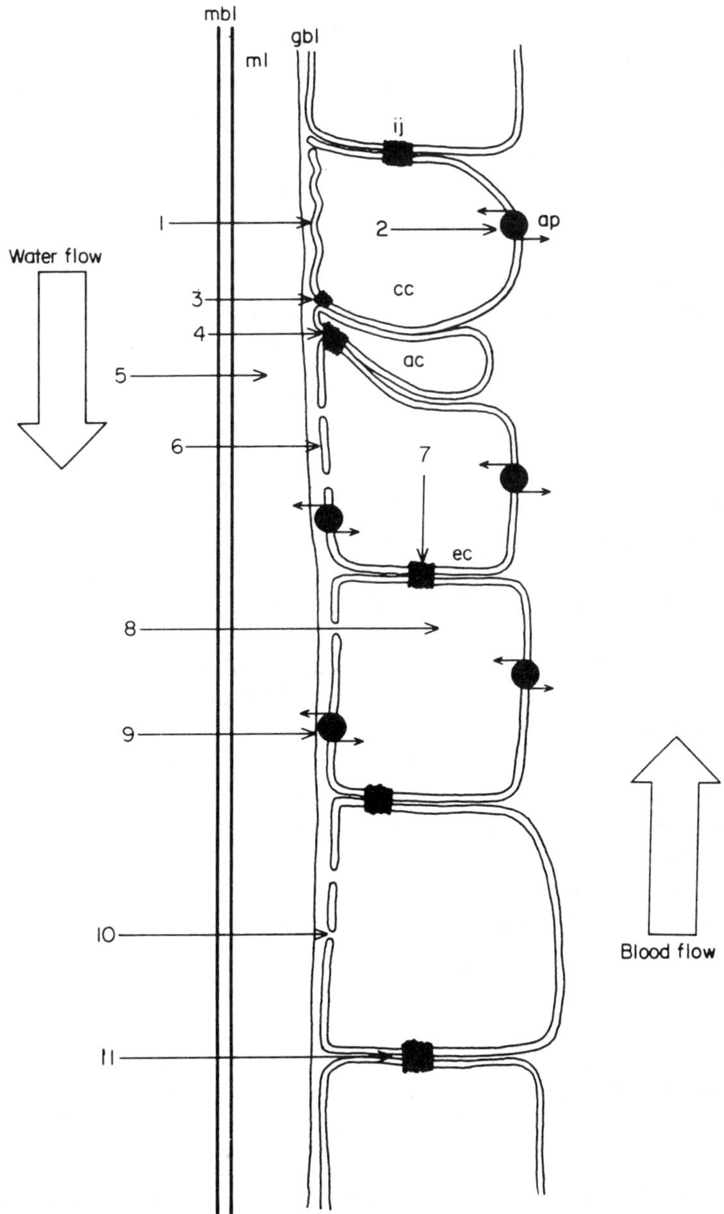
mbl
gbl
ml
ij
1
2
ap
Water flow
cc
3
4
ac
5
6
7
ec
8
9
10
11
Blood flow

sialic acid in mucous, the phosphate groups on membrane phospholipids, carboxylate groups on the membrane proteins, and the binding structure of membrane transport proteins (Exley and Phillips 1988; Exley et al. 1991). Of crucial importance is the fact that the gill boundary layer closest to the epithelium (see Figure 16.3) is both chemically and physically dissimilar to the water outside this layer. The chemistry of boundary layer is very much influenced by the organism itself. It is determined by differences in ion flux rates between transmembranal transport and diffusion into the water outside the layer, and "stabilized/protected" by the overlaying mucous, which acts as a hydrodynamic barrier and supports ion gradients between the boundary layer and the water outside (see Exley et al. 1991). The pH of the layer is determined by the hydration equilibrium of expiration products (mainly carbon dioxide and ammonia) and diffusion into and out of the boundary layer of acid or base equivalents from water and fish. An important property is the facility to maintain a circumneutral boundary layer pH, even at low environmental pH. This protects fish in acidic waters from deterious effects of acidity per se; normally, acidic water above a pH of 4.5 is not acutely toxic for salmonids (McDonald 1983). When fish are exposed to a pH that is not acutely toxic, there is initially a great increase in ion efflux, followed by a relatively rapid recovery of the efflux (McDonald 1983). The specific mechanism involved is unknown; however, some local adjustments in gill tissue have been suggested to contribute. It is also possible that a rapid decrease in the boundary layer pH may contribute to the initial efflux; however, the buffering that follows from the epithelial cells restores the boundary layer pH, thereby reducing Ca loss and membrane permeability. Since the boundary layer is the first surface to make contact with the environment, this boundary layer should receive more attention, so as to effect a better understanding of acclimation effects and acid tolerance, including species, strain, and life-history stage differences.

According to Figure 16.3, Al may interact with the gill function before it comes in contact with the epithelium itself (in mucous layer, site 5), at different sites on the apical epithelial cell surface, between adjacent cells (sites 1,6,9,10 and 3,4,11, respectively), or within the cell (sites 2 and 7). Toxicity of Al is caused by inorganic monomeric Al species, and the presence of Al-complexing ligands other than $OH^-$ (such as fluoride, silic acid, organic substances as humic acids and citrate) generally reduces the toxicity of Al (see review by Rosseland et al. 1990). At a pH of 5, the sialic acid in mucous is primarily negatively charged; positively charged Al hydroxides may therefore bind to mucous. The polymerization of Al may cause (a) irritation of the gill epithelium, (b) stimulation of mucous secretion that along with the Al precipitation causes interlamellar clogging, and (c) several of the gill lesions described in Figure 16.2. Since polymerization kinetics is temperature dependent, it is not surprising that Al toxicity is reduced at low temperatures (Poleo et al. 1991).

According to Exley et al. (1991), Al binds on the gill membrane via the trivalent cation ($Al(H_2O)_{6(aq)}{}^{3+}$), which is the only oxidation state available to biology. The nature of this ion requires it to bind to small electronegative species, such as the common biologically oxygen-based functional groups (phosphates, carboxylates, carboxyls and hydroxyls; see Exley et al. 1991). The circumneutrality of the boundary layer at moderate pH (4.5–5.5) favors binding to such oxygen-based functional groups. Thus, Al may bind to and neutralize the charge of both the phosphate groups on membrane phospholipids and the carboxylate groups on the membrane proteins, subsequently reducing the membrane fluidity. Exley et al. (1991) also suggested that Al may substitute for (a) metal co-factors in the functional domains of transport proteins or for (b) transport species per se. All of these effects on the apical cell surface may be summarized in the prediction of interference between both important trans-cellular diffusional and active processes, of which could be the uptake of ions ($Na^+$, $Cl^-$, $Ca^{2+}$), excretion of waste products ($NH_4^+$, $HCO_3^-$), or diffusion of respiratory gases ($O_2$, $CO_2$). Al has also been suggested to substitute for $Ca^{2+}$ in the intercellular cement. Increased paracellular permeability caused by $H^+$- and Al-induced removal of Ca from the anionic sites in the cement—especially between chloride cells and adjacent pavement epithelial cells—probably contribute to the increased efflux of ions in acidic water. The ameliorating effect of Ca on Al and the pH response is probably caused by a tightening of the junctions, which prevents the passive loss of ions (Wood and McDonald 1987). Prolactin production increases after chronic exposure to sublethal acidic waters, mainly as a response to a drop in plasma electrolytes (Wendelaar Bonga et al. 1987). Prolactin reduces the gill epithelium permeability; this time-dependent increase is clearly a mechanism of resistance.

What actually is the dominating or most important of these processes outside the cells: hydroxide precipitation in mucous, or the binding to epithelium? We could hypothesize that hydroxide precipitation in mucous or the binding to epithelium are primarily the result of reaction rates and concentrations, and that they are probably affected by temperature and situations such as the unstable Al chemistry of mixing zones, characterized by an already ongoing Al polymerization (Rosseland et al. 1992).

How Al actually comes into a cell is unknown. From the findings of *in vitro* experiments with phospholipid vesicles, however, Exley et al. (1991) suggest that apically bound Al alters membrane permeability to allow the intracellular accumulation of Al. The possibility for entrance via carrier systems should be investigated. Inside the cell, citrate might be an excellent ligand to Al and may act as an intermediate chelator, passing Al to groups with higher affinity. Gill activities of the enzymes carbonic acid anhydrase and Na-K-ATPase are inhibited in acid/Al-exposed salmonids (Staurnes et al. 1984; Rosseland et al. 1990, 1992). The interaction between Al and ATP (mainly the tendency for ATP to form stronger complexes with Al than with Mg) may imply that Al can affect many enzymes reactions where ATP is a substrate, possibly causing severe disturbances of the cells' energy metabolism (Ganrot 1986). Al also binds to calmodulin, which is a multifunctional, Ca-dependent protein that regulates a variety of cellular reactions, including the regulation of many enzymes (Ganrot 1986). In mammals, Al has been shown to interfere with the Ca regulatory system and Ca homeostasis. A similar interference might be suspected in the gill epithelium cells (discussed by Exley et al. 1991). The cytosolic plasma $Ca^{2+}$ concentration is very well regulated, and a higher concentration than $10^{-7}$–$10^{-8}$ causes the breakdown of cellular functions (Wiercinski 1989). Myocardial cell necrosis during heart muscle arrest seems to be caused by Ca "overload" in the cell (Wiercinski 1989). A similar cell necrosis could also be suggested to take place in gill epithelia. Therefore, the overall effects of the possible interference of Al with basic processes in the gill epithelial cells may include severe effects on the epithelial barrier properties (trans- and paracellular transport) and accelerated cell death (Exley et al. 1991).

## Hormones Related to Osmoregulation

Prolactin and cortisol are important hormones related to osmoregulation: prolactin reduces ion permeability and increases mucous production, while cortisol stimulates the onset of cellular proliferation and differentiation in the primary gill epithelium and increases the specific activity of Na-K-ATPase. Both hormones are affected by acidic waters (Wendelaar Bonga et al. 1987; Witters et al. 1991). Plasma cortisol increases in fish when (a) they are exposed to a low pH when Al is present at a high concentration (this is presumably a response to compensate for the $H^+$/Al toxicity; Witters et al. 1991) or (b) they are exposed to a low external concentration of NaCl (Perry and Laurent 1989) or general chronic stress (Pottinger and Pickering 1992). Prolactin production increases after chronic exposure to sublethal acidic waters, mainly as a response to a drop in plasma electrolytes (Wendelaar Bonga et al. 1987). Since this is a time-dependent rise, increased prolactin production is clearly a mechanism of resistance.

Although hormones play an important role in various resistance mechanisms (Exley and Phillips 1988), an important aspect involves the potentially negative effects of an increased level of cortisol as a response to prolonged (chronic) exposure to acidic Al-rich waters. Since a permanently increased level of cortisol has a negative effect on the immune system (Mazeaud and Mazeaud 1981; Pickering and Pottinger 1985), such a response might therefore have a generally negative effect on the health status of fish populations in acid lakes. Another important aspect is a possible post-episodic effect. The combination of a primary sublethal physiological stress (e.g., osmoregulatory and circulatory problems) and a secondary reduced immunity caused by a cortisol response might lead to an increased mortality over a prolonged period. Thus, the overall effect might be substantially greater than the direct observed mortality during and shortly after episode or an exposure to a "mixing zone chemistry." If such a relation exists, it might explain the phenomena of postspawning mortality and "juvenilization," where brown trout postspawners dissappear in some populations after their first spawning, resulting in lack of older-year classes in the population (see Rosseland 1986).

## Metabolism and Growth

Metabolic activity, measured as oxygen uptake, is not affected by moderate $H^+$ concentrations alone but increases as a response to Al in the water. The increased respiratory and heart rate observed in acidic waters are not believed to cause the increased energy expenditure per se, since the increased metabolism reflects rather the increased activity of the intrinsic compensatory mechanism trying to restore homeostasis. Hyperventilation in acidic waters seems to be a specific response to the labile Al concentration, since the addition of a chelator (such as citrate) depresses hyperventilation (see reviews by Wood and McDonald 1987; Rosseland et al. 1990).

In long-term experiments, even low concentrations of Al reduce growth (Sadler and Turnpenny 1986). Results from a study of stocking brown trout in the limed Lake Hovvatn, which was carried out to study growth during the reacidification period, have documented the relationship between reduced growth and increased mortality, and critical levels of pH, Al concentrations, and Ca concentrations (Barlaup et al. 1993). A decreased appetite was observed in brook trout exposed to acidic water (see Rosseland et al. 1990). Recent experiments indicate a reduction in the gastric evacuation rate in acidic waters (Åtland and Barlaup 1991). The growth reduction response seen in Lake Hovvatn might therefore have been caused by a combination of an increased metabolism, as the lake water reacidified, and a general decrease in appetite and food conversion rate.

## Reproduction and Early Life Stage

### *Oogenesis*

The oogenesis and fertilization period are sensitive to low pH. A reduced level of Ca in the serum and plasma of female fish from acidic lakes has been reported, indicating a probability of failure to produce viable eggs. A depletion of Ca from bone and increased numbers of females with unshed eggs have also been noticed. Some studies have also reported reduced vitellogenin levels under acid exposures (Roy et al. 1990; Mount et al. 1988). However, several studies indicate that females from acidic lakes develop eggs and spawn normally, even though the plasma Ca during oogenesis has been low (see reviews by Rosseland 1986; Muniz 1991).

### *Egg Stage*

After fertilization, the embryo seems to be susceptible to acidic waters throughout the whole period of development, although the period shortly after fertilization, and also prior to hatching, seems to be most critical (Rosseland 1986). For a long time, $H^+$ was considered alone to be the major toxicant affecting the egg stage, although some effects of Al at intermediate pH ranges were demonstrated. Low pH in the surrounding waters results in pH depression inside the egg (in the perivitelline fluid), which leads to either a prolongation of the hatching or to reduced hatching success. The low pH of the perivitelline fluid also depresses the activity of the hatching enzyme, chorionic dehydrogenase, which in turn reduces an effective breakdown of the eggshell (chorion). Species and strain differences in sensitivity to acidic waters at hatching may therefore reflect variation in levels of inactivation of the hatching enzyme. Chorion thickening, probably the result of protein denaturation due to the surrounding low pH, and a reduced activity of the embryo inside the egg (reduced mechanical breakdown), due to the low pH of the perivitteline fluid, enforces these negative effects.

More recently, it has been demonstrated that Al reduces both ion uptake at the eyed-egg stage and the activity of Na-K-ATPase in the embryo (see Rosseland et al. 1990). The actual mechanisms for Al uptake at the basal-membrane level have not been described; however, we suggest that they are similar to those described for the gill epithelium. Even at the embryo stage, waterborne Ca has been demonstrated to have an important protective role, thus factors related to permeability must be central.

### *Alevin Stage*

An increasing negative influence of Al with age occurs after hatching (Baker and Schofield 1982; Wood and McDonald 1982). The reason for this is still unclear; however, it might have something to do with the changes that take place in the respiratory system/organ. Shortly after hatching, alevins still have significant skin respiration; this gradually turns to gill respiration as the primary source of gas and ion exchange. The sensitivity related to the gill as a target organ thus gradually develops in the alevin. This might explain the gradual importance of Al as a toxicant after hatching.

Al and pH are known to interfere with whole-body mineral content and skeletal calcification at the embryo and fry stage (Sayer et al. 1991). Recent studies on various strains of brown trout having different sensitivity to acidic waters seem to indicate differences in the calcification rate at the alevin stage (Kroglund et al. 1992). Despite a comparable total-body Ca, the most resistant strains exhibited the lowest calcification rate for finrays and skeleton. This phenomenon might indicate an important resistant mechanism for embryo survival before swimup, thus giving priority to a high level of Ca in the plasma/serum in order to ensure Ca and electrolyte homeostasis. Development of screening techniques to investigate strain tolerance in fish, based on early calcification rates, is under way in the ReFish project (Kroglund et al. 1992).

A reduced metabolism, indicated by an increased number of degree-days from fertilization to hatching and through the yolksack period, has been suggested as a mechanism of resistance. By increasing the period until swimup, the possibility of avoiding acidic episodes in stream habitats will be reduced, and the ability to survive will be enhanced (Rosseland 1986).

## ENVIRONMENTAL IMPLICATIONS

The decline in fish populations is explained by what happens with critical life processes (e.g., reproduction, migrations), at critical life stages (e.g., embryo, alevin, smolt, spawners) during critical environmental situations (e.g., occurence of episodes, "mixing zones"). For inland fish populations, reproduction failure (i.e., reduced egg production or failure of eggs and larvae to survive) is still recognized as the main cause of extinction (Rosseland 1986; Muniz 1991). The main cause of death for the first life-history stages might therefore, from an ecological point of view, have the greatest impact on the environment.

For the anadromous fish population, the situation is more complex; effects on the older life stages are possibly of greater importance. The smolt stage is considered to be the most sensitive stage, especially for Atlantic salmon; however, spawning fish also have a low tolerance (see reviews by Rosseland 1986; Rosseland et al. 1990). During the short smolting period, when bottom-dwelling and territorial freshwater parr change to become smolt prepared for a pelagic marine life, the gill epithelium gradually change toward that of a marine fish. This change is characterized by an increase in membrane permeability, chloride cell number and structure complexity, and activity of the ion pump enzyme Na-K-ATPase (reviewed, e.g., by Hoar 1988). These changes render the smolt epithelium extremely sensitive to the deleterious effects from $H^+$ and Al complexes. Returning fish from sea (e.g., spawners) also have a seawater-like gill epithelium, which renders them very vulnerable to acidic/Al stress. The effects on Atlantic salmon smolts (Figure 16.4) demonstrate an inhibition of gill Na-K-ATPase activity to parr level, ionoregulatory failure, and a complete loss of seawater tolerance when exposed to acidic water with a low Al concentration. These results could also be used in the integration model (Figure 16.1); acid and Al (abiotic level) cause the inhibition of an enzyme (biochemical level) important for the chloride cells' ability to pump ions (cell level), which then contribute to the failure of the gill (organ level) to regulate the blood plasma ion concentrations in both fresh and sea water (organism level: homeostasis). The results also indicate an impaired ability to smolt, which could adversely affect seawater survival or cause fish not to migrate (organism level: life performance). This will certainly affect the salmon population and river community. Yet, since the results originate from a laboratory study, they can only *imply* what happens under natural conditions. There is thus a real need for future verification under natural conditions before interpretation of possible environmental implications is complete.

When we consider what actually constitutes critical processes or stages and what happens under natural conditions, we must not forget to include the ability or will to avoid or escape. Predictions of environmental impact on fish, however, are often based on laboratory or field tolerance data that do not take mobility of fish into consideration. A conceptual model to link avoidance and toxicological data for environmental assessment has been proposed by Gray (1990). This model (Figure 16.5) takes into account the fact that fish avoidance/escape unfavorable conditions (Figure 16.5 b), but that this avoidance/escape response is modified by, e.g., the initial shock effects when concentrations are very high (Figure 16.5c) or when intrinsic factors are involved (Figure 16.5d).

Intrinsic factors that can override the avoidance/escape reaction could include "positive drive forces," such as the urge to spawn (which affects fish migrating to their

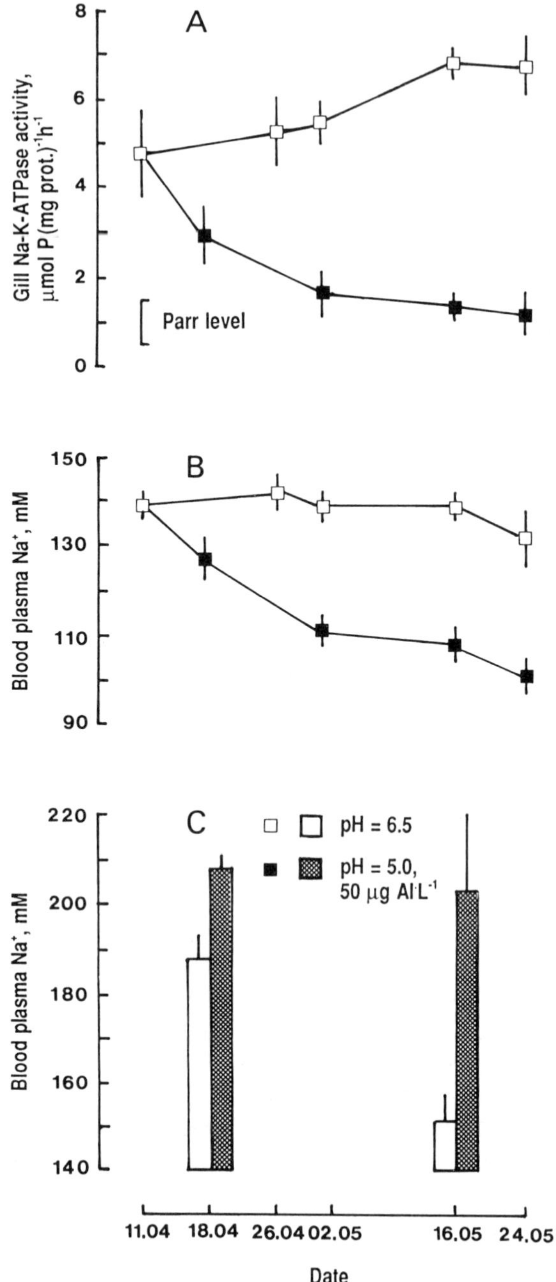


**Figure 16.4** A: Activity of gill Na-K-ATPase. B: Plasma $Na^+$ concentration in Atlantic salmon smolt kept in fresh water at pH 6.5 and pH 5.0 with 50 μg Al $L^{-1}$ (~2 μmol $L^{-1}$). C: Plasma $Na^+$ concentration 24 hrs after transfer to sea water. After Staurnes et al. (1993).

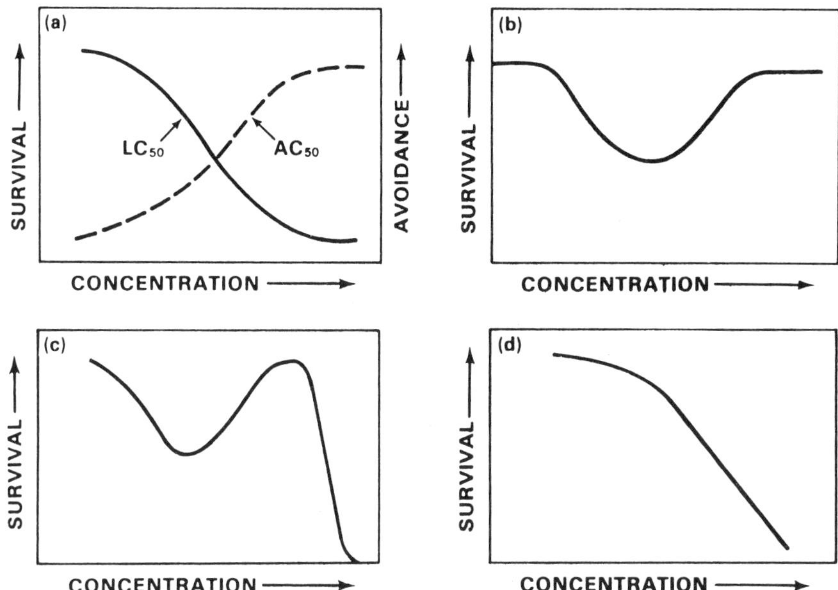


**Figure 16.5** Conceptual models for: (a) separate concentration/avoidance ($AC_{50}$) and concentration/mortality ($LC_{50}$) functions; (b) presumed joint concentration/ avoidance/mortality functions; (c) the initial schock effects when the concentrations is very high; (d) presumed effect of intrinsic factors as urge to spawn on joint function. After Gray (1990).

spawning grounds), or factors such as the reduced upstream swimming ability (which causes downstream-migrating smolts to drift more or less passively). The situation as seen in Figure 16.5c could be due to homeostatic disruption (such as uncontrolled ion loss) or by impairment of the olfactory epithelia functions. This might occur during episodes or in mixing zones. In nature, such combinations as seen in Figure 16.5c and Figure 16.5d may occur and enforce the negative effects of each effect alone. The massive fish kill of Atlantic salmon spawners in River Ogna (Skogheim et al. 1984) is possibly an example of such a situation.

In lake systems, there are more escape possibilities than in rivers. Accordingly, mass fish kills have been primarily reported from rivers and brooks. Most brown trout populations spend their early low-mobility stages in brooks and streams. Thus, the effects on these stages seem to be the most important for fish decline. River-dwelling species or stocks have very restricted escape possibilities; this renders them especially vulnerable for acidic waters. For territorial, river-dwelling species like Atlantic salmon or sea trout, their chance to behave according to the scenario in Figure 16.5b is restricted; during episodes, they are more likely to face situations comparable to Figure 16.5c than fish in lakes. Their territorial behavior may cause their concentration/avoidance reactions to fit more into a situation according to Figure 16.5d than Figure 16.5b. For anadromous fish, the situation is even worse since their smolt stage is extremely sensitive to acidic waters.

## CONCLUSION

In this chapter, we have tried to describe possible ways of explaining what happens during acidification under natural conditions. We conclude that the interpretation of potential environmental implications must be based on (a) proper understanding of the fundamental basic processes involved (cf. Figure 16.2 and 16.3), preferably discussed under a conceptual framework (as shown in Figure 16.1 and 16.5), and (b) from a best-possible knowledge base regarding natural conditions. Perhaps a more thorough review of the acidification literature in such an approach would be useful in providing a guideline for future research.

## REFERENCES


Åtland, Å., and B. Barlaup. 1991. Rate of gastric evacuation rate in brown trout (*Salmo trutta L.*) in acidified and non-acidified water. *Water, Air, Soil Pollut.* **60**:197–204.

Baker, J.P., and C.L. Schofield. 1982. Aluminum toxicity to fish in acidic waters. *Water, Air, Soil Pollut.* **18**:289–309.

Barlaup, B., Å. Åtland, and E. Kleiven. 1993. Stocking of brown trout (*Salmo trutta* L.) cohorts after liming: Effects on survival and growth during five years of reacidification. *Water, Air, Soil Pollut.*, in press.

Barlaup, B., Å. Åtland, G.G. Raddum, and E. Kleiven. 1989. Improved growth in stunted brown trout (*Salmo trutta* L.) after reliming of Lake Hovvatn, Southern Norway. *Water, Air, Soil Pollut.* **47**:139–151.

Calow, P. 1991. Physiological cost of combating chemical toxicants: Ecological implications. *Comp. Biochem. Physiol.* **100C**:3–6.

Exley, C., J.S. Chappell, and J.D. Birchall. 1991. A mechanism for acute aluminum toxicity in fish. *J. Theor. Biol.* **151**:417–428.

Exley, C., and M.J. Phillips. 1988. Acid rain: Implications for the farming of salmonids. In: Recent Advances in Aquaculture, ed. J.F. Muir and R.J. Roberts, vol. 3, pp. 225–341. London: Croom Helm.

Ganrot, P.-O. 1986. Biochemistry and metabolism of $Al^{3+}$ and similar ions. A review. *Environ. Health Persp.* **65**:363–441.

Gray, R.H. 1990. Fish behavior and environmental assesment. *Environ. Toxicol. Chem.* **9**:53–67.

Henriksen, A., O.K. Skogheim, and B.O. Rosseland. 1984. Episodic changes in pH and aluminum-speciation kill fish in a Norwegian salmon river. *Vatten* **40**:225–260.

Hoar, W.S. 1988. The physiology of smolting salmonids. In: Fish Physiology: The Physiology of Developing Fish. Viviparity and Posthatching Juveniles, ed. W.S. Hoar and D.J. Randall, vol. XIB, pp. 275–343. New York: Academic.

Klaprat, D.A., S.B. Brown, and T.J. Hara. 1988. The effect of low pH and aluminum on the olfactory organ of rainbow trout (*Salmo gairdneri*). *Environ. Biol. Fishes* **22**:69–77.

Kroglund, F., T. Dalziel, B.O. Rosseland, L. Lien, E. Lydersen, and A. Bulger. 1992. Restoring endangered fish in stressed habitats. ReFish Project 1988–1991. NIVA Acid Rain Research Report 30/1992. ISBN–82–577–2184–0. Oslo: NIVA, 43 pp.

Lemly, A.D., and R.J.F. Smith. 1987. Effects of chronic exposure to acidified water on chemoreception of feeding stimuli in fathead minnows (*Pimephalis promelas*): Mechanisms and ecological implications. *Environ. Toxicol. Chem.* **6**:225–238.

Mallat, J. 1985. Fish gill structural changes induced by toxicants and other irritants: A statistical review. *Can. J. Fish. Aquat. Sci.* **42**:630–648.

Mazeaud, M.M., and F. Mazeaud. 1981. The role of catecholamines in the stress response of fish. In: Stress and Fish, ed. A.D. Pickering, pp. 49–75. London: Academic.

McDonald, D.G. 1983. The effects of $H^+$ upon the fish gills of freshwater fish. *Can. J. Zool.* **61**:691–703.

Mount, D.R., J.R. Hockett, and W.A. Gern. 1988. Effects of long-term exposures to acid, aluminum, and low calcium on adult brook trout (*Salvelinus fontinalis*). 2. Vitellogenesis and osmoregulation. *Can. J. Fish. Aquat. Sci.* **45**:1633–1642.

Muniz, I.P. 1991. Freshwater acidification: Its effects on species and communities of freshwater microbes, plants and animals. *Proc. Roy. Soc. Edinburgh* **97B**:227–254.

Perry, S.F., and P. Laurent. 1989. Adaptional responses of rainbow trout to lowered external NaCl concentration: Contribution of the branchial chloride cell. *J. Exp. Biol.* **147**:147–168.

Pickering, A.D., and T.G. Pottinger. 1985. Cortisol can increase the susceptibility of brown trout, *Salmo trutta L.*, to disease without reducing the white blood cell count. *J. Fish Biol.* **27**:611–619.

Poleo, A.B.S., E. Lydersen, and I.P. Muniz. 1991. The influence of temperature on aqueous aluminum chemistry and survival of Atlantic salmon (*Salmo salar* L.) fingerlings. *Aquatic Toxicol.* **21**:267–278.

Pottinger, T.G., and A.D. Pickering. 1992. The influence of social interaction on the acclimation of rainbow trout, *Oncorhynchus mykiss* (Walbaum) to chronic stress. *J. Fish Biol.* **41**:435–447.

Rosseland, B.O. 1986. Ecological effects of acidification on tertiary consumers. Fish population responses. *Water, Air, Soil Pollut.* **30**:451–460.

Rosseland, B.O., I.A. Blakar, A. Bulger, F. Kroglund, A. Kvellestad, E. Lydersen, D. Oughton, B. Salbu, M. Staurnes, and R. Vogt. 1992. The mixing zone between limed and acid river waters: Complex aluminum chemistry and extreme toxicity for salmonids. *Environ. Pollut.* **78**:3–8.

Rosseland, B.O., T.D. Eldhuset, and M. Staurnes. 1990. Environmental effects of aluminum. *Environ. Geochem. Health* **12**:17–27.

Roy, R.L., S.M. Ruby, D.R. Idler, and D.R. Ying. 1990. Plasma vitellogenin levels in prespawning rainbow trout, *Oncorhynchus mykiss*, during acid exposure. *Arch. Environ. Contam. Toxicol.* **19**:803–806.

Sadler, K., and A.W.H. Turnpenny. 1986. Field and laboratory studies of exposure of brown trout to acid waters. *Water, Air, Soil Pollut.* **30**:593–599.

Sayer, M.D.J., J.P. Reader, and R. Morris. 1991. Embryonic and larvae development of brown trout, *Salmo trutta*: Exposure to trace metal mixtures in soft water. *J. Fish Biol.* **38**:773–787.

Segner, H., R. Marthaler, and M. Linnenbach. 1988. Growth, aluminum uptake and mucous cell morphometrics of early life stages of brown trout, *Salmo trutta*, in low pH water. *Environ. Biol. Fishes* **21**:153–159.

Skogheim, O.K., B.O. Rosseland, and I.H. Sevaldrud. 1984. Deaths of spawners of Atlantic salmon (*Salmo salar* L.) in River Ogna, SW Norway, caused by acidified aluminum-rich water. *Rep. Inst. Freshw. Res. Drottningholm* **61**:195–202.

Staurnes, M., P. Blix, and O.B. Reite. 1993. Effects of acid water and aluminium on parr-smolt transformation and seawater tolerance in Atlantic salmon, *Salmo salar. Can. J. Fish Aquat. Sci.*, in press.

Staurnes, M., T. Sigholt, and O.B. Reite. 1984. Reduced carbonic anhydrase and Na-K-ATPase activity in gills of salmonids exposed to aluminum containing acid water. *Experientia* **40**:226–227.

Wendelaar Bonga, S., G. Flik, and P.H.M. Balm. 1987. Physiological adaptation to acidic stress in fish. *Ann. Soc. R. Zool. Belg.* **117(1)**:243–254.

Wiercinski, J.F. 1989. Calcium, an overview. *Biol. Bull.* **176**:195–217.

Witters, H., S. Van Puymbroeck, and O.L.J. Vanderborght. 1991. Adrenergic response to physiological disturbances in rainbow trout, *Oncorhynchus mykiss*, exposed to aluminum at acid pH. *Can. J. Fish. Aquat. Sci.* **48**:414–420.

Wood, C.M., and D.G. McDonald. 1982. Physiological mechanisms of acid toxicity to fish. In: Acid Rain/Fisheries, ed. R.E. Johnson, pp. 197–226. Bethesda, MD: Am. Fisheries Society.

Wood, C.M., and D.G. McDonald. 1987. The physiology of acid/aluminum stress in trout. *Ann. Soc. R. Zool. Belg.* **117(1)**:399–410.

# 17

# Molecular Approaches for the Investigation of the Genetic Composition in Aquatic Ecosystems

A. HARTMANN
GSF-Forschungszentrum für Umwelt und Gesundheit, Institut für Bodenökologie, Neuherberg, Postfach 1129, 85758 Oberschleissheim, F.R. Germany

## ABSTRACT

Conventional methods of morphological and physiological characterization and the counting of (micro)organisms in freshwater ecosystems are not satisfactory for a comprehensive population analysis, because part of the microorganisms are not culturable and a morphologically based identification is often not possible or misleading. The present state of the art of alternative, molecular biological methods for the characterization of the genetic pool in environmental samples is reviewed. Phylogenetically relevant oligonucleotide gene probes derived from sequences of rRNA species as well as other diagnostically valuable DNA probes are used to assess the diversity of the gene pool or to identify organisms directly by *in situ* hybridization. For the specific *in vitro* multiplication of phylogenetic indicator gene sequences, the polymerase chain reaction (PCR) is used; its present potentials and limitations are discussed. An assessment of genetic diversity changes through acidification may be accomplished by reassociation studies of the bacterial DNA isolated from these habitats. Comparative restriction fragment length polymorphism (RFLP) and hybridization studies of specific subpopulations may provide further insight of genetic shifts in acidified aquatic ecosystems.

## INTRODUCTION

The characterization of the genetic pool, with respect to the (micro)biological population in complex ecological communities, like the freshwater ecosystem, and its alteration by environmental changes due to acidification, provide major challenges for several reasons. First, the community structure is very complex and is comprised of probably more than a thousand different taxonomic species of bacteria, fungi, and algae as well as micro- and mesofauna. The complete identification of the microflora

*Acidification of Freshwater Ecosystems: Implications for the Future*
Edited by C.E.W. Steinberg and R.F. Wright 

by direct counts using common fluorescent DNA stains, such as 4,6-diamidino-2-phenylindole (DAPI) or acridine orange and morphological characteristics, is not possible due to the mostly nondiscriminative simple shapes of bacteria and fungi. Only species that are morphologically distinct or distinguishable from the rest of the native microflora can be counted selectively, and this is only possible for a very limited number of organisms. Algae offer a greater variety of typical structures, which are used for identification; however, there is ample evidence that also in this group morphological features are not satisfactory for an unequivocal identification. This was, e.g., recently demonstrated for the green algal class *Ulvophyceae* and the genus *Chlamydomonas* (Buchheim et al. 1990). Based on phylogenetic studies of the nuclear 18S rRNA sequences, Buchheim et al. provided evidence that the current taxonomy, which is mainly based on morphological characteristics, does not reflect natural or monophyletic groups. The present *Chlamydomonas* taxa may include distinct genera and may ultimately be regarded as distinct families or even orders. Therefore, the morphologically based identification could be misleading.

Besides direct counting techniques, microorganisms need to be cultivated in different media prior to their characterization by physiological criteria in the conventional microbial analysis (Austin 1988). To assess community diversity, the Shannon index has been widely used, which measures both the occurrence of individual species and the evenness of distribution within the community. The community structure of sessile heterotrophic aquatic bacteria stressed by acid mine drainage, has been studied by Mills and Mallory (1987). However, only a minor portion of the microorganisms can be grown in laboratory growth media and conditions (Ward et al. 1990). Some microbes change from a culturable to a viable but nonculturable state as a result of environmental conditions, and these organisms need to be resuscitated before cultivation (Nilsson et al. 1991). In addition, some organisms are present in very low numbers, and their determination requires some specific means of amplification before identification and counting. Conventional methods provide no satisfactory solutions to these basic problems of comprehensive microbial community analysis.

Recent advances in the application of molecular biological methods in ecological studies (Olsen et al. 1986; Pickup 1991; Sayler and Layton 1990; Steffan and Atlas 1991) as well as progress in immunological methods, computer assisted microscopy (Caldwell and Lawrence 1989), and flow cytometry (Shapiro 1990) have greatly expanded the array of tools that can be applied in studies of microbial ecology. On the basis of ribosomal RNA sequences, molecular phylogeny has made the natural classification of microorganisms possible by using the molecular record for the evolutionary relatedness of organisms. Since the molecular genetic approach focuses on molecules rather than organisms, the method is not limited to species that are amenable to laboratory cultivation. In addition, the diversity of community structure can be studied using DNA reassociation kinetics of the total pool of isolated DNA from environmental samples. In this chapter, I discuss the potential and present limitation of the molecular approach and give examples for its application in studies of natural water ecosystems. The increasing application of molecular tools should also greatly advance our knowledge of genetic alterations and losses in acidified freshwater ecosystems.

## THE MOLECULAR GENETIC APPROACH

The availability of taxonomically relevant nucleic acid sequences is bringing an essential phylogenetic perspective into biology. There is no more fundamental or direct way to classify and relate organisms than by appropriate nucleic acid sequence comparisons or hybridization with "signature" gene probes. Since this identification approach does not necessarily need cultivation of the organisms, the major constraint for the characterization of nonculturable organisms is circumvented.

### Gene Probes

#### *Ribosomal RNA-Directed Probes*

For the analysis of natural microbial populations, in which unknown diversity must be anticipated, there are several reasons to focus on the rRNAs (Olsen et al. 1986). As key elements of the protein-synthesizing machinery, the rRNAs are functionally and evolutionarily homologous in all organisms. They are extremely conserved in overall structure and nucleotide sequences. Some sequence stretches are invariant across the primary kingdoms while others differ in a genus- and even species-specific manner in hypervariable regions. The conserved sequences and secondary structure elements allow the alignment of variable sequences so that only homologous nucleotides are employed in any phylogenetic analysis. The highly conserved regions also provide primer sequences to start specific amplification by polymerase chain reaction (PCR, see below) as well as convenient hybridization targets. Since active cells harbor about 10,000 ribosomes, the rRNAs constitute a significant component of the biomass, and they are readily recovered from all types of organisms for accumulation of data bases of reference sequences (for further references, see Olsen et al. 1986 ).

There are three size classes of rRNAs in bacteria and eukaryotes (in brackets): 5S (5,8S), 16S (18S), and 23S (26S). Because of the small size (about 120 nucleotides) and limited information content, the 5S rRNA can only be used for populations of limited complexity. The 5S rRNA is isolated from the environmental biomass and the various species-specific molecules are separated by high-resolution gel electrophoresis. In addition, unique 5S rRNA types can be sequenced and compared with a reference data base to confirm the phylogenetic affinities. Using the 16S and 23S rRNA species, even complex populations can be analyzed. Due to the very high information content of the 23S rRNA (about 3000 nucleotides), even closely related species can be discriminated (Höpfl et al. 1989; Kirchhof and Hartmann 1992).

As a first step, the total DNA isolated from the collected biomass can be cloned, with the recombinant DNA library then screened by hybridization with a general 16S

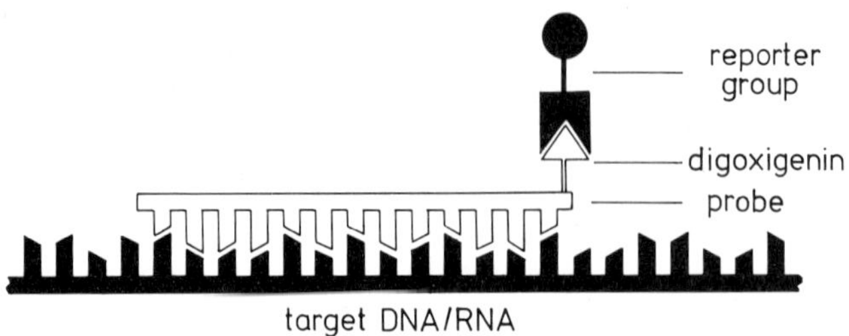


**Figure 17.1** Detection of specific DNA sequences with labeled, diagnostic oligonucleotide gene probes.

rRNA probe. Subsequently, the 16S rRNA sequences are determined using 16S rRNA-specific primers and finally compared with a reference data base. Using this approach, Giovannoni et al. (1990) described the genetic diversity of Sargasso Sea bacterioplankton and found forms that have not been described phylogenetically. Oligonucleotides (about 15 nucleotides in length) with phylogenetic signature character can be derived from the hypervariable 16S or 23S rRNA sequences and coupled with fluorescent dyes or another reporter function like digoxigenin (Zarda et al. 1991; Figure 17.1). These synthetic gene probes can be used as "phylogenetic stains" for the identification of single cells after fixation by *in situ* hybridization (DeLong et al. 1989; see below).

*DNA-Directed Probes*

DNA probes can be both double- or single-stranded. Single-stranded DNA probes, however, allow a more sensitive detection because they hybridize only to the target sequence and cannot reanneal to each other (Sayler and Layton 1990). Double-stranded probes are still very popular for environmental applications because they are easy to prepare and can be made from a wide variety of DNA sequences of interest. They can be obtained from poorly studied organisms, and the gene function of the DNA sequence need not be known. Total genomic DNA of different organisms have been used as gene probes. Species-specificity can be screened by hybridization for DNA fragments that bind to the desired species only (Sayler and Layton 1990). Gene probes for a specific phenotypic trait are useful for detecting organisms with common functions, such as certain biodegradative abilities or resistance factors.

## Environmental Recovery of DNA/RNA

The use of nucleic acid extraction of total DNA/RNA from communities *in situ* without cultivation of individual populations circumvents the limitations imposed by cultivation and achieves a more representative sampling of the genetic diversity in a given environment. In nucleic acid extraction from organisms in the water column of aquatic ecosystems, the cells usually need to be concentrated from large volumes of water by filtration. Single-unit constructions for filtration, cell lysis, and DNA extraction have been developed (for a review, see Sayler and Layton 1990). DNA obtained from water samples is usually of high purity and large in fragment size. The major problem in extracting DNA from sediments is to obtain DNA with high purity. Both clay and organic particles raise problems in DNA isolation and purification. Clay has a tendency to bind DNA adsorptively, whereas humic polymers tend to copurify with DNA. Two approaches for nucleic acid isolation have been used. In the cell extraction method, the microorganisms are extracted first from soil and sediments by several homogenization and centriguation steps in the presence of polyvinylpolypyrrolidone (PVPP). Then the DNA is isolated by standard cell lysis procedures. In the study of 16S rRNA sequences of uncultivated hot-spring cyanobacteria, even isolated ribosomes were used as starting material for RNA and subsequent cloning of the chromosomal DNA (Weller et al. 1991). In the alternative direct lysis method, the cells are first lysed in the environment using physical or chemical disruption. The DNA is subsequently extracted with alkali and purified using phenol, cesium chloride gradients, and/or hydroxyapatite chromatography. The yield and quality of DNA is improved when PVPP is added after cell lysis. A straightforward extraction and purification procedure of environmental DNA and RNA with subsequent molecular analysis was recently described by Selenska and Klingmüller (1992). For the general detection of a genotype in a microbial community, direct extraction followed by dot-blot hybridization analysis is usually sufficient (see below).

The gene expression in natural populations can be studied by extracting the RNA and probing with the genes of interest in dot-blot hybridization experiments (Pichard and Paul 1991).

## Amplification of Environmental DNA by Polymerase Chain Reaction

In many cases, the concentration of extracted DNA species in environmental samples is too low to be further analyzed. The development of the PCR was a major methodological breakthrough in molecular biology and permits the *in vitro* replication and amplification of defined gene segments. The PCR greatly enhances the probability of detecting rare sequences in heterologous mixtures of DNA (for a review, see Steffan and Atlas 1991). The basic steps are (a) the melting of DNA to convert double-stranded DNA to single-stranded DNA, (b) the annealing of specific primers to the target DNA, and (c) the synthesis of ds-DNA from the primers by the action of the thermotolerant *Taq* DNA polymerase (Figure 17.2). The oligonucleotide primers are designed to hybridize to regions of DNA flanking a desired target gene sequence. DNA strands without this sequence will not be replicated and will remain as single strands. The product DNA duplexes are melted by heating and the process is repeated many times, resulting in an exponential increase in the amount of the desired target sequence.

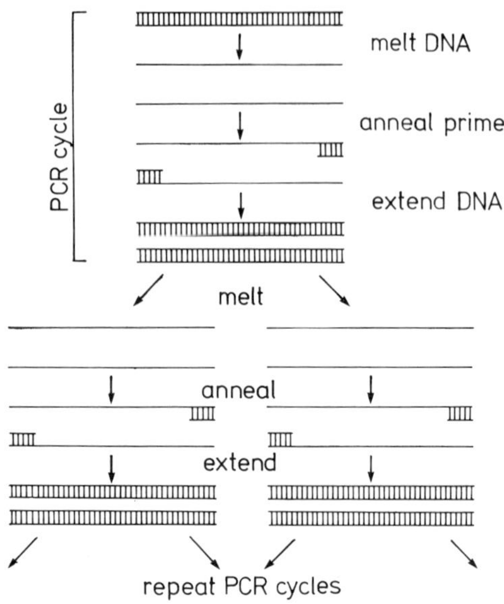


**Figure 17.2** Amplification of specific DNA sequences by the polymerase chain reaction using specific primers.

Several DNA segments can be simultaneously amplified using multiple pairs of primers. Bej et al. (1990) used this multiplex PCR to detect low levels of certain indicator bacteria in water. Because of the extremely sensitive setup of the PCR method, the DNA from environmental samples must be extensively cleaned. Standard purification schemes probably do not work with every environmental sample. Therefore, a proper protocol must be ascertained for each type of sample (Steffan and Atlas 1991; Selenska and Klingmüller 1992).

The gene-probe detection limit observed without amplification is generally in the range of $10^3$–$10^4$ cells/g of sediment. However, even single target cells can be detected using amplification by PCR. The quantitative estimation of a target population based on PCR reaction products is difficult because the amount of PCR products formed during the amplification cycles increases exponentially. A quantification is possible using a competitive PCR scheme in which target DNA is quantified by co-amplifying of target DNA in the presence of known quantities of a competitive DNA (Gilliland et al. 1990).

## Hybridization Techniques

DNA hybridization techniques can be powerful tools for the detection of organisms in environmental samples because they offer high sensitivity and specificity. Under optimal conditions, a 17-base pair (bp) oligonucleotide probe can detect a single gene in a large genomic background of, e.g., $3 \times 10^9$ base pairs (Sayler and Layton 1990). This specificity and sensitivity results from high-fidelity hydrogen bonding between complementary nucleotide sequences. In ecological studies, these tools can be applied in a more frequent real-time analysis of environmental samples and may provide knowledge of the absolute composition and structure of natural communities and the dynamics of individual populations or genes within this community.

Nucleic acid hybridization involves the detection of target nucleic acid sequences through the binding of nucleotide sequences with a homologous complementary probe sequence. Factors affecting the hybridization or reassociation of two complementary DNA strands include temperature, incubation time, salt concentration, the degree of mismatch between the base pairs, and the length and concentration of the target and probe sequences.

### *Filter Hybridization*

This technique is based on the hybridization of immobilized sample DNA on a solid phase, such as nitrocellulose or nylon filter. After the attachment of single-stranded sample, or target nucleic acids to the filter surface, the filters are prehybridized to block nonspecific nucleic acid-binding sites. As target nucleic acids, mixed nucleic acids extracted directly from an environment source and immobilized on a filter surface as well as purified and restricted DNA transfered to filters by southern transfer are possible. Labeled probe DNA is then added to the filters, and the probe is allowed to hydridize by forming a double-stranded molecule with the complementary target sequences. Finally, excess unbound labeled probe is washed off and the hybrid (target:probe) is detected. The specificity is largely controlled by the stringency of the hybridization and washing conditions. The advantages of filter hybridizations are that different types of nucleic acids with varying purity can be analyzed and that multiple samples can easily be processed and quantified simultaneously.

As probe-labeling methods, the radioactive $^{32}P$ label—introduced as $^{32}P$ nucleotide triphosphate via, e.g., nick translation—is the radioactive label of choice. It allows the most sensitive detection.

Nonradioactive methods mostly involve an indirect detection of the target DNA via enzyme-linked antibodies. For this purpose, haptens (biotin or digoxigenin) are incorporated into the DNA probes. After hybridization, enzyme-linked antibodies specific for the haptens are added and a colorimetric assay is used to detect the enzymes attached to the probe-target sandwich. Generally, alkaline phosphatase or horseradish peroxidase are used as the indicator enzymes.

Colony or plaque hybridization are also often used techniques because they can easily be coupled to the conventional microbiological analysis on culture plates. After transfer of bacterial colonies or phage plaques to filters, enzymatic or alkaline lysis treatments are perfomed prior to hybridization analysis (Sayler and Layton 1990).

### *Combination of Restriction Fragment Length Polymorphism Analysis with Hybridization Studies*

DNA extracted from a particular organism or environmental sample can be cut by site-specific nucleases to yield a characteristic so-called restriction pattern. This pattern is obtained after separating the restricted DNA fragments on a sodium dodecylsulfate polyacrylamide gel electrophoresis (SDS-PAGE) or other suitable separation techniques. Each microbial species or strain has a specific pattern, which can be used for a detailed characterization. To get an identification pattern with less complexity, gene probes with, e.g., phylogenetic relevance can be used in subsequent hybridization studies of the restriction pattern after transfer to filters. In eukaryotic organisms, such as algae and other protists, the RFLP of their mitochondrial DNA provides important taxomic information (Coleman and Goff 1991).

### *In situ Hybridization*

Information about the spatial distribution of individual organisms and their possible interaction with other organisms and the physical environment can be accomplished using the *in situ* single-cell hybridization method. Small oligonucleotides (Figure 17.1), e.g., targeting-selective 16S or 23S rRNA sequences, have been successfully used in this respect (DeLong et al. 1989; Amann, Krumholz et al. 1990; Amann et al. 1992). Even nonculturable bacteria, like the magnetotactic bacteria of freshwater and marine sediments, could be identified (Spring et al. 1992). Fluorescently labeled oligonucleotide probes yield results with excellent spatial resolution, and they can be evaluated using an epifluorescence microsopy or confocal laser scanning microscopy. Their sensitivity, however, is rather limited because the fluorescent probes used so far confer only a single fluorescent dye molecule per ribosome. Their application is therefore restricted to the detection of actively growing microbes that have a high rRNA content. Oligonucleotides, which were end-labeled with digoxigenin (DIG), could be detected in whole fixed cells by antibodies labeled with either alkaline phosphatase or horseradish peroxidase (Zarda et al. 1991). This technique resulted in a significant signal amplification as compared to the fluorescently labeled oligonucleotide and can be used in natural samples with autofluorescence.

### *DNA Reassociation Studies for Diversity Assessments*

The genetic diversity of environmental microbial communities was measured by Torsvik et al. (1990) using DNA reassociation kinetics. In this technique, the bacteria are first extracted from soil or sediment samples, and then the bacterial DNA is obtained and purified. After shearing the DNA by sonication, it is precipitated with ethanol. The rate of reassociation after thermal denaturation provides an estimation of the genetic diversity of the bacteria in the environmental samples using reference mixtures of DNA with known complexities: the greater the diversity of the DNA, the slower the reassociation. Using this approach, the presence of about 3000 different bacterial species in one gram of forest litter soil was estimated (Torsvik et al. 1990).

## IMMUNOLOGICAL TECHNIQUES

In addition to the genetic techniques described above, immunological tools can also be used in population analysis. Protein or polysaccharide structures on the cell surfaces provide specific targets (determinants) for a specific characterization using polyclonal or monoclonal antisera (Bohlool and Schmidt 1980). The specificity can be at different levels, but even a strain-specific detection of organisms is possible. For quantification, different enzyme-linked immunosorbent assays are performed. The sensitivity of detection could be improved using monoclonal antibodies and chemoluminescent detection (Schloter et al. 1992). However, for extremely low numbers of organisms, as in freshwater ecosystems, selective enrichment techniques are necessary for counting. A single cell identification with specific antibodies uses a sandwich technique with second antibodies labeled with, e.g., a fluorescent dye as indicator.

## CONCLUDING REMARKS

With the rapid advances of molecular genetic approaches in systematics (Stackebrandt and Woese 1981; Theriot 1989; Olsen 1990) and their beginning application in ecological studies, new research possiblities arise. Although the methods are rapidly becoming easier, they still need very experienced and careful handling to avoid final misinterpretion. Presently, the limitations are given by the still-limited number and accessibility of genetic probes and the relatively high costs per assay. Using present protocols, not all microorganisms are permeable to the labeled oligonucleotide probes. Organisms in the resting or "quiescent" state cannot yet be tested with these probes, because rRNA is present in these forms only in low copy numbers. Especially when PCR is applied, the experiments have to be performed and interpreted with great care. There is evidence that under certain conditions, PCR-derived artificial DNA is created (Liesack et al. 1991). Some of the present obstacles may soon be solved. After specifically labeling subpopulations with fluorescent dyes, flow cytometric analysis may be used as an automatic and specific quantification method (Amann, Binder et al. 1990). The great potential of molecular approaches in ecological studies should also provide answers concerning shifts and losses of genetic information in acidified freshwater ecosystems.

## REFERENCES

Amann, R.I., B.J. Binder, R.J. Olson, S.W. Chisholm, R. Devereux, and D.A. Stahl. 1990. Combination of 16S rRNA-targeted oligonucleotide probes with flow cytometry for analysing mixed microbial populations. *Appl. Environ. Microbiol.* **56**:1919–1925.

Amann, R.I., L. Krumholz, and D.A. Stahl. 1990. Fluorescent oligonucleotide probing of whole cells for determinative, phylogenetic, and environmental studies in microbiology. *J. Bacteriol.* **172**:762–770.

Amann, R.I., J. Stromley, R. Devereux, R. Key, and D. Stahl. 1992. Molecular and microscopic identification of sulfate-reducing bacteria in multispecies biofilms. *Appl. Environ. Microbiol.* **58**:614–623.

Austin, B. 1988. Methods in Aquatic Bacteriology. Chichester: Wiley.

Bej, A.K., M.H. Mahbubani, R. Miller, J.L. DiCesare, L. Haff, and R.M. Atlas. 1990. Multiplex PCR amplification and immobilized capture probes for detection of bacterial pathogens and indicators in water. *Mol. Cell. Probes* **4**:353–365.

Bohlool, B.B., and E.L. Schmidt. 1980. The immunofluorescence approach in microbial ecology. In: Advances in Microbial Ecology, vol. 4, ed. M. Alexander, pp. 203–241. New York: Academic.

Buchheim, M.A., M. Turmel, E.A. Zimmer, and R.L. Chapman. 1990. Phylogeny of *Chlamydomonas* (Chlorophyta) based on clastic analysis of nuclear 18S rRNA sequence data. *J. Phycol.* **26**:689–699.

Caldwell, D.E., and J.R. Lawrence. 1989. Microbial growth and behavior within surface microenvironments. In: Recent Advances in Microbial Ecology, ed. T. Hattori, Y. Ishida, Y. Maruyama, R.Y. Morita, and A. Uchida, pp. 140–145. Tokyo: Japan Scientific Societies Press.

Coleman, A.W., and L.J. Goff. 1991. DNA analysis of eukaryotic algal species. *J. Phycol.* **27**:463–473.

DeLong, E.F., G.S. Wickham, and N.R. Pace. 1989. Phylogenetic stains: Ribosomal RNA-based probes for the identification of single cells. *Science* **243**:1360–1363.

Gilliland, G., S. Perrin, and H.F. Bunn. 1990. Competitive PCR for quantitation of mRNA. In: PCR Protocols: A Guide to Methods and Applications, ed. M. Innis, D. Gelfand, D. Sninsky, and T. White, pp. 60–69. New York: Academic.

Giovannoni, S.I., T.B. Britschgi, C.L. Moyer, and K.G. Field. 1990. Genetic diversity in Sargasso Sea bacterioplankton. *Nature* **345**:60–63.

Höpfl, P., W. Ludwig, K.H. Schleifer, and N. Larsen. 1989. The 23S ribosomal RNA higher-order structure of *Pseudomonas cepacia* and other prokaryotes. *Eur. J. Biochem.* **185**:355–364.

Kirchhof, G., and A. Hartmann. 1992. Development of gene-probes for *Azospirillum* based on 23S rRNA sequences. *Symbiosis* **13**:27–35.

Liesack, W., H. Weyland, and E. Stackebrandt. 1991. Potential risks of gene amplification by PCR as determined by 16S rDNA analysis of a mixed-culture of strict barophilic bacteria. *Microb. Ecol.* **21**:191–198.

Mills, A.L., and L.M. Mallory. 1987. The community structure of sessile heterotrophic bacteria stressed by acid mine drainage. *Microb. Ecol.* **14**:219–232.

Nilsson, L., J.D. Oliver, and S. Kjelleberg. 1991. Resuscitation of *Vibrio vulnificus* from the viable but nonculturable state. *J. Bacteriol.* **173**:5054–5059.

Olsen, G.J., D.J. Lane, S.J. Giovannoni, and N.R. Pace. 1986. Microbial ecology and evolution: A ribosomal RNA approach. *Ann. Rev. Microbiol.* **40**:337–365.

Olsen, J.L. 1990. Nucleic acids in algal systematics. *J. Phycol.* **26**:209–214.

Pichard, S.L., and J.H. Paul. 1991. Detection of gene expression in genetically engineered microorganisms and natural phytoplankton populations in the marine environment by mRNA analysis. *Appl. Environ. Microbiol.* **57**:1721–1727.

Pickup, R. W. 1991. Development of molecular methods for the detection of specific bacteria in the environment. *J. Gen. Microbiol.* **137**:1009–1019.

Sayler, G.S., and A.C. Layton. 1990. Environmental application of nucleic acid hybridization. *Ann. Rev. Microbiol.* **44**:625–648.

Schloter, M., W. Bode, A. Hartmann, and F. Beese. 1992. Sensitive chemoluminescence-based immunological quantification of bacteria in soil extracts with monoclonal antibodies. *Soil Biol. Biochem.* **24**:399–403.

Selenska, S., and W. Klingmüller. 1992. Direct recovery and molecular analysis of DNA and RNA from soil. *Microb. Rel.* **1**:41–46.

Shapiro, H.M. 1990. Flow cytometry in laboratory microbiology: New directions. *AMS News* **56**:584–588.

Spring, S., R. Amann, W. Ludwig, K.H. Schleifer, and N. Petersen. 1992. Phylogenetic diversity and identification of nonculturable magnetotactic bacteria. *Sys. Appl. Microbiol.* **15**:116–122.

Stackebrandt, E., and C.R. Woese. 1981. The evolution of prokaryotes. In: Molecular and Cellular Aspects of Microbial Evolution, ed. M.J. Carlisle, J.R. Collins, B.E.B. Moseley, pp. 1–31. Cambridge: Cambridge Univ. Press.

Steffan, R.J., and R.M. Atlas. 1991. Polymerase chain reaction: Applications in environmental microbiology. *Ann. Rev. Microbiol.* **45**:137–161.

Theriot, E. 1989. Phylogenetic systematics for phycology. *J. Phycol.* **25**:407–411.

Torsvik, V., J. Goksoyr, and F.L. Daae. 1990. High diversity in DNA of soil bacteria. *Appl. Environ. Microbiol.* **56**:782–787.

Ward, D.M., R. Weller, and M.M. Bateson. 1990. 16S rRNA sequences reveal numerous uncultured microorganisms in a natural community. *Nature* **345**:63–65.

Weller, R., J.W. Weller, and D.M. Ward. 1991. 16S rRNA sequences on uncultivated hot spring cyanobacterial mat inhabitants retrieved as randomly primed cDNA. *Appl. Environ. Microbiol.* **57**:1146–1151.

Zarda, B., R. Amann, G. Wallner, and K.-H. Schleifer. 1991. Identification of single bacterial cells using digoxigenin-labeled, rRNA-targeted oligonucleotides. *J. Gen. Microbiol.* **137**:2823–2830.

# 18

# The Biological Effects of Acid Episodes

S.J. ORMEROD[1] and A. JENKINS[2]
[1]Catchment Research Group, School of Pure and Applied Biology, University of Wales College of Cardiff, Cardiff, U.K.
[2]Institute of Hydrology, Crowmarsh Gifford, Wallingford, Oxfordshire, U.K.

## ABSTRACT

In this chapter, we critically review the information available on factors influencing acid episodes and on their biological impact. We suggest that, whereas many factors influencing episodicity have been described across a range of temporal and spatial scales, few systematic data or analyses are available. In part this reflects the lack of a common format applicable to all surface waters, which describes the key features of different kinds of acid episode: duration, frequency, concentration, amplitude, rate of change. A full quantitative understanding of the biological importance of acid episodes has been prevented by this absence of a common hydrochemical currency through which to assess severity, but also because:

1. Biological systems vary in their response to acid episodes at different levels of organization (individual, species, population, community).
2. Data available on the impacts of acid episodes have involved either experiments (and hence artifacts) or field data (which lack control, precision, and cause-effect knowledge).
3. Biological impacts have been seen largely as ecotoxicological phenomena affecting individuals rather than ecological phenomena affecting populations or communities.
4. The effects of single acid episodes, multiple episodes, and chronic acidification are difficult to separate under real field conditions.

The extent to which further knowledge and research is required depends on information needs. At one extreme, we might suggest that acid episodes are damaging and thus should be prevented. At the other, quality standards for acid episodes might be set only if we can quantify precisely the components that are damaging to different organisms and in different habitats. Such quantification is not currently available.

"Upon those who step into the same rivers, different and ever different rivers flow down." —Heracleitus (ca. 540–480 B.C.).

*Acidification of Freshwater Ecosystems: Implications for the Future*
Edited by C.E.W. Steinberg and R.F. Wright 

## INTRODUCTION

In many branches of the study of water pollution, recent decades have seen a shift from approaches emphasizing chronic (or steady-state) conditions, to those in which brief exposures or episodes of pollutants are important ecotoxicological stressors (e.g., Seager and Maltby 1989; McCahon and Pascoe 1990). In part, this shift has resulted from the increasing frequency of intermittent relative to chronic sources of pollution and from the realization that ecotoxicological responses to fluctuating conditions can differ from responses to constant conditions with the same mean chemistry (see Siddens et al. 1986; Seager and Maltby 1989). Also germane to the context of this chapter, fluctuations in water chemistry are a key element of some kinds of pollution, among which surface water acidification is the prime example (Davies et al. 1992).

Most features of water chemistry that are affected by acidification—acid-neutralizing capacity (ANC), pH, sulfate, nitrate, base cations, aluminum, manganese, zinc, and other metals—fluctuate over time scales of hours (e.g., during rainstorms), weeks (e.g., during snow melt), months (e.g., following drought or forest clearance), or years (during chronic acidification). Other changes occur continually, for example with the varying contributions of water from different soil and groundwater sources. Changes in acid-base status at all of these time scales have been postulated to be important influences on the biology of streams and lakes. Short-term episodes of acidity have been seen as particularly important, and evidence of fish kills following such events is widespread (e.g., observations by Hultberg and others, reviewed by Muniz 1991). While they can occur through a range of natural processes, acid deposition has contributed to their occurrence and/or severity through direct effects at snowmelt and by conditioning catchments, so that acid episodes are more likely or more pronounced. Conditioning processes include an increase in the catchment pools of sulfate and $H^+$ ions available for transport into surface waters during high flow and also a reduction in the base saturation of soils. Further additions of acid from the atmosphere onto such base-depleted soils thus result in disproportionately large reductions in the pH of runoff.

In reality, however, quantifying and isolating the biological effects of acid episodes, as distinct from those of chronic acidification, have proven to be perplexing problems. Difficulties arise because episodic changes are largely a characteristic of surface waters that have relatively low ANC, and hence also tend to be chronically acidified (Weatherley and Ormerod 1991). Additionally, continual fluctuation in many chemical parameters over different time scales leads to problems in defining what we mean by acid episodes: markedly differing hydrological events of contrasting chemistry and duration are lumped together under this term by different authors. Difficulties also arise because acid episodes involve simultaneous changes in several important chemical features whose biological effects are difficult to distinguish (Weatherley et al. 1990), and because these changes have many characteristics that vary over many temporal and spatial scales (see Figure 18.1); all can potentially affect the severity of biological response (Table 18.1). Further difficulties arise because there has been a

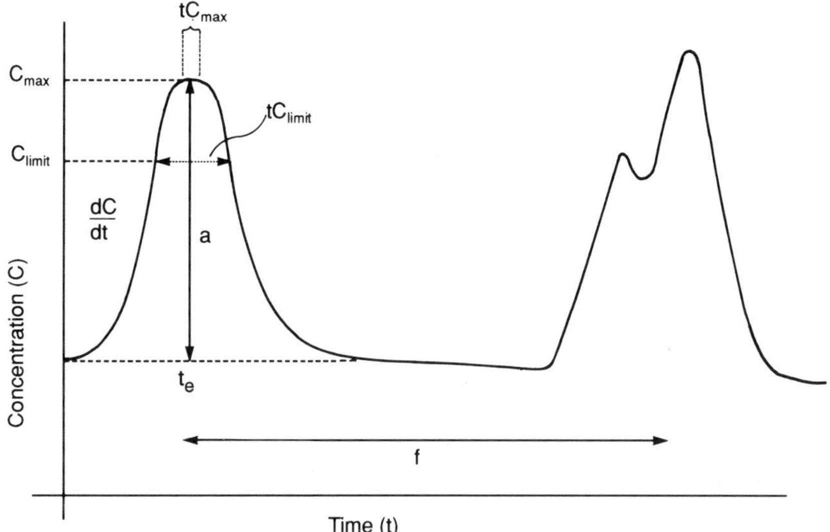


**Figure 18.1** Some variable features of acid episodes. They include amplitude (*a*), rate of change (*dC/dt*), duration of peak concentration ($C_{max}$) or over the threshold concentration for some toxic effect ($C_{limit}$), and frequency between episodes (*f*).

**Table 18.1** A selection of likely factors influencing the response of aquatic organisms or communities to episodic change in the acid-base status.

| Biological Factors (individual) | Biological Factors (ecological) | Water Quality |
|---|---|---|
| Species | Individual responses | Concentration |
| Age/life cycle stage | Population responses | Duration |
| Species interactions | Scope for recovery | Rate of change |
| Disease/parasitism | Scope for recolonization | Episode frequency |
| Reproductive state | Habitat quality | Synergisms |
| Nutritional status | Time of year | Mitigators |
| History/acclimation | Species interactions | |
| Genotype | Other ecological stressors | |
| Lethal or sublethal effect? | | |

tendency to see acid episodes as *toxicological* phenomena, causing lethal and sublethal responses in individual organisms. This is in contrast to their being seen and understood as *ecological* phenomena, effecting biological change at the population or community level (with a few notable exceptions). Together, these contrasting problems underscore a need to define, quantify, and characterize the biological effects of acid episodes more accurately than hitherto.

Here, we will review some of the problems that have confounded a clear understanding of the biological importance of acid episodes across all field conditions under which they occur. We draw, in part, on recent reviews by Davies et al. (1992) and Wigington et al. (1992), but go beyond these descriptive accounts to argue that, despite research into acidification over three decades, we have yet to develop:

- a common format, applicable to all surface waters, that describes the key features of different kinds of acid episode: duration, frequency, concentration, amplitude, rate of change;
- a systematic approach to assess how different features influence episodicity;
- a clear and unequivocal understanding of those features of acid episodes that are biologically most important, most damaging, and which are distinct from chronic acidification in their effects on populations and communities.

In other words, we call for a common currency of hydrochemical parameters that describes severity and frequency of acid episodes, a common currency of parameters that describes the severity of biological response, and we suggest a need for systematic studies which interrelate the two.

## A COMMON CURRENCY?

A theoretical and highly simplified rendering of the character of an acid episode is shown for one chemical parameter (let us say aluminum) in Figure 18.1. It has duration ($t_e$), amplitude ($a$), rate of change ($dC/dt$) in both ascending and descending limb, peak concentration ($tC_{max}$), and duration of peak concentration ($tC_{max}$). To these, other parameters might be added such as the mean concentration ($xC$) during the episode, or the amplitude and duration over which some known ecotoxic threshold ($C_{limit}$) is exceeded ($tC_{limit}$). Among these parameters, few are usually considered; Davies et al. (1992) reviewed largely duration, amplitude, and peak concentrations for $H^+$. They suggested that the most acidic episodes tended to occur in those waters where the pre-episode pH is already low, with the largest amplitude of change occurring in waters of higher pH.

We should also note that several chemical parameters will change, possibly intercorrelated ways during acid episodes, and will have the properties depicted in Figure 18.1; acid episodes, generated either by snowmelt or rainfall, often have increased sulfate and/or nitrate concentrations, loss of ANC, reduced pH, reduction in base cation concentration (in which dilution is important), and increased concentrations of other metals. In addition, other acid episodes will follow at a given site with a given frequency ($f$), although the subsequent episodes may differ in any of the listed parameters. It is possible also that we would wish to differentiate between features that are the properties of individual *episodes* and those that are properties of given *sites* (e.g., Harriman et al. 1990). In other words, under some circumstances, we may ask whether given episodes (or suite of episodes) have certain characteristics that exceed thresholds sufficient for biological impact. Alternatively, we may ask whether sites have properties that determine the character of the episodicity they exhibit. In some cases, this *site-specific* assessment might involve determining the average behavior of a given site using one or more of the parameters shown in Figure 18.1. We may wish, however, to parametrize the character of a site in a different way, e.g., as concentration/duration curves or frequency duration curves (Figure 18.2). However, these might not necessarily show discrete events, and so a further possibility is provided by event-frequency distributions (Figure 18.2c). As with all ecological phenomena, the properties of either given acid episodes or given sites will be influenced by a mix of deterministic and stochastic features (Figure 18.3).

## IN SEARCH OF SYSTEMATIC UNDERSTANDING: HYDROCHEMISTRY

Several hydrochemical features of acid episodes or sites, as depicted in Figures 18.1 and 18.2, vary on a range of spatial and temporal scales.

### Spatial Features

Freshwater systems are characterized by varying episodicity (as both frequency and intensity) at different spatial scales, ranging from the regional or national scale to the very precise and local scale, e.g., at the gill surface of a fish or insect. At the international or regional scale, climatic variability may mean that some areas are subject to different annual flow regimes driven by snowmelt, as opposed to orographic or frontal rainstorms. The many contrasts between rainfall and snowmelt episodes are important in this context. Snowmelt, particularly where it occurs over frozen ground, permits a more direct link between anthropogenic pollution and surface water chemistry; pre-event groundwater sometimes makes a smaller contribution to runoff under these circumstances and buffering processes are bypassed. This direct pathway means that nitrate can, under some conditions, make a more important contribution to acidity than in rainfall episodes, since there will be less chance for uptake into the terrestrial ecosystem. Temporal features also contrast between these two types of episode (see below).

Similarly, the pollution climate at a regional scale may be responsible for either conditioning catchments in different ways or by directly influencing runoff, e.g., at snowmelt (Davies et al. 1992). An important example here is the extent to which nitrate makes a greater contribution to the acidity of episodes in the northeast U.S. than elsewhere (Wigington et al. 1992). In part, this difference may reflect a greater contribution to deposition by nitrate, although it may also reflect the relatively large contribution to the available data from afforested catchments. However, we have little chemical information of sufficient temporal resolution through which these and other

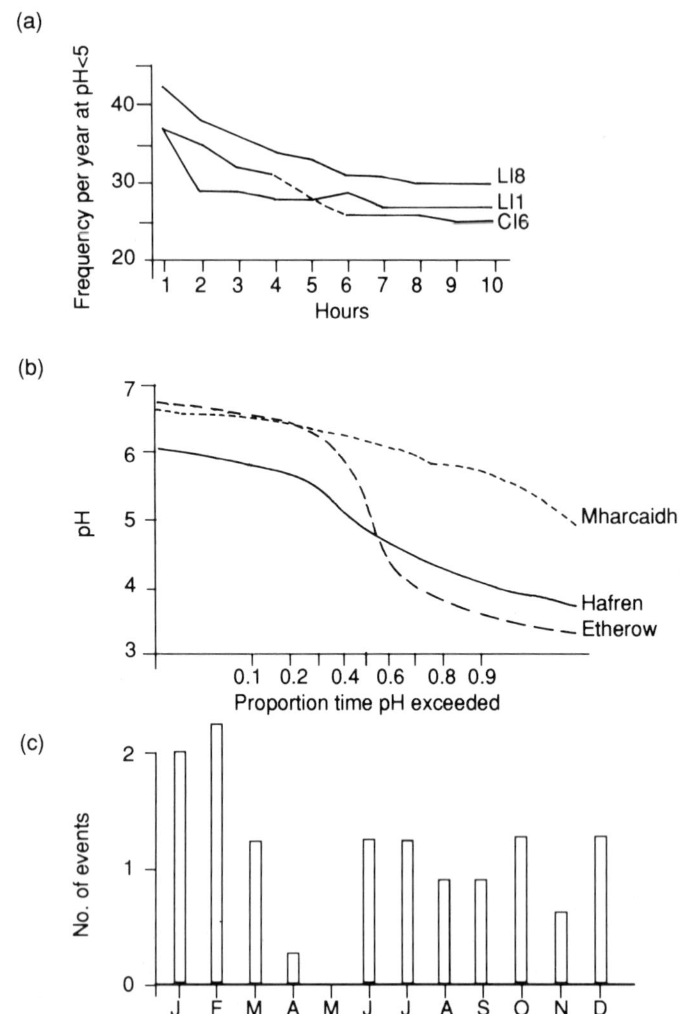


**Figure 18.2** Depicting the episodicity of sites: (a) event frequency/duration curves for three sites at Llyn Brianne, mid Wales; (b) pH duration curves for three sites in the U.K. acid waters monitoring network; (c) the mean monthly number of episodes with pH < 5.0 at Allt a Mharcadh, Scotland.

regional trends in episodicity might be quantified using some or all of the parameters suggested in Figures 18.1 and 18.2.

At the national scale, many countries now have networks of monitoring stations across which continuous monitors permit some assessment of episodic behavior. These networks already provide information on differences in episodicity between catchments (e.g., Figure 18.2), providing us with the opportunity to understand how

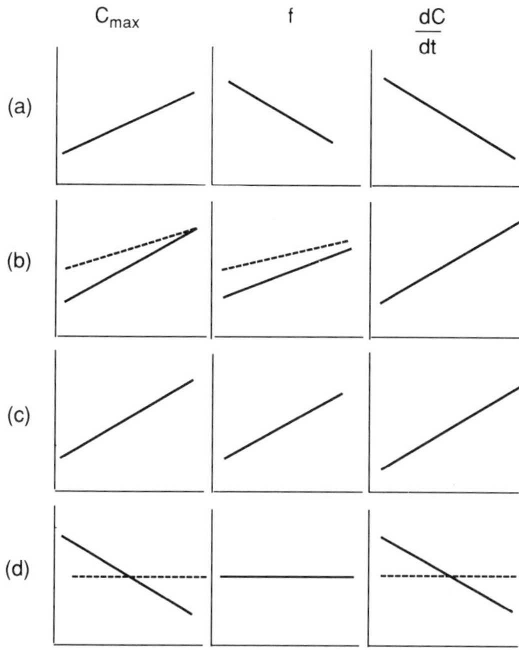


**Figure 18.3** Hypothetical expectations of a range of influences on the episodicity of streams. The y axis shows peak concentration ($C_{max}$), frequency (*f*) and rate of change (*dC/dt*) of $H^+$ or aluminium episodes, while the x axis shows s trend in: (a) sites influenced by frontal rainfall to those influenced by snowmelt; (b) from clean to polluted deposition (the solid line represents an inland site, and the hatched line a coastal site influenced by marine salts); (c) a shift from grassland to plantation forestry; (d) a shift from surface to interstitial water in instances with poor mixing (solid) and rapid mixing (hatched).

both the location of a catchment and its attributes influence episodicity. These attributes include soil type, soil depth, vegetation type, altitude, basin form, topography, and drainage density. Davies et al. (1992) suggest that the relative mixing of "event" (i.e., storm runoff or meltwater from the snowpack) versus "pre-event" (soil or ground) water contributing to runoff is a central and important issue in episode character. The mix is affected by the flowpaths through which runoff reaches the stream or lake, so that catchment characters can be important, as are antecedent conditions (e.g., soil wetness in different horizons, temperatures leading to speed of snowmelt). Catchment features and location can also profoundly affect the causative mechanisms behind acid episodes. Davies et al. (1992) suggest that acid episodes can occur due to natural factors, which include base-cation dilution, effects by organic acids, and the deposition of sea salts; the last of these, with chloride acting as the driving anion, can be relatively frequent in areas dominated by deposition from marine sources. They cite an example from western Scotland, where as many as six such acid episodes occur annually.

At the reach scale, subtle spatial effects on episodicity might arise where tributaries of different chemical quality are mixed (e.g., Rosseland et al. 1992), because additions from the streambed affect episode quality with progression downstream (Norton et al. 1992), or because local flow paths alter such features as hydrological residence time and the mix of pre-event versus event water. Examples of the latter effect include differences between the surface and interstitial water during acid episodes (e.g., Weatherley et al. 1989) or differences between the littoral and profundal zone of lakes. The net results of these reach effects are habitat-specific differences in several of the parameters depicted in Figure 18.1. Mixing waters can sometimes create areas of increased toxicity (Rosseland et al. 1992); however, in other cases, reach-scale effects during acid episodes can lead to the presence of refuges for organisms either located in certain microhabitats or mobile enough to change their location. At least some age classes of fish are known to show such responses to downstream changes in water quality during acid episodes (Gunn 1986; Carline et al. 1992), although data from invertebrates are more equivocal (Ormerod et al. 1987a). Nevertheless, there is some evidence that invertebrates in the interstitial meiofauna may be less affected by acid episodes than those in surface waters (Giberson and Hall 1988).

Even more local effects on the character of acid episodes are likely to occur at the highly intimate and chemically active boundary between a freshwater organism and its surroundings. For example, at the gill surface of a fish, the loss of microquantities of plasma salts (sodium, chloride, potassium) and excretory products across the gill during acid stress almost certainly affects chemistry at an extremely small scale; buffering reactions, with increased pH, are possible, with consequent effects on aluminium chemistry similar to those described at the interface between aluminum- and lime-rich water (Rosseland et al. 1992). These reactions are unlikely to affect the outcome to the fish of exposure to acid conditions; however, it may mean that general measurements of episodicity in lakes or rivers do not necessarily represent the conditions intimately experienced by the fish. Measuring and understanding the importance of these fine-scale effects is a particular problem.

Clearly, while the effects on episodicity of some of these spatial factors have been described and reviewed (e.g., Davies et al. 1992), few systematic data are available. Potential relationships are suggested in Figure 18.3 and might be regarded as just some of the hypotheses requiring further testing. An important influence in all these cases will be stochasticity.

## Temporal Factors

Several features governing acid episodes have a temporal dimension, partly because time is the central feature of the y axis of Figure 18.1. For example, rainfall-induced episodes occur over time scales of hours to days, while snowmelt occurs over days to weeks. Annual frequencies in these types of events also differ by about 1–2 orders of magnitude. While snowmelt, in general, instigates the most pronounced depressions in pH, particularly in the near-surface layers of lakes, Davies et al. (1992) point out that rainfall-induced episodes may be more "important" because they are more frequent on an annual basis. Unfortunately, we have no real criteria with which to determine what is "important" in this context.

Temporal elements will also affect the antecedent and ambient conditions governing attributes of the x axis. These might include trends over time scales of minutes to hours (e.g., the rate of snowmelt or rainfall intensity), to years (e.g., changes in deposition acidity). As with the spatial factors governing episodicity, temporal elements will have a large component of stochasticity, e.g., many of the factors governing antecedent and ambient conditions: the back-trajectory of air masses influencing pollution content; the likelihood of sea-salt deposition; soil hydraulic conductivity in different horizons; meteorological influences on rainfall or snowmelt. Moreover, temporal elements impinging on acid episodes will vary across all of the spatial scales outlined above, from the reach to the region. Thus, for example, y axis features in Figure 18.1 might vary between regions (e.g., the duration of acid episodes generated by snowmelt versus rainfall) or between habitats in the same reach (because the shape of Figure 18.1 will change). As with spatial factors, few systematic data are available on which to assess such variation.

## Chemical Factors

Chemical change is clearly a salient feature of acid episodes but there are surprisingly few instances where all coincident changes have been described. Davies et al. (1992) drew attention to this dearth of information, concentrating their account largely on changes in pH. This is clearly unfortunate given that other aspects of acid episodes, notably aluminum speciation, are more likely to be important biologically. Detailed chemical examination is difficult; however, the unusually precise assessment by Goenaga and Williams (1988), of changes in pH and aluminum speciation in a Welsh hillstream, involved substantial laboratory work to characterize just one episode. As expected from this and other studies, many chemical parameters change during snowmelt or rainstorms, so that biological responses in the field correlate equally strongly with a range of potential stressors (e.g., Weatherley et al. 1990). Weatherley et al. showed that peaks in several aluminum species occurred over the period preceding the death of exposed fish. As a result, no one aluminum species or concentration could be held responsible for death. Moreover, this study (in common with most others) examined correlations between lethality and mean concentrations of each aluminum species rather than any of the other potentially important biological changes during acid episodes (see Figure 18.1). Systematic information will be necessary for all potentially important parameters if we are to understand the true influences of acid episodes on biology.

## IN SEARCH OF SYSTEMATIC UNDERSTANDING: BIOLOGY

Whereas the hydrochemical problems, preventing a fuller understanding of acid episodes, stem from a need for systematic information and a common currency, a different set of problems prevent a fuller understanding of their biological importance. These might be summarized as follows.

### Biological Variability

With little exception (e.g., Hall et al. 1980; Ormerod et al. 1987a; Steinberg and Putz 1991), much of the emphasis on the biological impact of acid episodes has been on fish; their response to episodes governs much of our thinking on the whole subject area. However, the biotic components of freshwater ecosystems also include primary producers among bacteria, algae and macrophytes; consumers among the micro- and macro-invertebrates; and vertebrates which include Amphibia, birds, and mammals. At least for those groups that are wholly aquatic, the response of any individual organism to chronic or episodic acidification is likely to vary due to a range of features (Table 18.1). Only a few of those factors likely to influence response have been systematically studied. For example, different species of fish vary in their toxic response to chronic acid conditions or to acid episodes (Ormerod et al. 1987a; Muniz 1991), as do different invertebrate species (Chmielewski and Hall 1992). Differences between the toxic responses of different life stages of fish are also well known (e.g., eggs vs. fry vs. smolts vs. adults). Thus, some organisms may be killed by single acid episodes (e.g., Ormerod et al. 1987a), some by repeated acid episodes in the same location (e.g., Merrett et al. 1991), while others show sublethal effects only during exposure (e.g., Hall et al. 1988). In many cases, patterns of toxic response during acid episodes are consistent with patterns of distribution in relation to stream chemistry. For example, salmonids or the mayfly *Baetis rhodani* are generally scarce or absent from acid streams (e.g., Rutt et al. 1990), and are also sensitive to acid episodes (e.g., Ormerod et al. 1987a).

However, many difficulties arise in interpreting such information:

1. *Data quality/context*: Partly because of the stochastic and uncontrolled nature of acid episodes, most assessments of their biological effects have been made either in laboratory or field experiments. Both approaches share the difficulty that artifacts and a lack of realism are likely, not least that chemical conditions may be atypical; the complex chemistry of aluminum means that test solutions or induced conditions seldom truly mimic those in real acid episodes. In experiments at Llyn Brianne, the additions of aluminum sulfate to streams often resulted in unexpectedly large fractions of aluminum that would not pass through filters of 0.45 or 0.22 μm pore size; they were likely to be somewhere between polymeric and "dissolved" (S.J. Ormerod, unpubl. data). Other factors

lacking realism, even in field experiments, include the absence of confounding features that accompany acid episodes, such as increased flow. In addition, organisms are often constrained in exposure vessels and thus lack the opportunity of avoidance reactions. Furthermore, the scale of such field or ecosystem scale experiments means that replication is seldom achieved (Hurlbert 1984).

At the other extreme, observations of the effects of real acid episodes in the field have realism but lack control: pre-episode conditions are seldom known, so that biological responses can be judged only from the collection of dead animals (e.g., fish). Population consequences are liable to be unknown, and in any case the strong intercorrelation of many chemical features of acid episodes means that features responsible for mortality cannot be assessed easily.

2. *Difficulties of extrapolation from ecotoxicological to ecological responses*: Most information on the biological effect of acid episodes, either from laboratory experiments, field experiments, or field observations, is in the form of a lethal or sublethal ecotoxicological response. Organisms exposed and affected by acid episodes in this way may represent:
   a) dead organisms that have the potential to be replaced by "surplus" organisms from other locations unaffected by acid episodes,
   b) only a part of the population at any given location,
   c) partially affected organisms which may subsequently recover,
   d) organisms apparently unaffected during the duration of exposure which subsequently show mortality or sublethal responses.

   In cases (a), (b), and (c), population- or community-level responses may not be apparent following exposure to acid episodes, particularly where populations are controlled by some other factor (e.g., predation, food abundance, habitat availability). In case (d), population-level responses may be greater than anticipated, although even here there is room for recovery. Much will depend on the time scale over which population- or community-level change is being assessed. In instances where attempts are made to assess population or community responses to acid episodes in the field, a further problem arises:

3. *The difficulty of separating the effects of chronic acidification, single episodes, and multiple episodes under field conditions:* At least in areas where acid episodes are governed by frontal rainstorms, surface waters are characterized annually by many episodes. Examples would be Wales and Scotland. Moreover, indices of episodicity (e.g., minimum pH, mean –2 pH standard deviations, coefficient of variation in pH, maximum aluminum concentration) all correlate with mean chemical conditions. This would be expected since:
   a) samples taken during episodes contribute to the calculation of mean chemistry, and
   b) streams with low base status are anyway more susceptible to the occurrence of acid episodes.

Attempts to assess fish populations and invertebrate communities in these same areas show that empirical relationships between biological status and mean chemistry are not significantly improved by incorporating measures of episodicity (Weatherley and Ormerod 1991). In other words, either acid episodes or chronic acidity might be equally responsible for site-to-site population change among fish and invertebrates. This simple idea becomes even more robust when we consider that multiple acid episodes are required for cumulative toxic responses among some invertebrates (Merrett et al. 1991), and also because acid-base status probably affects invertebrates through a mix of direct physiological effects and indirect trophic effects (Ormerod et al. 1987a, 1987b). The latter are just as likely to reflect chronic conditions as acid episodes.

Because areas subject to snowmelt may have acid episodes of greater duration and peak concentrations, and might therefore be of greater severity, we need similar survey-scale exercises to those carried out by Weatherley and Ormerod (1991). These would ascertain whether:

a) mean chemical conditions can still represent the likelihood of episodes in these areas, or
b) biological status correlates as closely with mean chemistry as episodicity.

Parameters from Figure 18.1 might be used in such an analysis. Further difficulties in separating the effects of chronic acidification from those of acid episodes might arise if chronic exposure resulted in either:

a) organisms becoming weakened by chronic acid stress, hence more susceptible to episodic exposure, or
b) acclimated, and hence less susceptible to acid episodes; in this case, "acclimation" reflects responses at either the genetic (i.e., population) or phenotypic (i.e., individual) level.

## A WAY FORWARD?

In part, the ways forward depend on the aims of knowledge. These might include:

1. A simple "need to know" that acid episodes are biologically important, so that they can be prevented by appropriate management strategies (e.g., liming).
2. A need to parametrize relationships between biological response and episodicity for modeling purposes. This may not require detailed understanding, with chemical-biological links represented by "black boxes." In this context, Weatherley and Ormerod (1991) showed that even mean chemistry was a suitable predictor of biological status.
3. The need to understand the importance of different aspects of episodicity (e.g., damaging concentrations, frequencies, durations) so that water quality standards might be set. Both critical load assessments or statutory water quality objectives might require such standards.
4. The expansion of fundamental information about biology and hydrochemistry.

Applications such as Pt. 1 may require only acceptance and consolidation of existing knowledge or data, with an assumption that acid episodes are liable to be damaging and therefore should be prevented. By contrast, for Pt. 3, we might set accurate quality standards for acid episodes only if we know fundamentally and precisely the frequency, duration, amplitude, rate of change, and extreme concentrations that are damaging to different organisms and in different habitats. Such a need is likely to be only satisfied if we develop a common currency similar to that recommended here, which better represents episodicity at all temporal scales, at spatial scales from the reach to the region, and which can be more clearly related to biological response.

## ACKNOWLEDGEMENTS

We thank Dahlem Konferenzen for the opportunity to present this paper. Ron Harriman and Gunnar Raddum made valuable comments on the manuscript.

## REFERENCES

Carline, R.F., D.R. Dewalle, W.E. Sharpe, B.A. Dempsey, C.J. Gagen, and B. Swistock. 1992. Water chemistry and fish community response to episodic stream acidification in Pennsylvania, U.S.A. *Environ. Pollut.* **78**:45–48.

Chmielewski, C.M., and R.J. Hall. 1992. Responses of immature blackflies (Diptera: Simuliidae) to experimental pulses of acidity. *Can. J. Fish. Aquat. Sci.* **49**:833–840.

Davies, T.B., M. Tranter, P.J. Wigington, and K.N. Eshleman. 1992. "Acidic episodes" in surface waters in Europe. *J. Hydrol.* **132**:25–69.

Giberson, D.J., and R.J. Hall. 1988. Seasonal variation in faunal distribution within the sediments of a Canadian Shield stream, with emphasis on response to spring floods. *Can. J. Fish. Aquat. Sci.* **45**:1994–2002.

Goenaga, X., and D.J.A. Williams. 1988. Aluminium speciation in surface waters from a Welsh upland area. *Environ. Pollut.* **52**:131–149.

Gunn, J.M. 1986. Behavior and ecology of salmonid fishes exposed to episodic pH depressions. *Environ. Biol. Fishes* **17**:241–252.

Hall, R.J., R.C. Bailey, and J. Findeis. 1988. Factors affecting survival and cation concentration in the blackflies *Prosimulium fuscum/mixtum* and the mayfly *Leptophlebia cupida* during spring snowmelt. *Can. J. Fish. Aquat. Sci.* **44**:2123–2132.

Hall, R.J., G.E. Likens, S.B. Fiance, and G.R. Hendrey. 1980. Experimental acidification of a stream in the Hubbard Brook experimental forest, New Hampshire. *Ecology* **61**:976–989.

Harriman, R., E. Gillespie, D. King, A.E. Watt, A.E.G. Christie, A.A. Cowan, and T. Edwards. 1990. Short-term ionic responses as indicators of hydrochemical processes in the Allt A' Mharcaidh catchment, Western Cairngorms, Scotland. *J. Hydrol.* **116**:267–285.

Hurlbert, S.H. 1984. Pseudoreplication and the design of ecological experiments. *Ecol. Monogr.* **45**:187–211.

McCahon, C.P., and D. Pascoe. 1990. Episodic pollution: Causes, toxicological effects and ecological significance. *Funct. Ecol.* **4**:375–383.

Merrett, W.J., G.P. Rutt, N.S. Weatherley, S.P. Thomas, and S.J. Ormerod. 1991. The response of macroinvertebrates to low pH and increased aluminium concentrations in Welsh streams: Multiple episodes and chronic exposure. *Arch. Hydrobiol.* **121**:115–125.

Muniz, I. 1991. Freshwater acidification: Its effects on species and communities of freshwater microbes, plants and animals. *Proc. Roy. Soc. Edinburgh* **97B**:227–254.

Norton, S.A., J.C. Brownlee, and J.S. Kahl. 1992. Artificial acidification of a non-acidic and an acidic headwater stream in Maine, USA. *Environ. Pollut.* **77**:123–128.

Ormerod, S.J., P. Boole, P. McCahon, N.S. Weatherley, D. Pascoe, and R.W. Edwards. 1987a. Short-term experimental acidification of a Welsh stream: Comparing the biological effects of hydrogen ions and aluminium. *Freshwater Biol.* **17**:341–356.

Ormerod, S.J., K.R. Wade, and A.S. Gee. 1987b. Macro-floral assemblages in upland Welsh streams in relation to acidity, and their importance to invertebrates. *Freshwater Biol.* **18**:545–557.

Rosseland, B.J., I.A. Blakar, A. Bulger, F. Kroglund, A. Kvellstad, E. Lyderson, D.H. Oughton, B. Salbu, M. Staurnes, and R. Vogt. 1992. The mixing zone between limed and acidic river waters: Complex aluminium chemistry and extreme toxicity for salmonids. *Environ. Pollut.* **78**:3–8.

Rutt, G.P., N.S. Weatherley, and S.J. Ormerod. 1990. Relationships between the physicochemistry and macroinvertebrates of British upland streams: The development of modelling and indicator systems for predicting fauna and stream acidity. *Freshwater Biol.* **24**:463–380.

Seager, J., and L. Maltby. 1989. Assessing the impact of episodic pollution. *Hydrobiologia* **188/189**:633–640.

Siddens, L.K., W.K. Seim, L.R. Curtis, and G.A. Chapman. 1986. Comparisons of continuous and episodic exposure to acidic, aluminium-contaminated waters of brook trout (*Salvelinus fontinalis*). *Can. J. Fish. Aquat. Sci.* **43**:2036–2040.

Steinberg, C., and R. Putz. 1991. Epilithic diatoms as bioindicators of stream acidification. *Verh. Internat. Verein Limnol.* **24**:1877–1880.

Weatherley, N.S., and S.J. Ormerod. 1991. The importance of acid episodes in determining faunal distributions in Welsh streams. *Freshwater Biol.* **25**:71–84.

Weatherley, N.S., A.P. Rogers, X. Goenaga, and S.J. Ormerod. 1990. The survival and early life stages of brown trout (*Salmo trutta* L.) in relation to aluminium speciation in upland Welsh streams. *Aquat. Toxicol.* **17**:213–230.

Weatherley, N.S., S.P. Thomas, and S.J. Ormerod. 1989. Chemical and biological effects of acid, aluminium and lime additions to a Welsh hill-stream. *Environ. Pollut.* **56**:283–298.

Wigington, P.J., T.B. Davies, M. Tranter, and K.N. Eshleman. 1992. Comparison of episodic acidification in Canada, Europe, and the United States. *Environ. Pollut.* **78**:29–35.

Standing, left to right:
Steve Ormerod, Rainer Putz, Magda Havas, David Schindler, Björn Rosseland, Helmut Segner, Jürgen Böhmer, Toni Hartmann

Seated, left to right:
Carol Kelly, Joan Baker, David Brakke, Alan Jenkins, Tomas Paces

# 19

# Group Report: Physiological and Ecological Effects of Acidification on Aquatic Biota

D.F. BRAKKE, Rapporteur

J.P. BAKER, J. BÖHMER, A. HARTMANN, M. HAVAS, A. JENKINS, C. KELLY, S.J. ORMEROD, T. PACES, R. PUTZ, B.O. ROSSELAND, D.W. SCHINDLER, H. SEGNER

## INTRODUCTION

Acidification affects all components of biological communities in lakes and streams: microbes, algae, macrophytes, invertebrates, fish, amphibians, and other vertebrates that rely on aquatic ecosystems for habitat or food. Mechanisms of effect are both direct (toxic responses to changes in chemistry) and indirect (e.g., expressed through the food chain or caused by changes in habitat), and the responses may be immediate or delayed. In turn, many biological processes, especially microbial processes, can influence surface water acid-base chemistry. Thus, chemical and biological changes are intricately linked and complex, with extensive feedbacks.

Research on the effects of acid deposition and acidification on aquatic biota has been ongoing in Europe and North America for over the last 15 years, and many comprehensive reviews have been published (e.g., Leivestad and Muniz 1976; Haines 1981; Dillon et al. 1984; Havas 1986; J.P. Baker et al. 1990; Schindler 1988). Even though research has been conducted in several areas and it has concerned many different organisms, it has not been exhaustive and there are gaps in our understanding of the response of organisms to acidic conditions. These gaps often have occurred because funding has focused on chemical mechanisms and modeling response of systems rather than in making resource inventories or resolving uncertainties in biological responses to acidification. In addition, research has been conducted piecemeal, and coordinated regional studies at varying scales have been few.

*Acidification of Freshwater Ecosystems: Implications for the Future*
Edited by C.E.W. Steinberg and R.F. Wright 

Our objective was not to detail current understanding or to provide a complete analysis of published information but rather to summarize some aspects of what is known, consider emerging research findings, to highlight remaining uncertainties, and evaluate needs for future research. We organized our consideration of the effects of acidification on biota into six major topic areas: (1) mechanisms of direct effects on aquatic organisms; (2) mechanisms of indirect effects; (3) biological processes that affect chemistry; (4) extent, directions, and rates of change in biological communities; (5) identification of biologically relevant chemical reference points, and (6) improved approaches for evaluating biological effects of acidification.

## PHYSIOLOGICAL MECHANISMS OF DIRECT EFFECTS ON AQUATIC ORGANISMS

The biological effects of acidification have been studied using a variety of approaches including laboratory and field bioassays, whole-ecosystem and smaller-scale field experiments, field surveys, and analyses of temporal trends by monitoring or paleolimnological analyses. By drawing on this diverse data base, we can define with confidence the *basic* changes in biological organisms and communities in response to increasing surface water acidity caused by acid deposition. The effects listed could have resulted from direct or indirect effects as detailed below. We have summarized many of the observed changes in Table 19.1, without attempting to provide a complete review for all groups of organisms or comprehensive citations for any one country (see also Schindler, this volume).

In waters containing low concentrations of Ca, pH and Al are the two major chemical variables of concern. In general, acidic aluminum-rich water appears to affect fish and invertebrates through similar physiological processes. For gill-breathing animals, the primary target organ for the various physiological effects of acidic water is the gill, due to interruption of gas and ion transport across respiratory membranes. Five major functions, all performed by the gill organ, are affected: (1) ion regulation; (2) osmoregulation; (3) acid-base balance; (4) N-excretion; and (5) respiration. Except for the influence on respiration, all of the other effects on gills lead to impairment of osmoregulation, which is the primary cause of mortality.

When aluminum concentrations are elevated, aluminum can play a major, and often, overriding role in toxicity. In these environments, and especially in clear, low calcium waters ($< 1$ μeq $L^{-1}$ Ca), understanding the toxic mode of aluminum is key to understanding the gross effects on organisms and populations. Dissolved organic carbon (DOC) mediates the toxic influence of Al, apparently due to the complexation of Al. As a consequence, DOC is an extremely important influence on the survival of organisms exposed to acidic waters. Silica can also form complexes with Al (see Rosseland and Staurnes, this volume, and Böhmer and Rahmann 1990).

Evidence for uptake mechanisms of Al through the gill membrane has been produced as well as new theories for specific sites of direct binding of Al to cell structures. These theories might link the enzymatic impairments observed and explain fundamental impacts on calcium homeostasis. In our opinion, the trans-boundary layer between the mucus and the gill epithelial membrane is of primary importance. This layer is both chemically and physically dissimilar to the water outside, and the organism can influence the gross chemistry through its acid-base regulation. A direct effect of $H^+$ and Al from the environment is the stripping of Ca bound to intercellular junctions in the membrane, thus changing Na and Cl transport. Any mechanism reducing this primary effect on permeability and ion loss will be of vital importance. In this context, secretion of mucus at low environmental pH is a highly appropriate physiological response. For example, it has been shown that swim-up brown trout fry, which have a high density of epidermal mucus cells, are more tolerant to low pH than are juveniles, which have fewer mucus-producing cells (Segner et al. 1988).

When we compare species of invertebrates or fish that are closely related taxonomically, but differ in their acid sensitivity, we find that the acid-sensitive organism loses Na more quickly than the acid-tolerant organism.The extreme differences in ion-regulation ability (Na transport) of acid-tolerant species may reflect a difference in origin, whether marine or fresh water. The functional properties of gills or activity of different enzyme systems associated with gills may be a useful way to examine geographic distributions and classify species for sensitivity to acids. Because the relationship between both acid and aluminum sensitivity and ion balance, represented by net sodium loss, appears to be robust for many species and groups (crustaceans, insects, fishes), we may be able to predict the acid sensitivity of different species, strains, and life-history stages based on an understanding of ion regulation under normal (non-stressful) conditions. For example, under normal conditions, organisms that have a high exchange rate of Na (more than 10% of total body Na exchanged per hour) may be more sensitive to acidic episodes than those having a lower exchange rate. Similarly, during certain life-history stages (molting, smoltification, etc.) changes in membrane permeability, water uptake, and sodium regulation may protect from or predispose the organisms to acid stress.

We gave some consideration to the question of whether there are any meaningful impacts on metabolism at sublethal levels of exposure. Based on the evidence examined from several studies, it was suggested that any sublethal effects would be expressed more through modification of behavior than by reductions in growth or declines in the production of eggs. At sublethal exposures, sensory recognition may be weakened, which could result in decreased avoidance of critical water quality episodes. Some studies indicate reduced appetite (Muniz and Leivestad 1979) and evacuation rate of food (Åtland and Barlaup 1991). As indicated by a majority of studies, however, the metabolic changes caused by sublethal acidic stress are not expressed by reductions in growth or declines in the production of eggs (Jacobsen 1977; Hill et al. 1988; Mount et al. 1988; McCormick et al. 1989). The absence of consistent consequences of metabolic changes on growth and fecundity may be explained by the acclimation of fish to acidic stress, even though this is the subject of

**Table 19.1** Evidence for effects of acidification on aquatic biota.

| Observed Effect | Type of Study | Selected References |
|---|---|---|
| Fish population loss | Temporal trends | Schofield 1976; Muniz 1981; Sevaldrud et al. 1980; Rosseland et al. 1980, 1986; Rosseland 1986; Sevaldrud & Skogheim 1986; Harriman et al. 1987; Beamish et al. 1975; Henriksen et al. 1989; Berger et al. 1992 |
| | Field experiments | Schindler et al. 1985; Mills et al. 1987 |
| Reduction in species richness and loss of acid-sensitive species: | | |
| Algae | Field surveys | Almer et al. 1974; Siegfried et al. 1989 |
| | Paleolimnology | Charles 1984; Battarbee et al. 1985 |
| | Field experiments | Yan & Stokes 1978; Schindler et al. 1991 |
| Macrophytes | Field surveys | Grahn 1977; Jackson & Charles 1988 |
| Zooplankton | Field surveys | Hobæck & Raddum 1980; Confer et al. 1983 |
| | Field experiments | Malley et al. 1982; Schindler et al. 1991 |
| Benthic invertebrates | Temporal trends | Mossberg 1979; Hall & Ide 1987 |
| | Field surveys | Økland 1980; Stoner et al. 1984 |
| | Field experiments | Davies 1989; Weatherley et al. 1988; Hopkins et al. 1989 |
| Loss or inhibition of microbial activities: | | |
| Inhibition of decomposition | Whole-lake experiments | Kelly et al. 1984 |
| | Lab studies | McKinley & Vestal 1982 |
| Inhibition of nitrification | Whole-lake experiments | Rudd et al. 1988 |
| | Lab studies | Suzuki et al. 1974 |
| Development of benthic algal mats/clouds | Field surveys | Lazarek 1985; Howell et al. 1990 |
| | Field experiments | Schindler et al. 1991 |
| *Sphagnum* invasion into profundal zone | Field observation | Melzer & Rothmeyer 1983 |
| Reduced bird density, reproductive success, survival | Field surveys | Ormerod & Tyler 1987; Nyholm 1981; McNicol et al. 1987 |

*(continues)*

**Table 19.1** *continued*

| Observed Effect | Type of Study | Selected References |
|---|---|---|
| Direct mortality: | | |
| Fish | Field observations (fish kills) | Jensen & Snekvik 1972; Leivestad & Muniz 1976; Skogheim et al. 1984; Hesthagen 1986 |
| | Lab and field bioassays | Grande et al. 1978; Leivestad et al. 1980; Baker & Schofield 1982; Brown 1983; Fivelstad & Leivestad 1984; Rosseland & Skogheim 1984; Rosseland et al. 1992 |
| Amphibians | Lab and field bioassays | Freda & Dunson 1985; Andren et al. 1988 |
| | Lab bioassays | Böhmer & Rahmann 1990, 1992 |
| Zooplankton | Lab bioassays | Havas 1985; Price & Swift 1985 |
| Benthic invertebrates | Lab bioassays | France & Stokes 1987; Matthias 1983 |
| Disturbance of osmoregulation: | | |
| Fish | Lab bioassays | see Rosseland & Staurnes, this volume |
| Amphibians | Lab bioassays | Freda & Dunson 1984 |
| Zooplankton | Lab bioassays | Havas & Likens 1985 |

debate (Orr et al. 1986; Audet and Wood 1988). A direct effect on energy balance is possible and it should be given attention in future work.

Other sublethal impacts have been observed. Profound endocrine changes (prolactin, cortisol, urotensin II, thyroid hormones), alterations of carbohydrate metabolism, and inhibition of protein synthesis have been observed both in adult and larval acid-exposed fish (Brown et al. 1984; Tam et al. 1988; Segner et al. 1991). We know from several studies that hormones play an important role in tolerance to acidic waters (see Rosseland and Staurnes, this volume). However, the potential negative effects of prolonged, elevated cortisol due to chronic exposure to acidic aluminum-rich water have not been examined (Witters et al. 1991). Elevated cortisol levels, which are a primary stress response, depress the immune system of fish, thereby increasing the risk of pathogenic infections in acid-exposed fish populations (Mazeaud and Mazeaud 1981). Chemical stress can weaken other biota as well and make them more susceptible to infections and parasitism. Such interactions have been observed for *Daphnia* (Havas and Hutchinson 1983), but have not been explored extensively. Studies on the effects of low pH and Al in predisposing biota to parasitism and disease are clearly needed.

Behavioral responses of organisms were considered to be important determinants of survival of organisms exposed to deleterious water quality. For example, we know from several studies and observations in nature that fish seek refugia based on better water quality, often in spite of large population densities or shortages of food at these sites (Muniz et al. 1978; Rosseland et al. 1986; Rosseland and Staurnes, this volume). The driving force for the active search for refugia is probably a mixture of processes disturbing the homeostasis of the fish; however, the sensory organs related to olfaction or taste must play an important role.

The existence of refugia in a certain habitat might be crucial for survival, both under chronic or fluctuating acidic conditions. Lakes will normally provide more refugia for fish than streams, although in the latter environment fish are more dependent upon them for protection during episodes and chemical conditions that represent mixing zones of two different types of water quality. The interstitial and hyporheic environment in streams provide excellent refugia for invertebrates. Organisms also can influence their chemical environment. Large static schools of fish and invertebrates in low flow environments may enhance water quality and create their own refugia in acidic environments. In enclosures and static experiments, fish and invertebrates increase the pH within their experimental chambers by excreting $NH_3$ and organic compounds (Havas 1987). The degree to which this happens in nature and its biological significance are unknown.

When water of different chemistry mixes, very unstable chemical conditions can result. Associated with these mixing zones have been very acute toxic conditions for fish. The detrimental effects of mixing zone waters are caused by the ongoing transformation processes changing aluminum from low to high molecular weight species, which leads to precipitation onto the gills of fish (Rosseland et al. 1992). As the kinetics of these processes are highly temperature dependent, the toxic zones in streams due to such conditions will have a much larger extension during periods of low temperature (autumn to spring). Mixing zones and temperature effects need much more extensive study because they represent critical conditions for fish populations.

Behavioral responses of fish to acidic episodes are not well understood. Aluminum has been found to reduce or block olfaction in fish at acute exposures in laboratory experiments (Klaprat et al. 1988). Avoidance and/or escape reactions are important resistance mechanisms, especially under unstable chemical conditions such as are found during episodes or in mixing zones. Impairments of the ability to seek refugia under such conditions might therefore be lethal. An olfactory organ able to detect changes in water chemistry, but still tolerant of the negative effects of pH and aluminum, would be an important adaptation.

The disruption of olfaction has the potential for impairing the inprinting mechanism in seaward-migrating salmonid smolts, thus increasing their chance of straying on return from the ocean. The effect of low pH on feeding and food search behavior has been examined, but so far no negative effects have been observed. No direct effects on spawning behavior have been described. Nonetheless, because of the great economic significance of salmon, additional research is needed on the behavior of salmon stocks in acidified areas.

Good relationships have been developed to predict the distribution of fish populations as a function of pH, Ca, and Al (Gunn, this volume). Recent research in Norway also indicates that several approaches can be used to predict fish status from water chemistry (Cosby et al., this volume). For example, acid-neutralizing capacity (ANC), defined by charge balance, is an excellent predictor of the conditions when fish populations are healthy and also when populations become extinct (Henriksen et al. 1992; Lien et al. 1992). The fish do not respond directly to changes in ANC, however, the charge balance definition clearly represents the chemical conditions determining response. While the relationships are fairly tight, within the range of ANC –20 to +20 $\mu eq\ L^{-1}$, there is variation in the status of individual species of fish. Variations in DOC and the presence of refugia in certain habitats produce variability in the success of fish populations near critical levels of pH, Ca, and Al. One must be aware that the relationships summarize the influence of many factors and do not represent the chemistry of inlet brooks and streams and may underestimate the critical levels of water chemistry. Defining relationships between fish populations and water chemistry requires further expansion, especially when modeling changes related to recovery.

Post-episodic effects on fish populations have not been examined. A combination of primary sublethal physiological stresses (e.g., osmoregulatory and respiratory) and a secondary reduced immunity caused by a response to cortisol might lead to increased mortality in populations over longer periods. Longer-term effects might be substantially greater than those observed during and shortly after an episode or exposure to lethal mixing zones.

Much of our discussion centered on mechanisms of fish response, while recognizing that some of the physiological responses of invertebrates and other vertebrates are similar as noted above. Two major areas for further research involve plants and bacteria. We know more about the effects of acid on bacterial processes than for the responses of individual genera or species. Bacterial processes are affected as follows: decomposition of organic matter is not affected significantly until the pH drops below 5 (McKinley and Vestal 1982; Kelly et al. 1984); denitrification is not affected until 4.5, the lowest pH at which it has been measured in lake sediments (Rudd et al. 1986); sulfate reduction occurs at pH 4.5, but because of changes in sulfur oxidation it is difficult to determine whether it is inhibited or not (Herlihy and Mills 1985; Rudd et al. 1986); nitrification is inhibited in some lakes at pH 5.4–5.6 (Rudd et al. 1988), but not in others (Gerletti and Provini 1978).

There have been few taxonomic studies done on the effects of acidification on bacteria, fungi, yeasts, and protozoans. What is known is that there are more acid-tolerant fungal species than bacterial species and there is an apparent shift from bacterial to fungal populations as some lakes acidify. The critical pH for microbial acidity appears to be between 5.0 and 5.5. Above this range abundance and the rates of decomposition and nitrification are relatively unaffected. Microbial populations also respond to concentrations of organic matter. Microbial populations are more abundant in acidic, humic lakes than in acidic, clearwater lakes presumably due to the higher concentrations of DOC.

The interaction of Al and $H^+$ on microbes is poorly understood, but it may be important. Some of the problems with previous experiments include: (a) incompletely characterized substrates, making the determination of Al availability virtually impossible; (b) pH was either not provided or was unrealistically low (< 3) or high (> 7); and (c) unrealistically high Al concentrations.

The mechanisms of direct effects on algae and aquatic macrophytes are not well known. These groups provide the primary food base for consumers and they can influence elemental cycles. The richness of macrophyte taxa is lower in lower pH lakes (Jackson and Charles 1988). This empirical relation suggests the possible direct response of macrophytes to declining pH. Likewise, shifts in the composition of phytoplankton communities have been observed in many studies (see Table 19.1). There is general agreement that acidic lakes have lower species diversity than neutral or slightly alkaline lakes and that one or two dominant species in acidic lakes may make up as much as 80% or more of the total community biomass. The effects of acidification on phytoplankton biomass and productivity are inconsistent because acid-tolerant organisms appear to exploit vacant niches, and nutrient availability and zooplankton herbivory can mask the effect of pH. There appears to be a shift from planktonic to benthic or periphytic forms in acidic lakes.

The influence of low pH and elevated Al on algae and aquatic macrophytes has been reviewed (Havas 1986; Farmer 1990) but is poorly understood. Some studies show that Al accumulates in the nucleolus of cells and may therefore interfere with cell division. Aluminum-tolerant species, such as *Mougeotia*, appear to have a mechanism to prevent Al uptake. Critical areas identified for future research are (a) metal toxicity, particularly to Al; (b) the role of phosphatase activity in determining the survival of species exposed to low pH conditions; (c) the direct effect of hydrogen ions on algae; and (d) changes in the ratio of nutrients and forms of carbon available for photosynthesis.

Calcium concentrations may be more important than $H^+$ concentrations in influencing the distribution of macrophytes, particularly in the low Ca waters of Scandinavia. In the acidic lakes of Europe, there has been a general shift from *Lobelia dortmanna* and *Littorella uniflora* to *Sphagnum* spp. and *Juncus bulbosa*, presumably because of their ability to use $CO_2$ instead of $HCO_3^-$ for photosynthesis. *Sphagnum* also has a high cation exchange capacity and can efficiently remove divalent cations from the water.

The effect of Al on macrophytes is incompletely understood. A few studies have shown that mosses have a higher Al concentration than monocots, which have higher concentrations than dicots; however, in none of the species is there a clear correlation between Al in water and Al in plant tissue. Some species, especially mosses, accumulate surprisingly large concentrations of Al. Some species sequester Al in cell walls with no apparent toxicity. However, Al is known to affect terrestrial plants and affect the uptake of Ca and P. Studies are required to determine: (a) under what conditions Al is harmful to macrophytes; (b) whether macrophytes influence the water quality experienced by other organisms; and (c) whether Al interferes with Ca and P uptake by macrophytes.

## INDIRECT EFFECTS OF ACIDIFICATION ON AQUATIC ORGANISMS

Acidification can also have significant indirect effects on organisms because of the importance of ecosystem linkages. We discussed these indirect effects separately for lakes and streams due to the differences in the physical and chemical environments and the composition of biological communities.

### Lakes

As systems are stressed, sensitive species are lost and more tolerant species may fill unoccupied niches (Odum 1985; Schindler 1990). In general, it has been found that sensitive parts of the community are more responsive than the whole community. Acid-sensitive species are lost due to direct effects of low pH and related chemical conditions on organisms. Additional effects of acidification on lacustrine communities appear to be indirect. For example, increased water transparency could be a result of the loss of DOC, which would allow phytoplankton growth at greater depth. Changes in communities can also result from disruption of the trophic cascade. As the result of elimination of a top predator or other keystone organism, changes in the distribution and abundance of species can occur (Schindler, this volume). However, our detailed knowledge of these effects stems from a few systems, most notably Lake Gårdsjön in Sweden, Lake Hovvatn in Norway, and experimental whole-lake manipulations (Lakes 223 and 302) in Ontario and Little Rock Lake in Wisconsin; it is unclear how widely results from these systems can be extrapolated.

The changes in DOC with acidification are still unclear and the paleolimnological record shows inconsistent responses of DOC to acidification (Cumming et al. 1992; Hemond, this volume). This situation is due in part to the fact that some experimental treatments have manipulated a drainage lake, but not the watershed, while other treatments have dealt with seepage lakes. Therefore, any potential changes in the flux or forms of DOC leached from the watershed to surface waters are not known. Current research underway at the HUMEX project site in Norway should help evaluate the degree to which acidification alters the flux of DOC from watersheds.

In one experimental lake (Lake 223), increased transparency appeared to open or expand a new "niche" for phytoplankton below the thermocline. This expansion in turn corresponded with increased concentrations of *Daphnia* at subthermocline depths

(Kettle et al. 1987). In general, higher transparency, as might occur in acidic lakes, is known to cause increased predation pressure in zooplankton, causing changes in the patterns of vertical migration. The direct linkage between increased transparency and changes in the distribution of phyto- or zooplankton has not been studied at other sites. Increases in light penetration may be especially significant if they result in a change in stratification pattern that shrinks the volume of hypolimnion available for coldwater fishes.

Changes in water clarity and littoral communities, due to the appearance of algal clouds or mats, have also probably altered habitats for aquatic organisms. In Lake 223, development of littoral mats on the spawning grounds of lake trout caused them to spawn at less favorable sites (Mills et al. 1987). Littoral algal mats are also known to disrupt habitats for sulfate-reducing bacteria, thereby reducing internal alkalinity generation (Kelly et al., in review). The significance of algal mats in disrupting other chemical cycles is unknown because detailed studies have not been made.

Because aquatic macrophytes are important structural and functional components of littoral communities, significant biotic changes in littoral and possibly pelagic regions could result from the loss of or change in macrophyte communities. Macrophytes provide a refuge for small fish and habitat for invertebrates, and they can influence nutrient cycling from sediments. Do macrophyte communities respond directly to pH or are they influenced more by the changing environment represented by metaphytic algal clouds and algal mats? In addition, the influence of changes in macrophyte communities on other organisms is not known.

Most acid-sensitive lakes are phosphorus-limited. Any changes in phosphorus loading from watersheds or alterations of P cycling within lakes could result in changes in productivity and species composition. However, we have little information to address the important question of P cycling (Schindler, this volume). Phosphorus concentrations are extremely low and at the detection limit in many lakes in southern Norway (Wright et al. 1977). Even small reductions in phosphorus would have major impacts on productivity; however, it is not known whether concentrations have declined. We do not have long-term data bases available to address the problem. Effective monitoring of changes in phosphorus would require special procedures to stabilize the samples; however, some effort should be directed to resolving this question.

Elimination of predatory fishes and their replacement by corixids or *Chaoborus* appear to have changed the zooplankton communities of Lake Gårdsjön from domination by small species to domination by large ones due to changes in predation pressure (Stenson and Eriksson 1989). Studies on trophic interactions elsewhere indicate that this change would have an important effect on the composition of phytoplankton; however, the productivity of acidic lakes is generally very low and such a response has not been observed. The change from piscine to insect predators also would be expected to cause decreased abundance of rotifers in acidified lakes as it has elsewhere (Neill 1984), but this linkage has not been studied.

Schindler et al. (1985) described how elimination of acid-sensitive organisms from the food chain of Lake 223 caused the decline of lake trout at pH values higher than those directly toxic to trout. Such effects have not been documented at other sites; however, only three or four lakes have been studied comprehensively enough for such phenomena to be detected, so that it is not known how common such reductions in fish populations due to food chain effects might be.

The food web of many lakes in acid-sensitive areas is often very simple. For example, higher elevation lakes in southern Norway often have only one or two species of fish and a restricted set of benthic invertebrates. In such simple systems, we might expect indirect effects to be most important and a potential limitation to the survival of fish populations. However, hundreds of fish populations examined in Norway have not indicated food shortage as being critical for fish survival and development (Rosseland et al. 1980; Rosseland et al. 1986). Such a response may occur only in situations where the primary food organism is more sensitive to low pH than is the fish predator; even in those cases, fish may switch to other prey items without a major impact on growth. We conclude that food chain effects are not likely to be a major influence restricting fish populations in most acidified waters.

Similarly, the possible role of changes in littoral grazers in causing the development of algal mats has not been studied. Studies in Lake 302 South show that the filamentous Zygnematales, which prevail in the littoral zones of many acidified North American lakes, are better able to photosynthesize at low pH than the original diatom and bluegreen communities (Turner et al. 1991); however, the possible role of eliminating the crustaceous *Hyalella azteca, Orconectes virilis*, and other large grazers has not been considered. So far no study has calculated the extent to which acid-induced changes in littoral and pelagic communities have changed the overall production of lakes.

## Streams

There is clear evidence of direct physiological effects by acid-related factors on stream organisms, but some organisms are affected indirectly through ecological processes. Stream ecologists have separated such effects into those operating in "top-down" and "bottom-up" directions.

The scarcity of insectivorous vertebrates, such as cottid and salmonid fish or cinclid birds, has been proposed as a feature that permits the proliferation of large-bodied invertebrate predators into acidic streams. Two mechanisms through which such effects might occur include the removal of size-selective predation and freedom from diffuse competition for their smaller invertebrate prey. Schofield et al. (1988) have tested, and supported, the predation hypothesis experimentally in streams of the Ashdown Forest, England. Other examples of such effects have yet to be suggested and tested, in part because foodweb structures in streams are still poorly quantified.

Changes in the trophic base of streams with reduced pH include qualitative changes in epilithic algal communities, from diatoms to chlorophytes, and a change in diatom species composition. In addition, changes in the availability of energy inputs to streams from allochthonous litter occur as altered hyphomycete communities lead to reduced rates of decomposition. This altered energy base is paralleled in the invertebrate community, because species present in acidic streams are most commonly leaf shredders or litter collectors. Filter-feeders and epilithic grazers, such as many mayflies, are absent at low pH but are abundant at a higher pH. Two hypotheses proposed to explain these invertebrate patterns are: (a) they result from altered trophic conditions; and (b) tolerance of acidic conditions varies systematically between feeding groups (perhaps as an evolutionary consequence of the conditions under which their favored trophic conditions occur).

Dobson and Hildrew (1992) showed how the densities of shredders increase in acidic streams when the input and standing crop of leaf litter are artificially increased. Less evidence exists for the trophic exclusion of grazers from acidic streams. In North America, Rosemond et al. (1992) recently showed that some epilithic grazers were absent from acidic streams despite abundant food supplies. This response would suggest that physiological factors, and not trophic interactions, are responsible for their scarcity at low pH. One corrolary to this feature is a hitherto unexplored possibility that grazers in streams might exert a top-down effect on algal communities once stream chemistry is suitable for their survival. For example, some of the diatoms characteristic of circumneutral streams (e.g., *Achnanthes minutissima*) have small cells and are known to proliferate under increased grazing pressure.

Bottom-up effects of acidification on fish in streams are unlikely, in part because reductions in invertebrate production at low pH are not pronounced. Moreover, allochthonous inputs of terrestrial insects to stream systems can more than equal instream production. Changing invertebrate quality does, however, have consequences for a riparian bird, the dipper *Cinclus cinclus*. Not only is the scarcity of preferred prey accompanied by reduced density and productivity, but the scarcity of calcium-rich prey is accompanied by reduced plasma calcium concentrations in birds attempting to breed on acidic streams. This, in turn, is accompanied by reduced eggshell thickness, egg size, and clutch size. Such an example of physiological effects arising through an indirect trophic pathway is an unusual, but infrequently sought, consequence of acidification (Tyler and Ormerod 1992).

## BIOLOGICAL PROCESSES THAT AFFECT CHEMISTRY

There are a number of biological processes that make the chemical composition of water in streams and lakes different from what it would be if only geochemical processes were operating. Even processes that we think of as geochemical, such as weathering, may be affected strongly by biological processes. We agreed that the state of knowledge of aquatic biological processes that affect water chemistry is at a state where we can predict certain responses, but that there are many uncertainties in the role and significance of processes in terrestrial systems.

For a drainage system, plant and microbial communities in the watershed represent the first site of contact with acidic inputs from the atmosphere. The collection efficiency of different tree types (e.g., conifers vs. deciduous) may or may not be considered a "biological" effect, as it is probably determined by the physical dimensions of the trees. Nevertheless, the existence of different types of vegetation and their characteristics must be considered in understanding effects on receiving waters. This understanding is particularly important in areas where forests are managed and where decisions might be made to change management practices and choices of tree species (Harriman et al., this volume).

In upland soils, the degree to which sulfate is retained in the watershed is primarily determined by regional soil adsorption characteristics rather than by $SO_4^{2-}$ reduction or plant uptake. Wetlands, however, take up $SO_4^{2-}$ efficiently. In addition, wetter soils that are not strictly wetlands might retain sulfate more efficiently than drier soils. Therefore, as soil moisture changes in response to climatic factors, $SO_4^{2-}$ retention could be affected.

We know that nitrification, nitrogen fixation, N mineralization, denitrification, and plant uptake of nitrate and ammonia all occur; however, the rates and relative importance of each need to be quantified under different loadings and for different types of soils and plant communities. Increased yields of nitrate from watersheds have been known for many areas (Henriksen and Brakke 1988). When inputs of nitrate exceed the uptake of primary producers, nitrate contributes to acidity, and denitrification is the only removal process (Kelly et al. 1990). An important question relevant to both acidification and global warming is the effect of increased atmospheric N inputs on $N_2O$ releases. $N_2O$ is an intermediate product of both denitrification and nitrification and its production affects acidity within soils and receiving waters. In addition, $N_2O$ is more radiatively active than $CO_2$ and even more than $CH_4$.

Dissimilatory sulfate reduction can be a major contributor to lake ANC. Internal alkalinity generation by denitrification is much faster than that by sulfate reduction, but at present, sulfate reduction is more important. The magnitude of each is expected to change as the flux of nitrate and sulfate varies. Within lakes, models that predict sulfate and nitrate removal, and associated acid neutralization, are based on the consistent relationships seen between concentrations and the activities of the sulfate reducing and denitrifying bacteria (Baker et al. 1986; Kelly et al. 1987, 1990). These models work in a variety of acidified lakes. An important exception recently seen is the change in net sulfate reduction that occurred in an experimental lake when the pH reached 4.5. These results suggest that lakes at this very low pH may have much less internal alkalinity generation by sulfate reduction and thus may be particularly slow to recover as inputs are decreased. This change was apparently caused by the influence of metaphytic algae that developed at pH 5 and below. Because nitrate and sulfate are also taken up by algae, and because algal sedimentation is the major source of organic C for the denitrifying and sulfate-reducing bacteria, the addition of phosphorus must increase internal alkalinity generation. Small additions of P might reduce nitrate in lakes and significantly assist recovery. These additions should not be detrimental to the lake in the long term because experiments have shown that the effect of nutrient additions is short-lived after they are ceased (Shearer et al. 1987).

As discussed earlier, DOC is also important in affecting not only chemical properties (acidity from organic acids) but also physical properties, such as light penetration and depth of the thermocline, that affect biota. It also affects metal transformations, which are important in toxicity to organisms or to fish consumers (accumulation of mercury). The production of DOC is carried out by organisms in watersheds (root excretions, litter decomposition), and in lakes (algal excretions, decomposition, and algal remains). Removal of DOC in lakes is by bacteria, photochemical reactions, and flocculation. Differences in water retention time can significantly affect the removal of DOC. We need to know much more about the biological processes that determine the amount and character of DOC exported. The whole-ecosystem experiments for the RAIN and HUMEX projects in Norway are providing information on the changes in DOC exported following manipulations of acid loading (Wright et al. 1993).

The passage of material through the guts of important aquatic organisms, such as shredders, zooplankton, and fish, may also influence water chemistry. For example, it has been observed that introduction or removal of fish affects the rate at which nutrients turn over. Changes in species composition may thus affect phosphorus and nitrogen concentrations as well as other chemicals. This factor has not been examined in detail.

An especially important metal transformation carried out by microorganisms is mercury methylation. Inorganic mercury is not well retained by organisms; however, methyl mercury, especially the monomethyl form, is retained efficiently. In addition, it is much more toxic to fish and to consumers of fish (fish-eating birds and mammals, including humans) than is inorganic mercury. Conditions of elevated mercury in aquatic biota are especially prevalent in areas with low ANC lakes and are related to acidification, because lower pH and lower DOC are known to enhance mercury methylation (Xun et al. 1987; Miskimmin et al. 1992).

Mercury inputs from watersheds are positively correlated with DOC inputs (Mierle and Ingram 1991; Bodaly et al. 1992). Therefore, terrestrial processes associated with canopy capture and incorporation of mercury into organic litter may be significant to mercury loading to aquatic systems. Changes in DOC export from watersheds, which might be influenced by climatic change or by acidic precipitation, should also affect mercury concentrations in lakes and mercury methylation. Both atmospheric and terrestrial processes of methylation require further quantification, especially those processes mediated by terrestrial microorganisms. In addition, wetlands and peatlands are important sites for methylation and should receive further study.

An additional factor, the creation of impoundments, can also represent a situation where mercury methylation is greatly enhanced. In these situations, bacteria have abundant carbon substrates available (in dead vegetation and flooded peat and soils) and methylation is stimulated (Johnston et al. 1991). Fish from reservoirs have been shown to have higher concentrations of mercury than fish from other surface waters in the same area. This conclusion applies to many different situations and regions. Flooding of wetland or peatland areas would be expected to result in greatly enhanced incorporation of mercury into fish and waterfowl. Much greater attention and research should be devoted to resolving the interaction of land-use practices, reservoir creation, and operation; because low pH enhances methylation, reservoirs in areas with low ANC waters and acidic precipitation might be particular problems.

Elevated mercury in fish tissue is an increasing problem in many areas, including the U.S., Canada, and Scandinavia. Elevated mercury has also been reported for birds, especially loons. Mercury accumulation in fish is a significant issue for human health, and human consumption advisories have been issued for fish from lakes in Minnesota and Wisconsin in the U.S. and from other countries. Some models have been developed to predict mercury concentrations in fish; however, these models apply mainly to clearwater, seepage lakes and require further development to extend predictability beyond the range of calibrated conditions, especially in areas where watershed sources of DOC and mercury are significant. The issue of elevated mercury also extends to one of concern for the health and integrity of aquatic biota. We do not know the extent of direct toxicity of mercury to aquatic organisms or potential sublethal effects on behavior. Because of the significance of mercury to aquatic organisms and their consumers, greatly accelerated research is warranted.

In general, we have much less understanding of the extent to which biological processes influence chemistry within streams than within lakes. The scarcity of soft sediments at high flow may provide less potential for the microbial generation of alkalinity, however, this topic requires assessment and quantification. Nevertheless, reach-scale processes in streams are known to have effects on chemistry. Epilithic biofilms and bryophytes are involved in the flux of metals through streams, sequestering aluminum at higher pH (Caines et al. 1985), and releasing it as pH declines (Henriksen et al. 1988). Moreover, pulse experiments and artificial enrichments indicate that algal production in many temperate streams is phosphate, rather than nitrate, limited (D'Angelo et al. 1991). These observations may mean that streams, like lakes, will not remove increased nitrate except by denitrification, but this process has not been examined.

## EXTENT, DIRECTIONS, AND RATES OF BIOLOGICAL CHANGE

In our analysis of biological changes associated with acidification, we discussed the procedures that could be used to estimate damage to biological populations and considered results from several areas. We also discussed the magnitude and directions of changes in communities that have occurred during acidification and could occur during recovery. Several lines of evidence are required to evaluate fully the extent of changes in biological populations and to estimate possible recoveries in communities. Our discussions focused on five issues needing further attention and research that relate to the extent, directions, and rates of change in response to acidification:

1. the need for additional monitoring and modeling for regional-scale assessments;
2. opportunities for new insights on biological responses through paleolimnology;

3. biological recovery, in particular evidence from whole-system manipulations that can improve our understanding of how biological communities are likely to respond to reductions in acid loadings;
4. potential differences in the biological effects of acidification in warmwater and coldwater systems, and potential interactions with climatic change; and
5. effects of acidification on genetic diversity.

## The Need for Monitoring and Model Development for Regional Assessments

Trends in biotic communities through time can be assessed best through integrated long-term monitoring programs. Such monitoring programs were not in place prior to acidification, which made evaluation of changes in communities much more difficult. Moreover, many programs established monitoring of chemical conditions but failed to conduct any significant ongoing monitoring of biological status. As a result, with some exceptions, changes in biological communities could not be evaluated during the course of major research efforts on acidification. It is essential that existing long-term monitoring of biological communities be continued and expanded, not only to evaluate effects of acid deposition but to monitor trends in overall ecosystem status. Examples of new and effective integrated monitoring programs include the European Integrated Watershed Monitoring Program, the Norwegian Monitoring Programme for Long-Range Transported Air Pollutants, the Environmental Monitoring and Assessment Program (EMAP) in the U.S., the UN ECE Integrated Monitoring of Small Catchments, and the UN ECE ALPE-1 program. These programs need to be continued and expanded.

Greater emphasis should be placed on developing statistically valid regional evaluations and monitoring. A major strength of the lake surveys conducted in the U.S. and Finland was their sound statistical base (Landers et al. 1988; Forsius et al. 1990). Unfortunately, parallel chemical and biological information was not accomplished for all of the lakes. The lack of adequate design of other surveys has hampered evaluations of resource damage in many areas and was a major weakness of some programs. Only in a few cases has chemical and biological monitoring and evaluation been done in a coordinated fashion, with the Norwegian Thousand Lake Survey being an excellent example (Henriksen et al. 1988, 1989). Data on fish, invertebrates, and water chemistry are being used in a coordinated way to evaluate and monitor chemical and biological changes.

Monitoring and regional surveys provide a record of current status, an assessment of effects to date, and a means to evaluate past trends. To contribute to informed decisions regarding emission controls, predictions of future trends (effects and recovery) are also required. Such predictions must be based on a clear understanding of the full range of biological responses to acidification, and they must be based on conditions and biological populations representative of each area. The predictions also must consider recovery in terms of the effective reinvasion of species to systems as chemical conditions improve.

Some models have been developed to aid in projecting future trends in aquatic biota, in response to changes in chemistry (Schindler et al. 1989; Reckhow et al. 1987; Weatherly 1988; Minns et al. 1990, 1992; J.P. Baker et al. 1990; Cosby et al., this volume). Most of the models consider only the effects on fish, while other biological populations are ignored, and none of the models treat ecosystem processes. More research is needed on modeling biological responses and in making projections for other components of the biological community other than fish. Many other species respond to declining pH or elevated concentrations of Al before fish are lost from a system. As a consequence, resource damage estimates based solely on fish underestimate the total loss of populations. Further modeling should also be extended to include important ecosystem processes.

Some linked models have been developed. In Wales, for example, empirical models have been developed to relate pH, Al, Ca, and ANC to macroflora, invertebrates, fish, and riverine bird distribution (Ormerod et al. 1988). In Norway, empirical models have been used to estimate the recovery of fish populations (Henriksen et al. 1988, 1989), and fish population response has been predicted using MAGIC model estimates of changes in chemistry (Cosby et al., this volume).

All of the models developed to date rely on the assumption that biological response is immediate and in steady-state with any changes in chemistry. This assumption may be invalid because of delays in responses of organisms to changing chemical environments. For example, although the pH of lakes in southern Norway changed little from the mid-1970s to 1986 (Henriksen et al. 1988), there was an increasing number of lakes that had lost fish populations over the same period (Henriksen et al. 1989). These time lags in fish population response have been well documented (Berger et al. 1992). Delayed responses of other organisms may also occur (Webster et al. 1992). Such time lags become even more important in predicting recoveries because in addition to a delayed response to chemical changes, there will be lags due to the abilities of organisms for reinvasion. Moreover, some organisms or lake types may not recover completely because of barriers to reintroduction.

Making regional resource damage estimates is an important step. However, to date there has been an unfortunate tendency to infer that if a lake is not acidic (e.g., defined by Gran ANC $< 0$ or pH $< 5.3$), no biological damage has occurred even though there might be evidence that lakewater pH has declined. However, we know that species are lost from lakes if pH declines but remains $> 5.3$. More realistic evaluations of resource damage are required. As an example, preliminary models of resource damage have indicated that the damage caused by acidic precipitation may be widespread in lakes in the northeastern U.S.A. (Schindler et al. 1989) and eastern Canada (Minns et al. 1990, 1992). These models indicate that some lakes may have lost 30% or more of their overall taxa. For eastern Canada, this estimate may mean that millions of species populations may have been lost from hundreds of thousands of lakes. Modeled projections indicate that recent reductions in sulfur oxide emissions will allow a 50% recovery of the lost populations.

None of the resource damage models have included changes in nitrogen deposition or nitrate export from watersheds. Given the demonstrated significance of increased contributions of nitrate to the acidity of systems in many areas (Henriksen and Brakke 1988), consideration of nitrogen is necessary in future modeling efforts.

Further development of resource damage models is warranted for both lakes and streams. In this work, more attention should be given to streams and to evaluating the responses of the full range of hydrologic lake types. Resource damage estimates for streams are few. Estimates of changes in fish communities in trout streams in Pennsylvania suggest that fish species diversity has declined in response to acidification and 3000 km of streams have been impacted (Carline et al. 1992).

Resource damage models should be linked geochemical-biological models designed to apply to regional-scale evaluations. In addition, further modeling should be extended to represent the loss of all species, changes in community structure, and disruptions of ecosystem processes.

## Evidence from Paleolimnology and the Opportunity to Examine Biological Populations

The remains of many organisms accumulate in lake sediments. Paleolimnology is the study of historical conditions in lakes using the biological, chemical, and physical records of lake sediments. The direct use of this information to assess historical changes in biological communities and ecosystem processes in response to acidification has been relatively overlooked. In addition, to be able to infer changes over long time periods, one of the great advantages of the paleolimnological approach is that it can be used to examine any time lags in response and reveal trends that can be buried in shorter time frames (cf. Magnuson 1990). Increased emphasis on the analysis of sediment cores to estimate biological trends was identifed as an important research need.

The accumulation of sediments in lakes represents an historical record of atmospheric deposition and processes occurring in lakes and their watersheds. The sediment of a lake is usually deposited continuously and therefore represents a chronology of the lake. Unfortunately, not all organisms leave identifiable remains that are preserved in sediment, and thus the record is an incomplete representation of biological populations and lake processes. In addition, biological communities are shaped by physical and chemical characteristics of the environment as well as biological interactions. Consequently, paleolimnological interpretations are dependent upon our understanding of modern distributions of organisms relative to environmental factors. Moreover, changes in communities preserved in sediments could be in response to changes in physical or chemical conditions, biotic factors, or some combination thereof.

Numerous, recent paleolimnological studies (cf. Charles et al. 1989) have demonstrated their applicability in assessing the rates and directions of change in lake acidification. Thus far, nearly all of the focus of paleolimnological studies has been on the reconstruction of chemical changes in lakes. These reconstructions have been based almost solely on diatom remains, with more recent application of chrysophytes. As a result, the full range of changes in biological populations has not been described completely. For example, basic information on the number of species of various taxa present throughout a core has rarely been discussed (even for the diatoms), and only in a few situations (cf. Charles et al. 1990; Steinberg et al. 1988) have additional groups of organisms been considered. To address the questions of whether community structure or rates of production have changed in response to acidification, additional analyses of existing and new cores is essential.

Paleolimnological studies of change in the total accumulation rates of diatoms with acidification are inconclusive. However, they do not suggest any widespread reduction in primary productivity. There are demonstrated changes in taxa and in the types of habitats. For example, at a pH < 5.5–6.0 inferred from diatom remains, there is a loss of planktonic diatoms and a shift to benthic diatoms. Diatoms represent only a fraction of the productivity in lake systems, and the influence of the declining contribution of planktonic forms on the amount of primary production is not known. A leading hypothesis for the loss of diatom taxa is their lack of an effective acid phosphatase system that becomes active at pH values < 5.5, and they are out-competed by other species of algae (Smith 1990). The effect of this change in planktonic assemblages on grazer populations is not known, but it has not been examined.

Evidence also exists for shifts in the species composition of chrysophytes and zooplankton. Paleolimnological data often indicate an overall decline in the number of species of diatoms and cladocerans in response to acidification. However, changes in diversity in the sediment record are often difficult to interpret (Smol 1981). Whether similar reductions in species number occur for all algal groups or for all of the zooplankton is not known, yet modern observations from experimental manipulations suggest that species number should decline. Similarly, no consistent pattern of changes in overall primary or secondary productivity is known from paleolimnological studies.

Some new approaches represent particularly exciting possibilities for more complete reconstructions of changes in lakes with acidification. The identification of spores of a group of algae (Zygnematales) might be used as an indication of when metaphytic algal clouds or an algal mat developed. Further analysis of benthic diatoms might also be used to identify the development of metaphytic algae or an algal mat. Some evidence of increases in littoral Cladocera may indicate colonization of the additional new habitat associated with algal mats (Charles et al. 1990) or with *Sphagnum* (Steinberg et al 1988). Chrysophytes might be used to represent situations when episodic events began to occur. Recent improvements in techniques of pigment analysis represent an excellent opportunity to address the questions of changes in system productivity and possible shifts in the composition of algae (Leavitt 1989). Some of the new techniques allow for the recognition of pigments specific to particular algal taxa, but they are best done on fresh sediment following proper handling to prevent pigment degradation.

We are often interested in what happens when fish populations were lost from a system. In some cases, reliable historical records are available (e.g., the hytte-books in Norway), but most often they are lacking. Fish do not leave adequate remains in lake sediments to reconstruct changes in their populations; however, changes in the composition of the biological community recorded in lake sediments might be used to indicate time periods when fish were lost. Large invertebrates, such as *Chaoborus*, may invade or expand populations when fishes are lost (Von Ende and Dempsey 1981) and therefore may indicate the time marking the loss of fish populations (Kingston et al. 1992). Changes in the size distribution of the zooplankton, when carefully interpreted, can also be used to represent altered predation pressure.

For zooplankton and other invertebrates, the potential direct effects due to chemical changes must be separated from those related to food chain interactions. Given the information summarized in Table 19.1 from a variety of sources and approaches, and other sources described below, it appears that changes in many species, e.g., *Daphnia*, are due to direct effects of changes in pH. Nonetheless, indirect effects can play a role in shaping community structure and inconsistent reconstructions from lake to lake could partly be due to indirect effects. The expansion of habitat for littoral organisms represented by algal clouds or mats could also lead to signficant indirect effects on the structure of communities. Greater attention should be placed on the separation of direct and indirect effects in paleolimnological analyses.

## Evidence from Experimental Manipulations and Implications for Recovery

In general, experimental manipulations of whole lakes in Ontario and Wisconsin and surveys of acidified lakes in eastern Canada indicate that similar changes in species occurred at similar pH thresholds. These results suggest that rates of acidification in experimental lakes were not unrealistically fast (Schindler et al. 1991) and that they can be used to represent changes in communities occurring over longer time scales and at slower rates of change. There are, however, several reasons to believe that biological recovery from acidification may be delayed and not follow with reintroductions of species at the same pH values they were lost during acidification. At present, no study has documented complete recovery of biological communities from acidification, although a number of recovering ecosystems have regained several acid-sensitive species (reviewed by Schindler et al. 1991; see also Keller et al. 1992). The process of dispersal can apparently lead to significant lags in response.

An important question for the recovery of biota is whether the initial stages of chemical recovery might include periods when chemical conditions deteriorated before they got better. For example, as sulfur deposition declines, base cation concentrations would decrease. In addition, MAGIC would predict that base cations also could decline further as base saturation of soils was restored following reductions in sulfur deposition. The result could be a period when Ca concentrations in surface waters might decline to very low levels so that even a higher pH and a lower inorganic monomeric Al concentration would represent toxic conditions for fish. Work on the initial recovery phase is extremely important and continued observations of long-term monitored sites, such as the 100 lake subset of the Norwegian Thousand Lake Survey, is critical.

Once a species is lost from a lake, its reinvasion depends on whether it has resting eggs, spores, or other dormant life stages in the lake, whether the lake is interconnected with waters from which the species may reinvade, and whether the species has life stages that might be carried by aquatic birds or mammals or if it has flying life stages. In general, zooplankton and phytoplankton have a good ability to disperse and can also reinvade from resting stages. Insect dispersal should also be relatively rapid. Some acid-sensitive invertebrates such as amphipods, lacking flying stages, and fish are especially dependent on interconnected watercourses for dispersal (Keller et al. 1992). As a result, even though chemical conditions might improve, biological recovery of seepage lakes or isolated alpine lakes, e.g., in southern Norway, may be incomplete and require reintroduction by stocking.

In many acidified lakes, the habitats once occupied by acid-sensitive organisms are invaded by acid-tolerant ones, and it is not clear whether the natural occupants can reinvade once less-acidic conditions return. Evidence does suggest that some species that develop with acidification, such as metaphytic algae, are lost immediately when pH increases (Jackson et al. 1990). Therefore, some niches may be vacated.

Even when reversal of acidification allows the remaining natural species to resume reproduction, population instability can delay recovery of the whole community. For example, white suckers (*Catostomus commersoni*) in an artificially acidified lake in Canada (Lake 223) did not reproduce for three years while the lake was kept at pH $< 5.1$, but adults remained in good condition (fat). Reproduction resumed in the first year that pH was raised, with the production of an abnormally large year-class. As a result, populations were several times higher than before acidification. Because of insufficient food, suckers became thin and ceased reproducing at higher pH. High predation by suckers also caused chironomid emergence to be lower than during the period of greatest acidity. Restabilization of the sucker population may take several years, as it did following cessation of nutrient addition to another experimental lake.

The potential for food resources to limit the success of fish reintroductions is not indicated from stocking experiments with strains of brown trout in Norway (ReFisk project). In addition, the lack of food is not clearly demonstrated during acidification prior to the loss of fish populations. Both lines of evidence suggest that the food base is unlikely to limit the recolonization of fish populations, which we expect will be limited by barriers to dispersal. In summary, while it is clear that some parts of aquatic communities will recover rapidly from acidification, complete recovery will be delayed by recovery of habitats, dispersal, competition, and restabilization of population age structures. In some cases, particularly for seepage lakes, dispersal processes are likely to be slow. However, the rate and total degree of recovery of biological populations is poorly understood and cannot be predicted at present (see also Dise et al., this volume).

## Acidification of Warmwater Systems and Acidification and Climatic Change in Northern Latitudes

Much less is known about how acidification affects biological communities and how biological processes affect the chemistry of warmwater systems than for lakes in boreal regions. In the lakes in Florida, rates of sulfate reduction appear to occur at rates similar to those in northern coldwater systems (Baker et al. 1986). Denitrification, which occurs much more rapidly in summer at northern latitudes (Rudd et al. 1990), may be more effective in warmer climates. Similarly, mercury methylation, which is very temperature-dependent (Bodaly et al. 1992), might be more effective in warmer climates. Such studies appear to be important, both because of the anticipated increases in acidification as industrial development occurs in tropical and subtropical developing countries, and as predictors of how climate warming might modify rates and effects of acidification and recovery in cold temperate regions.

Recent studies of lakes in Florida also highlight the need to consider hydrologic flowpaths. Most of the acidic lakes in Florida are of the seepage lake type (Eilers et al. 1988), and many of these can have very complex flowpaths (Pollman and Canfield 1991). Greater emphasis should be placed on examining seepage lakes in general, because they are widespread and significant resources in many areas including the Upper Midwest of the U.S., Alaska, and the Oregon Cascades. Special emphasis should be placed on seepage lakes in warmer regions.

Observations at the Experimental Lakes Area during two decades of increasing temperature and decreasing moisture suggest that climate change may have a number of important effects on biota, some exacerbate the effects of acidification, while others mitigate acidification. Warmer, drier conditions increase water residence time of lakes, causing higher concentrations of elements, including sulfate, calcium, and nitrogen (Schindler et al. 1990). This is expected to cause higher rates of internal alkalinity generation, which is strongly dependent on water renewal (Kelly et al. 1987). Lower water renewal and higher temperature also cause DOC concentration to decrease, causing higher transparency. This effect is similar to that caused by acidification. Forest fires caused by drought allow greater wind exposure, causing thermoclines to deepen, which may also enhance internal alkalinity generation because of better contact of the higher concentration epilimnitic water with sediments. Warmer temperatures also cause increased decomposition. These combine to reduce summer refuges for cold stenothermic, oxygen-sensitive organisms like lake trout and *Mysis,* two organisms that also are very acid-sensitive. As mentioned above, mercury methylation and denitrification may increase under warmer conditions.

For seepage lakes in the Upper Midwest of the U.S., ANC has been related directly to the fraction of groundwater inputs. As an illustration of the potential for climatic impacts on these seepage lakes, during a period of recent drought, lake levels declined dramatically. Morever, as groundwater inputs declined, rapid declines were found in lakewater ANC (Webster et al. 1990). These observations suggest that seepage lakes may be very responsive to a changing climatic environment and that the chemical environment in the lakes will be a complex interaction of groundwater inputs and in-lake processes.

Arctic ecosystems may be particularly susceptible to disruption by acid deposition, especially in parts of the industrialized circumpolar arctic where significant point-sources of $NO_x$ and $NO_y$ (e.g., Jaffe 1991) are overlain on a background of low levels of sulfate deposition associated with arctic haze (Barrie 1986; Barrie and Bottenheim 1991). Generally speaking, high-latitude/high-altitude ecosystems are species poor and have little functional redundancy. In base-poor regions, surface water acidification leading to species loss or dysfunction may therefore be expected to have significant consequences at the ecosystem level. The vulnerability of alkalinity-generating processes in these systems is not known. However, because alkalinity-generating mechanisms are mostly carried out anaerobically, one might expect that an increased photoperiod in summer, with associated photosynthetic oxygen production, would decrease the degree of anoxia in sediments and possibly decrease alkalinity generation. Finally, the lack of a dark period allows ongoing photolysis of organic compounds, which is of particular concern for those toxicants that increase with photolysis (see also Landers et al., this volume).

## Potential Changes in Genetic Diversity

The major concern for maintaining genetic diversity is the potential for *irreversible* loss of unique strains or endangered species. The threat to genetic or biological diversity from acidification and other impacts has largely been ignored even though significant losses have been reported due to various land-use practices (Hughes and Ness 1992). The application of molecular genetic tools opens up new horizons to study changes in aquatic biota at different trophic levels (Hartmann, this volume). In Norway, a gene bank of regional fish populations is being built to conserve genetic diversity related to specific local or regional adaptations. We do not know what, if any, important strains have already been lost. Molecular techniques might be useful in determining what strains will be best suited to the conditions during recovery and survive to establish permanent, reproducing populations.

Several other approaches are possible using genetic tools. For example, DNA fingerprinting of different zooplankton species is being used to study the recolonization process during the recovery of lakes in Ontario. In this way we can determine whether a system recovers to its original condition or whether some new combination of genetic strains is involved in reinvasion. Using genetic markers it should also be possible to study in detail the process of recolonization and to examine the rates at which reinvasion might occur. We might also evaluate the source of organisms that reinvade. Is recolonization dependent on chance dispersal from various spatial scales or are local source areas most important?

Microbial communities, such as those responsible for decomposition, denitrification, and sulfate reduction, also represent an excellent opportunity to make use of new genetic tools. For example, these communities continue to function as the pH of a lake decreases to 5, and in some cases even lower. However, it is not known whether the species composition of the communities changes greatly, which could be examined using recently developed genetic techniques (Hartmann, this volume). An especially interesting question concerns the nitrifying bacteria. Nitrification ceases in Canadian Shield lakes at about pH 5.4–5.6. However, it occurs in some very acidic lakes and also some acidic soils. Are different bacteria responsible in these different systems, or has some adaptation occurred? These and other questions argue for accelerated work with new molecular techniques in conjuction with effective monitoring of species losses or reintroductions.

## CHEMICAL AND BIOLOGICAL INDICES OF ECOSYSTEM HEALTH

We have made many advances in understanding the response of organisms to acidification, but can we use this information to develop effective measures of the status or health of biological communities? Are there ways to evaluate and monitor ecosystem health? There are many important functional and community characteristics to represent in any description of the status of a system. It is important to recognize and separate the direct toxic responses leading to the loss of genetic or species diversity from ecosystem-level responses. In order to represent ecosystem health and monitor changes in response to some perturbation or during a phase of recovery, we must use a variety of approaches and focus on a range of ecosystem attributes.

Appropriate indicators of ecosystem health can include biological and chemical variables as well as combinations of both. We can examine the distribution of biota in lakes and streams in relation to chemical conditions based on surveys, monitoring, and other field-based approaches as well as by bioassay procedures. Some chemical variables can have significance controlling the distribution of individual species and be used accordingly. Other chemical measures might serve as useful reference values for ecosystem processes. In both cases, it is important to identify the concentration, the time period when it occurs, and the length of the exposure to any organisms. Developing useful ecological indicators, particularly any that might be used to reflect ecosystem health, is extremely difficult, and there is little relevant literature to serve as background. Some work has been done to develop an index of biotic integrity for streams by Karr (1987), and it has been extended and applied in several areas. The index focuses on biological populations but does not include genetic diversity or reflect ecosystem processes. We are not aware of similar approaches related to acidification. Nonetheless, concerns for biological diversity and the health or integrity of aquatic systems appear to be increasing (Cairnes and Lackey 1992; Hughes and Noss 1992).

Based on recent research, some levels can be identified that relate to the occurrence or health of populations in relationship to aquatic organisms in acidified areas. All of these relationships are dependent on the response of biological populations, whose individual species often vary tremendously in their sensitivities to change in particular parameters. Moreover, the distribution of species often varies greatly among regions. Therefore, it is essential to recognize and account for regional distributions of different types of environment and of biota in applying any relationship to chemical conditions.

Acidification of fresh waters can produce a chemical environment leading to direct toxic effects on the populations of biological organisms. As an example, pH, low Ca, and high inorganic Al can produce extremely toxic conditions for fish. Nonetheless, fish species vary considerably in their tolerance to various levels of exposure. Even so, there has been a tendency to focus on the point when all fish populations were lost from a system and to equate that point with damage. Dead bodies are relatively easy to count, and limits based on death are easier to set than any based on the health of a population. However, given data from flow-through bioassays, with confirmation from regional surveys relating fish presence or absence to chemical levels, it is possible to determine biologically relevant chemical reference points that represent (a) the loss of any fish population or (b) the health of all fish populations. For example, a defined measure of ANC (charge balance) was used by Lien et al. (1992) to examine the distribution and health of populations of fish species in lakes and Atlantic salmon (*Salmo salar L.*) in rivers. Their results indicated that it was possible to establish for each fish species the relevant limits that reflected the ANC value when all fish populations were extinct and when the populations of each species were all healthy. Even though the fish do not respond directly to charge balance ANC, these data could be used to set an ANC limit for the protection of fish populations in Norway that could be incorporated into critical loading estimates. Similar approaches were used to examine the distribution of invertebrates in streams and lakes (Lien et al. 1992).

"Critical thresholds" for pH have been defined for many organisms and processes. "Acid stress indices" that combine pH, Al, and Ca have been proposed for some fish species (J.P. Baker et al. 1990). The development and use of such indices is only beginning, and we suggest that further development of chemical and biological indices of biological status is a high priority for future research. These indices can provide a useful composite measure by which to estimate the conditions of biological populations under various chemical conditions, evaluate scenarios of recovery, and to monitor changes in systems over time.

Additional factors have been recognized as important variables controlling the distribution and occurrence of organisms as a function of acidic conditions. These factors include DOC, nitrate, and temperature. As these other factors change in response to acidification, recovery, or climatic change, new conditions for biological populations result. Some additional variables should be explored in relationship to their influence on the occurrence of organisms. For example, there is some evidence that Si forms complex molecules or ions with Al, which might protect fish populations. Therefore, some combination of Si and nutrient additions might be used to inactivate Al and slowly titrate a lake to higher pH without resorting to liming as a mitigation strategy. We note, however, that mitigation of acidity with nutrients must maintain greater nitrate than ammonium in order to avoid even further acidification. Other variables might be explored as they relate to the occurrence, type, and distribution of benthic or metaphytic algal mats. The presence of such mats is poorly described and would be useful to predict.

Other chemical variables have potential usage for indicating the overall health or status of ecosystem processes. Some microbial processes, such as S-reduction and N-fixation, are pH dependent and can represent additional reference points indicating ecosystem status. The appearance of detectable nitrate in the epilimnia of lakes during mid-summer may indicate leakage of nitrate from the catchment, which might reflect the health of the terrestrial ecosystem within the watershed. The appearance of ammonia in surface waters might be explored as an indicator of the effectiveness of microbial processes. The accumulation of mercury and other toxicants at higher trophic levels may also represent an important indication of ecosystem health.

Further development of chemical and biological indicators of ecosystem health should be pursued. Effective monitoring of changes in acidity, especially in conjunction with response to other stressors, cannot be accomplished without careful examination of the appropriate measures. We note the development of programs, e.g., EMAP in the U.S. and the Thousand Lakes Survey in Norway, that are designed to monitor environmental health using chemical and biological indicators. We emphasize that such programs must be soundly designed and conducted with adequate extent and frequency so that meaningful trends can be identified. Work to establish such programs and the improvement of indicators for use in evaluating ecosystem health is a high priority.

## IMPROVED APPROACHES IN ENVIRONMENTAL RESEARCH

Not only has the acidification debate provided a framework for a broad range of scientists to understand a major environmental problem on a wide range of scales and from many disciplines, it has also provided a focus and forum for debate about basic aspects of the implementation and philosophy of the scientific method, especially as it applies to regional-scale evaluations. This result arises in part from our recognition that individual approaches (experiment, survey or monitoring, and modeling) have their own particular flaws. It arises also, however, because we recognize that in combination these approaches have great value: consistency between experiments, surveys, and models gives us an indication that results are robust, but inconsistencies give us the opportunity to examine critically the methods we adopt.

### The Problems of an Isolated Approach

For any ecologist working in isolation, the methodological problems of assessing the impact of acidity on biota might be summarized by suggesting that:

1. Experiments are limited by artifacts, unrealistic conditions, and difficulties in design. At the laboratory-scale, realism is likely to be the major problem, while

at the ecosystem-scale, design difficulties, problems of replication (and hence problems of generalization) are more likely. Artifacts, almost as a definition of the act of experimentation, are likely at all scales. Examples include the caging of animals for toxicology, the absence of confounding stresses, and inappropriate and artificial chemical conditons. The latter include not only inappropriate DOC and $Ca^{2+}$ concentrations but also the extreme difficulties of recreating aluminum speciation to mimic those seen in real acidic conditions.

2. Surveys and monitoring exercises (and here we might include paleoecology) are limited by confounding factors, a lack of precision, and a limited understanding of cause-effect relationships. The examples here are legion, but an example might include the difficulties during monitoring of detecting when a signal in any given parameter emerges from background variation, and then in interpreting its cause.
3. Models are, by definition, imperfect representations of the real world. Again, potential problems are too numerous to detail, but examples might include the tendency of some acidification models to simulate water quality accurately despite the absence or inadequate representation of some important processes known to occur in real environments. We note the frequent situation where a model is used to infer understanding and the unfortunate tendency, often expressed by decision-makers, to conclude that because a model is more complex it must be a better representation of the real world. An even more extreme abuse of models is to say that a process measured in the natural world but perhaps not included in a complex model, cannot be very important because the simulations from the complex model explain how nature operates. These faults lie not with the models, but with those who apply them.
4. All approaches have suffered from the unfortunate tendency of investigators to repeat mistakes already made by others. Scientists have a great tendency to report positive results and rarely communicate those results that do not fit a current paradigm or are simply not understood. We note many situations where similar observations were made over several years by different investigators, but the results were unknown to the others. Perhaps there is a need for a "*Journal of Puzzling Results*" and maybe another "*Journal of Failed Approaches*" to facilitate real communication of research experiences and results. Science requires complete communication of results, not just those supporting current views or the findings of experiments with clearly demonstrated effects.

## The Value of a Combined Approach

Rather than being characterized by an isolated approach, the acidification debate has provided a prime example of the value of collaboration between the experimental, survey, and modeling approaches. Field or laboratory experiments have reproduced, or aided, the results of surveys; models have been tested (and modified) following field or laboratory experimentation; surveys have generated hypotheses tested by experiments or models. In part, these combined approaches have been permitted not only by use of the literature but by the institutional provision of an appropriate forum for collaboration. We suggest that other large-scale environmental problems will benefit from the same style of work.

We recommend that combined approaches be institutionalized. Integrated research programs ensure better communication and cooperation among scientists expert in different methods and approaches. In addition, we strongly urge that such integrated programs be conducted at a variety of scales. Some work must be site-specific; however, scaling to landscape or regional levels is a critically important step that should be accomplished. As a consequence, integrated research programs will require substantial effort from the start to ensure adequate design.

We have a second recommendation targeted to educators who should use the experience of acidification research and the lesson of a combined approach as a basis for the design of courses, or as an example of how effective the combined approach can be in addressing specific problems. Thus, the education of scientists working on environmental problems needs to include the development of *skills* in interacting with other scientists using different approaches, and *appreciation* of the value of these different approaches. Environmental scientists need experience in the field and hands-on experience working at a variety of scales. The problems encountered encompass many fields, situations, and scales. It is essential that environmental scientists be prepared to be conversant with a range of topics and conditions and be flexible in their approach.

## RECOMMENDATIONS FOR COORDINATED FUTURE RESEARCH

We have identified a number of directions for future research on the biological response to acidification that have important implications for the recovery phase and the interaction of acidification with other stresses. We recognize that some of the research requires specific mechanistic approaches; however, we emphasize that the work should be coordinated with other efforts to lead to a comprehensive view of responses to acidification and recovery.

We encourage ongoing efforts in integrated monitoring of ecosystem status. Such monitoring programs should be continued and expanded. These programs should have a sound statistical base and operate at a variety of scales. A key component of integrated monitoring should be the extrapolation of results to make resource damage estimates. The status and health of biological populations should be evaluated and monitored to determine present and future extent of damage and rates of recovery. These evaluations must represent a linkage of geochemical and biological processes, and they should consider the full range of biological organisms and processes. Modeling biological responses to acidification should be expanded to include ecosystem processes.

The monitoring and evaluation of resource status is a measure of ecosystem health. There are many aspects of ecosystem health, including biological diversity at various levels (genetic, species, habitat, landscape) and the integrity and efficiency of ecosystem processes. Ecosystems often respond to many different stresses. We feel that evaluating and monitoring ecosystem health is a major direction for future research. This research should not be focused solely on any one stress, but attempt to evaluate system response to different stressors. Our discussions of biologically relevant chemical reference points suggested that while we could suggest some endpoints that might reflect changes in certain parts of a system, much greater attention should be placed on understanding the full range of biological response and looking for indicators of ecosystem health. This work is in its infancy and should be greatly expanded, because we believe there are great possibilities to develop effective measures of ecosystem health and integrity.

Specific toxicological studies are required for organisms and conditions to develop a sound, mechanistic basis to explain why species are lost from acidic environments. Uncertainties remain particularly in our understanding of the response of algae and aquatic plants, sublethal and chronic effects on organisms, identification of the most sensitive life stages, and indirect effects on communities. While we pursue mechanistic understanding, we should also develop useful measures to evaluate the status of organisms and populations.

New molecular techniques show particular promise in understanding why some strains are more or less responsive to changes in pH and other consituents. These techniques should also be used to examine and track sensitivities of strains to various chemical conditions and to explore ways to resolve uncertainties in the response of microbes to acidification. They also represent an excellent opportunity to document the source of species reinvading recovering systems.

The chemistry of zones where waters mix can be unstable and represent highly toxic conditions for biota. Much greater research is required to determine the significance of these zones on the success of biological populations. In particular, because of the great economic significance of Atlantic salmon and their frequent exposure to mixing zones, further research should emphasize delayed or sublethal effects of exposure on their populations.

Paleolimnology has been very important in evaluating the changes in lakes in response to acidic inputs. There should be increased emphasis on the analysis of sediment cores to reflect more completely the changes in biological communities and processes associated with acidification. This work will document a broader range of responses of biological communities to acidification, but importantly, it represents an opportunity to have other views on how recovery might operate. Further work in paleolimnology can be extended by using additional groups of organisms in making inferences of past conditions, and by developing and employing new techniques to examine questions that could not be addressed previously. We encourage futher work to document time periods when fish were lost from systems, when metaphytic algal blooms or mats occurred, and to address questions of changes in lake productivity and community structure. We note that all of this work should keep in mind the physical and chemical environment, and the biological interactions involved in order to separate direct and indirect effects on organisms and processes. Reconstructions of changes in communities should identify the initial biological communities present and aspects of the physical environment on which chemical changes are imposed.

There is much uncertainty associated with two important variables in the chemical environment, total organic carbon (TOC), and nitrate. We require much greater understanding of the role of TOC in mitigating toxicity, the production and flux of TOC, and the potential differences in the type of TOC as they relate to the processes of acidification and recovery. The contributions of nitrate to acidity are not accounted for in present models, and as nitrate is associated with maintenance of high Al concentrations, recovery of biological populations might be affected.

Elevated mercury in fish and fish-eating birds has been reported from many areas having low ANC surface waters. Low pH and changes in TOC can increase the methylation of mercury by microorganisms. Because of the significance of this issue to human health, work on processes of mercury transfer from terrestrial to wetland or wetland to aquatic systems should be accelerated. This research should also be expanded to examine mercury methylation during recovery phases and interactions with climatic change that might vary the flux of TOC and mercury from terrestrial environments.

Acidification research has most often been concentrated in the northern temperate and boreal regions. It also has been focused primarily on drainage lakes. There is a need to expand research to warmwater systems, conduct further work on a range of lake types and settings, and do more basic research on streams.

In all future work, we suggest the great strength of coordinated research efforts on a range of scales that involve a variey of approaches. Only by coming at a complex problem, such as acidification, from many different directions, approaches, and perspectives can we develop a comprehensive view of the response of biological communities and ecosystems to stress.

## REFERENCES

Almer, B., W. Dickson, C. Ekstrom, E. Hornstrom, and U. Miller. 1974. Effects of acidification of Swedish lakes. *Ambio* **3**:30–36.

Andren, C., L. Henriksen, M. Olsson, and G. Nilson. 1988. Effects of pH and aluminum on embryonic and early larval stages of Swedish brown frogs *Rana arvalis, R. temporaria,* and *R. dalmatina. Holarctic Ecol.* **11**:127–135.

Åtland, Å., and B. Barlaup. 1991. Rate of gastric evacuation rate in brown trout (*Salmo trutta* L.) in acidified and nonacidified water. *Water, Air, Soil Pollut.* **60**:197–204.

Audet, C., and C.M. Wood. 1988. Do rainbow trout (*Salmo gairdneri*) acclimate to low pH? *Can. J. Fish. Aquat. Sci.* **45**:1399–1405.

Baker, J.P., D.P. Bernard, S.W. Christensen, M.J. Sale, J. Freda, K. Heltcher, D. Marmorek, L. Rowe, P. Scanlon, G. Suter, W. Warren-Hicks, and P. Welbourn. 1990. Biological effects of changes in surface water acid-base chemistry. Report 13. Acid Deposition: State of Science and Technology. Washington, D.C.: Natl. Acid Precipitation Assessment Program.

Baker, J.P., and C.L. Schofield. 1982. Aluminum toxicity to fish in acidic waters. *Water, Air, Soil Pollut.* **18**:289–309.

Baker, L.A., P.L. Brezonik, and E.S. Edgerton. 1986. Sources and sinks of ions in a soft water acidic lake in Florida. *Water Resour. Res.* **22**:715–722.

Baker, L.A., P.R. Kaufmann, A.T. Herlihy, and J.M. Eilers. 1990. SOS/T 9: Current status of surface water acid base chemistry. Report 9. Acid Deposition: State of Science and Technology. Washington, D.C.: National Acid Precipitation Assessment Program.

Barrie, L.A. 1986. Arctic air pollution: An overview of current knowledge. *Atmos. Environ.* **20**:643–663.

Barrie, L.A., and J.W. Bottenheim. 1991. Sulfur and nitrogen pollution in the Arctic atmosphere. In: Pollution of the Arctic Atmosphere, ed. W.T. Sturges, pp. 155–183. New York: Elsevier.

Battarbee, R.W., R.J. Flower, A.C. Stevenson, and B. Rippey. 1985. Lake acidification in Galloway: A paleoecological test of competing hypotheses. *Nature* **314**:350–352.

Beamish, R.J., W.L. Lockhart, J.C. Van Loon, and H.H. Harvey. 1975. Long-term acidification of a lake and resulting effects on fishes. *Ambio* **4**:98–102.

Berger, H.M., T. Hesthagen, I.H. Sevaldrud, and L. Kvenild. 1992. Forsuring av innsjøer i Sør-Norge: Fiskestatus innen geografiske rutenett. (Mapping of fish community status in relation to acidification in Norwegian lakes). NIVA Forskrapport 32. 12 p. Oslo: NIVA. (In Norwegian).

Bodaly, R.A., J.W.M. Rudd, R.J.P. Fudge, and C.A. Kelly. 1992. A relationship between mercury concentrations in fish and lake size in remote Canadian Shield lakes. *Can. J. Fish. Aquat. Sci.*, in press.

Böhmer, J., and H. Rahmann. 1990. Influence of surface water acidification on amphibians. In: Progress in Zoology, ed. W. Hanke, pp. 287–309. New York: Gustav Fischer Verlag.

Böhmer, J., and H. Rahmann. 1992. Limnologische Unterschungen zur Versauerung stehender Gewasser im Nordschwarzwald unter besonderer Berücksichtigung der Amphibienfauna. Umweltforschung in Baden-Württemberg, vol. 1. Langenau: Ecomed Verlag, 230 pp.

Brown, D.J.A. 1983. Effect of calcium and aluminum concentration on the survival of brown trout (*Salmo trutta*) at low pH. *Bull. Environ. Contam. Toxicol.* **30**:582–587.

Brown, S.B., J.G. Eales, R.E. Evans, and T.J. Hara. 1984. Interrenal, thyroidal, and carbohydrate responses of rainbow trout (*Salmo gairdneri*) to environmental acidification. *Can. J. Fish. Aquat. Sci.* **41**:36–45.

Caines, L.A., A.W. Watt, and D.E. Wells. 1985. The uptake and release of some trace metals by aquatic bryophytes in acidified waters in Scotland. *Environ. Pollut. B* **10**:1–18.

Cairnes, M.A., and R.T. Lackey. 1992. Biodiversity and management of natural resources: The issues. *Fisheries* **17**:6–10.

Carline, R.F., W.E. Sharpe, and C.J. Gagen. 1992. Changes in fish communities and trout management in response to the acidification of streams in Pennsylvania. *Fisheries* **17**:33–38.

Charles, D.F. 1984. Recent pH history of Big Moose Lake (Adirondack Mountains, New York, U.S.A.) inferred from sediment diatom assemblages. *Verh. Int. Verein Limnol.* **22**:559–566.

Charles, D.F., R.W. Battarbee, I. Renberg, H. van Dam, and J.P. Smol. 1989. Paleoecological analysis of lake acidification trends in North America and Europe using diatoms and chrysophytes. In: Acid Precipitation, ed. S.A. Norton, S.E. Lindberg, and A.L. Page, vol. 4, pp. 207–267. Soils, Aquatic Processes, and Lake Acidification. New York: Springer.

Charles, D.F., M.W. Binform, E.T. Furlong, R.A. Hites, M.J. Paterson, J.P. Smol, A.J. Uutala, J.R. White, D.R. Whitehead, and R.J. Wise. 1990. Paleoecological investigation of recent lake acidification in the Adirondack Mountains, N.Y. *J. Paleolimnol.* **3**:195–241.

Confer, J.L., T. Kaaret, and G.E. Likens. 1983. Zooplankton diversity and biomass in recently acidified lakes. *Can. J. Fish. Aquat. Sci.* **40**:36–42.

Cumming, B.F., J.P. Smol, J.C. Kingston, D.F. Charles, H.J.B. Birks, K.E. Camburn, S.S. Dixit, A.J. Uutala, and A.R. Selle. 1992. How much acidification has occurred in Adirondack region lakes (New York, U.S.A.) since preindustrial times? *Can. J. Fish. Aquat. Sci.* **49**:128–141.

D'Angelo, D.J., J.R. Webster, and E.F. Benfield. 1991. Mechanisms of stream phosphorus retention: An experimental study. *J. N. Am. Benth. Soc.* **10**:225–237.

Davies, I.J. 1989. Population collapse of the crayfish *Orconectes virilis* in response to experimental whole lake acidification. *Can. J. Fish. Aquat. Sci.* **46**:910–922.

Dillon, P.J., N.D. Yan, and H.H. Harvey. 1984. Acidic deposition: Effects on aquatic ecosystems. *CRC Crit. Rev. Environ. Control* **13**:167–194.

Dobson, M., and A.G. Hildrew. 1992. A test of resource limitation among shredding detrivores in low order streams in Southern England. *J. Anim. Ecol.* **61**:69–77.

Eilers, J.M., D.H. Landers, and D.F. Brakke.1988. Chemical and physical characteristics of lakes in the Southeastern U.S. *Environ. Sci. Technol.* **22**:172–177.

Farmer, A.M. 1990. The effect of lake acidification on aquatic macrophytes: A review. *Environ. Pollut.* **65**:219–240.

Fivelstad, S., and H. Leivestad. 1984. Aluminum toxicity to Atlantic salmon (*Salmo salar* L.) and brown trout (*Salmo trutta* L.): Mortality and physiological response. *Rep. Inst. Freshw. Res. Drottningholm* **61**:69–77.

Forsius, M., J. Kämäri, P. Kortelainen, J. Mannio, M. Verta, and K. Kinnunen. 1990. Statistical lake survey in Finland: Regional estimates of lake acidification. In: Acidification in Finland, ed. P.Kauppi, P. Antilla, and K. Kenttämies, pp. 759–780. Berlin: Springer.

France, R.L., and P.M. Stokes. 1987. Influence of manganese, calcium, and aluminum on hydrogen ion toxicity to the amphipod. *Hyallela azteca. Can. J. Zool.* **65**:3071–3078.

Freda, J., and W.A. Dunson. 1984. Sodium balance of amphibian larvae exposed to low environmental pH. *Physiol. Zool.* **57**:435–443.

Freda, J., and W.A. Dunson. 1985. Field and laboratory studies of ion balance and growth rates of ranid tadpoles chronically exposed to low pH. *Copeia* **2**:415–423.

Gerletti, M., and A. Provini. 1978. Effect of nitrification in Orta Lake. *Prog. Water Tech.* **10**:839–851.

Grahn, O. 1977. Macrophyte succession of swedish lakes caused by deposition of airborne acid substances. *Water, Air, Soil Pollut.* **7**:295–305.

Grande, M., I.P. Muniz, and S. Andersen. 1978. Relative tolerance of some salmonids to acid water. *Verh. Inter. Verein Limnol.* **20**:2076–2084.

Haines, T.A. 1981. Acidic precipitation and its consequences for aquatic ecosystems: A review. *Trans. Am. Fish. Soc.* **110**:669–707.

Hall, R.J., and F.P. Ide. 1987. Evidence of acidification effects on stream insect communities in central Ontario between 1937 and 1985. *Can. J. Fish. Aquat. Sci.* **44**:1652–1657.

Harriman, R., B.R.S. Morrison, L.A. Caines, P. Collen, and A.W. Watt. 1987. Long-term changes in fish populations of acid streams and lochs in Galloway South West Scotland. *Water, Air, Soil Pollut.* **32**:89–112.

Havas, M. 1985. Aluminum bioaccumulation and toxicity to *Daphnia magna* in softwater at low pH. *Can. J. Fish. Aquat. Sci.* **42**:1741–1748.

Havas, M. 1986. Effects of aluminum on aquatic biota. In: Aluminum in the Canadian Environment, ed. M. Havas and J.F. Jaworski, pp. 79–127. NRCC #24759. Ottawa: Natl. Research Council of Canada.

Havas, M. 1987. Does hemoglobin enhance the acid-tolerance of Daphnia? *Ann. Soc. R. Zool. Belg.* **117(1)**:151–164.

Havas, M., and T.C. Hutchinson. 1983. Effect of low pH on the chemical composition of aquatic invertebrates from tundra ponds at the Smoking Hills, N.W.T., Canada. *Can. J. Zool.* **61**:241–249.

Havas, M., and G.E. Likens. 1985. Changes in Na influx and outflux in *Daphnia magna* (Straus) as a function of elevated Al concentrations in soft water at low pH. *Proc. Natl. Acad. Sci. U.S.A.* **82**:7345–7349.

Henriksen, A., and D.F. Brakke. 1988. Increasing contributions of nitrogen to the acidity of surface waters in Norway. *Water, Air, Soil Pollut.* **42**:183–201.

Henriksen, A., D.F. Brakke, T.S. Traaen, L. Lien, and S. Taubøll. 1992. A method for estimating and mapping critical loads of acidity to surface waters in Norway. *Sci. Total Environ.*, in press.

Henriksen, A., L. Lien, B.O. Rosseland, T.S. Traaen, and I.S. Sevaldrud. 1989. Lake acidification in Norway: Present and predicted fish status. *Ambio* **18**:314–321.

Henriksen, A., B.M. Wathne, E.J.S. Røgeberg, S.A. Norton, and D.F. Brakke. 1988. The role of stream substrates in aluminum mobility and acid neutralization. *Water Res.* **22**:1069–1073.

Herlihy, A.T., and A.L. Mills. 1985. Sulfate reduction in freshwater sediments receiving acid mine drainage. *Appl. Environ. Microbiol.* **49**:179–186.

Hesthagen, T. 1986. Fish kills of Atlantic salmon (*Salmo salar*) and brown trout (*Salmo trutta*) in an acidified river of southwest Norway. *Water, Air, Soil Pollut.* **30**:619–628.

Hill, J., R.E. Roley, V.S. Blazer, R.G. Werner, and J.E. Gannon. 1988. Effects of acidic water on young-of-the-year smallmouth bass (*Micropterus dolomieu*). *Environ. Biol. Fish.* **21**:223–229.

Hobæk, A., and G.G. Raddum. 1980. Zooplankton communities in acidified lakes in South Norway. SNSF Report IR 75/80. Oslo: SNSF Project, 132 pp.

Hopkins, P.S., K.W. Krants, and S.D. Cooper. 1989. Effects of an experimental acid pulse on invertebrates in high altitude Sierra Nevada stream. *Hydrobiologia* **171**:45–58.

Howell, E.T., M.A. Turner, R.L. France, M.B. Jackson, and P.M. Stokes. 1990. Comparison of sygnematacean (Chlorophyta) algae in the metaphyton of two acidic lakes. *Can. J. Fish. Aquat. Sci.* **47**:1085–1092.

Hughes, R.M., and R.F. Noss. 1992. Biological diversity and biological integrity: Current concerns for lakes and streams. *Fisheries* **17**:11–19.

Jackson, M.B., E.V. Vandermeer, N. Lester, J.A. Booth, L. Molot, and I.M. Gray. 1990. Effects of neutralization and early reacidification on filamentous algae and macrophytes in Bowland Lake. *Can. J. Fish. Aquat. Sci.* **47**:432–439.

Jackson, S.T., and D.F. Charles. 1988. Aquatic macrophytes in Adirondack (New York) lakes: Patterns of species composition in relation to environment. *Can. J. Bot.* **66**:1449–1460.

Jacobson, J.O. 1977. Brown trout (*Salmo trutta* L.) growth at reduced pH. *Aquaculture* **11**:81–84.

Jaffe, D.A. 1991. Local sources of pollution in the Arctic: From Prudhoe Bay to the Taz Peninsula. In: Pollution of the Arctic Atmosphere, ed. W.T. Struges, pp. 255–287. New York: Elsevier.

Jensen, K.W., and E. Snekvik. 1972. Low pH levels wipe out salmon and trout populations in southernmost Norway. *Ambio* **1**:223–225.

Johnston, T.A., R.A. Bodaly, and J.A. Mathias. 1991. Predicting fish mercury levels from physical characteristics in boreal reservoirs. *Can. J. Fish. Aquat. Sci.* **48**:1468–1475.

Karr, J.R. 1987. Biological monitoring and environmental assessment: A conceptual framework. *Environ. Manag.* **11**:249–256.

Keller, W., J.M. Gunn, and N.D. Yann. 1992. Evidence of biological recovery in acid-stressed lakes near Sudbury, Canada. *Environ. Pollut.* **78**:79–85.

Kelly, C.A., J.W.M. Rudd, A. Furutani, and D.W. Schindler. 1984. Effects of lake acidification on rates of organic matter decomposition in sediments. *Limnol. Oceanogr.* **29**:687–694.

Kelly, C.A., J.W.M. Rudd, R.H. Hesslein, D.W. Schindler, P.J. Dillon, S.T. Driscoll, S.A. Gherini, and R.E. Hecky. 1987. Prediction of biological and neutralization in acid-sensitive lakes. *Biogeochemistry* **3**:129–140.

Kelly, C.A., J.W.M. Rudd, and D.W. Schindler. 1990. Lake acidification by nitric acid: Future considerations. *Water, Air, Soil Pollut.* **50**:49–61.

Kettle, W.D., M.F. Moffett, and F. Denoyelles. 1987. Vertical distribution of zooplankton in an experimentally acidified lake containing a metalimnetic phytoplankton peak. *Can. J. Fish. Aquat. Sci.* **44**:91–95.

Kingston, J.C., H.J.B. Birks, A.J. Uutala, B.F. Cumming, and J.P. Smol. 1992. Assessing trends in fishery resources and lake water aluminum from paleolimnological analyses of siliceous algae. *Can. J. Fish. Aquat. Sci.* **49**:116–127.

Klaprat, D.A., S.B. Brown, and T.J. Hara. 1988. The effect of low pH and aluminum on the olfactory organ of rainbow trout (*Salmo gairdneri*). *Environ. Biol. Fish.* **22**:69–77.

Landers, D.H., W.S. Overton, R.A. Linthurst, and D.F. Brakke. 1988. Eastern Lake Survey. Regional estimates of lake chemistry. *Environ. Sci. Technol.* **22**:128–135.

Lazarek. S. 1985. Epiphytic algal production in the acidified Lake Gardsjon, SW Sweden. *Ecol. Bull. Stockholm* **37**:213–218.

Leavitt, P.R., S.R. Carpenter, and J.F. Kitchell. 1989. Whole-lake experiments: The annual record of fossil pigments and zooplankton. *Limnol. Oceanogr.* **34**:700–717.

Leivestad, H., G. Hendrey, I.P. Muniz, and E. Snekvik. 1980. Effects of acid precipitation on freshwater organisms. In: Impact of Acid Precipitation on Forest and Freshwater Ecosystems in Norway, ed. F.H. Brække, pp. 87–111. Oslo: SNSF Project.

Leivestad, H., and I.P. Muniz. 1976. Fish kill at low pH in a Norwegian river. *Nature* **259**:391–392.

Leivestad, H., I.P. Muniz, and B.O. Rosseland. 1980. Acid stress in trout from a dilute mountain stream. In: Ecological Impact of Acid Precipitation., ed. D. Drabløs and A. Tollan, pp. 318–319. Oslo: SNSF Project.

Lien, L., G.G. Raddum, A. Fjellheim, D.F. Brakke, and A. Henriksen. 1992. The relationship of fish and invertebrates to critical limits of water chemistry in Norwegian lakes. *Sci. Total Environ.*, in press.

Magnuson, J.J. 1990. Long-term ecological research and the invisible present. *BioScience* **40**:495–501.

Malley, D.F., D.L. Findlay, and P.S.S. Chang. 1982. Ecological effects of acid precipitation on Zooplankton. In: Acid Precipitation: Effects on Ecological Systems, ed. F.M. D'Itri, pp. 297–327. Ann Arbor: Ann Arbor Science.

Matthias, U. 1983. Der Einfluss der Versauerung auf die Zusammensetzung von Bergbachbiozonosen. *Arch. Hydrobiol. Suppl.* **65**:407–483.

Mauzeaud, M.M., and F. Mauzeaud. 1981. The role of catecholamines in the stress response of fish. In: Stress and Fish, ed. A.D. Pickering, pp. 49–75. London: Academic.

McCormick, J.H., K.M. Jensen, and L.E. Anderson. 1989. Chronic effects of low pH and elevated aluminum on survival, maturation, spawning, and embryo-larval development of the fathead minnow in soft water. *Water, Air, Soil Pollut.* **43**:293–307.

McKinley, V.L., and J.R. Vestal. 1982. Effects of acid on plant litter decomposition in an arctic lake. *Appl. Environ. Microbiol.* **43**:1188–1195.

McNichol, D.K., B.E. Bendell, and K. Ross. 1987. Studies of the effects of acidification on aquatic wildlife in Canada: Waterfowl and trophic relationships in small lakes in northern Ontario. Occasional paper No. 62. Ottawa: Canadian Wildlife Service, 31 pp.

Melzer, A., and E. Rothmeyer. 1983. Die Auswirkung der Versaerung der beiden Arberseen im Bayerischen Wald auf die makrophytenvegetation. *Ber. Bayer. Bot. Bes.* **54**:9–18.

Mierle, G., and R. Ingram. 1991. The role of humic substances in the mobilization of mercury from watersheds. *Water, Air, Soil Pollut.* 56:349–357.

Mills, K.H., S.M. Chalanchuk, L.C. Mohr, and I.H. Davies. 1987. Responses of fish populations in lake 223 to 8 years of experimental acidification. *Can. J. Fish. Aquat. Sci.* **44(1)**:114–125.

Minns, C.K., J.E. Moore, D.W. Schindler, P.G.C. Campbell, P.J. Dillon, J.K. Underwood, and D.M. Whelpdale. 1992. Expected reduction in damage to Canadian lakes under legislated decreases in sulfur dioxide emissions. Report 92–1, Committee on Acid Deposition. ISSN1188–911X37. Ottawa: Royal Society of Canada, 37 pp.

Minns, C.K., J.E. Moore, D.W. Schindler, and M.L. Jones. 1990. Assessing the potential extent of danger to inland lakes in eastern Canada due to acidic depostion. III. Predicted impacts on species richness in seven groups of aquatic biota. *Can. J. Fish. Aquat. Sci.* **44(1)**:3–5.

Miskimmin, B.M., J.W.M. Rudd, and C.A. Kelly. 1992. Influence of dissolved organic carbon, pH, and microbial respiration rates on mercury methylation and demethylation in lake water. *Can. J. Fish. Aquat. Sci.* **49**:17–22.

Mossberg, P. 1979. Benthos of oligotrophic and acid lakes. *Rep. Inst. Fresh. Res. Drottningholm* **11**:1–40.

Mount, D.R., J.R. Hockett, and W.A. Gern. 1988. Effect of long-term exposure to acid aluminum, and low calcium on adult brook trout (*Salvelinus fontinalis*). 2. Vitellogenesis and osmoregulation. *Can. J. Fish. Aquat. Sci.* **45**:1633–1642.

Muniz, I.P. 1981. Acidification and the Norwegian salmon. In: Acid Rain and the Atlantic Salmon, ed. S. Lee, pp. 66–71. International Atlantic Salmon Foundation Special Publication 10. St. Andrews, New Brunswick, Canada: IASF.

Muniz, I.P., H. Leivestad, and B.O. Rosseland. 1978. Stressmålinger på fisk i sure vassdrag. (Stress measurements on fish in acidic rivers). *Nordforsk, publ.* **2**:223–247. (In Norwegian).

Muniz, I.P., and H. Leivestad. 1979. Langtidseksponering av fisk til surt vann. Forsok med bekkerye, *Salvelinus fontinalis* Mitchill (Long-term exposure of brook trout to acid water). IR 44/79. 33 p. Oslo: SNSF Project. (In Norwegian).

Neill, W.E. 1984. Regulation of rotifer densities by crustacean zooplankton in an oligotrophic mountain lake in British Columbia. *Oecologia (Berlin)* **61**:175–181.

Nyholm, N.E.I. 1981. Evidence of involvement of aluminum in causation of defective formation of eggshells and of impaired breeding in wild passerine birds. *Environ. Res.* **26**:363–371.

Odum, E.P. 1985. Trends expected in stressed ecosystems. *BioScience* **35**:419–422.

Økland, J. 1980. Environment and snails (Gastropoda): Studies of 1000 lakes in Norway. In: Ecological Impact of Acid Precipitation, ed. D. Drabløs and A. Tollan, pp. 322–323. Oslo: SNSF Project.

Ormerod, S.J., and S.J. Tyler. 1987. Dippers (*Cinclus cinclus*) and Grey Wagtails (*Motacillia cinerea*) as indicators of stream acidity in upland Wales. *ICBP Tech. Publ.* **6**:191–208.

Ormerod, S.J., N.S. Weatherley, P.V. Varallo, and P. Whitehead. 1988. Preliminary empirical models of the historical and future impact of acidification on the ecology of Welsh streams. *Fresh. Biol.* **20**:127–140.

Orr, P.L., R.W. Gradley, J.B. Sprague, and N.J. Hutchinson. 1986. Acclimation-induced change in toxicity of aluminum to rainbow trout (*Salmo gairdneri*). *Can. J. Fish. Aquat. Sci.* **45**:243–246.

Pollman, C.D., and D.E. Canfield, Jr. 1991. Florida. In: Acidic Deposition and Aquatic Ecosystems. Regional Case Studies, ed. D.F. Charles, pp. 367–416. New York: Springer.

Price, E.E., and M.C. Swift. 1985. Inter- and Intra-specific variability in the response of zooplankton to acid stress. *Can. J. Fish. Aquat. Sci.* **42**:1749–1754.

Reckhow, K.H., R.W. Black, T.B. Stockton, Jr., J.D. Vogt, and J.G. Wood. 1987. Empirical models of fish response to lake acidification. *Can. J. Fish. Aquat. Sci.* **44**:1432–1442.

Rosemond, A.D., S.R. Reice, J.W. Elwodd, and J. Mulholland. 1992. The effects of acidity on benthic invertebrate communities in the south-eastern United States. *Freshwater Biol.* **27**:193–210.

Rosseland, B.O. 1986. Ecological effects of acidification on tertiary consumers. Fish population responses. *Water, Air, Soil Pollut.* **30**:451–460.

Rosseland, B.O., I.A. Blakar, A. Bulger, F. Kroglund, A. Kvellstad, E. Lydersen, D.H. Oughton, B. Salbu, M. Staurnes, and R. Vogt. 1992. The mixing zone between limed and acidic river waters: Complex aluminum chemistry and extreme toxicity for salmonids. *Environ. Pollut.* **78**:3–8.

Rosseland, B.O., I. Sevaldrud, D. Svalastog, and I.P. Muniz. 1980. Studies on freshwater fish populations: Effects of acidification on reproduction, population structure, growth and food selection. In: Ecological Impact of Acid Precipitation, ed. D. Drabløs and A. Tollan, pp. 336–337. Oslo: SNSF Project.

Rosseland, B.O., and O.K. Skogheim. 1984. A comparative study on salmonid fish species in acid aluminum-rich water. II. Physiological stress and mortality of one and two year old fish. *Rep. Inst. Freshw. Res. Drottningholm* **61**:186–194.

Rosseland, B.O., O.K. Skogheim, and I. Sevaldrud. 1986. Acid deposition and effects in nordic Europe. Damage to fish populations continue to space. *Water, Air, Soil Pollut.* **30**:381–386.

Rudd, J.W.M., C.A. Kelly, V. St. Louis., R.H. Hesslein, A. Furutani, and M.H. Holoka. 1986. Microbial consumption of nitric and sulfuric acids in acidified north temperate lakes. *Limnol. Oceanogr.* **31**:1267–1280.

Rudd, J.W.M., C.A. Kelly, D.W. Schindler, and M.A. Turner. 1988. Disruption of the nitrogen cycle in acidified lakes. *Science* **240**:1515–1517.

Rudd, J.W.M., C.A. Kelly, D.W. Schindler, and M.A. Turner. 1990. A comparison of the acidification efficiencies of nitric and sulfuric acids by two whole-lake addition experiments. *Limnol. Oceanogr.* **35**:663–679.

Schindler, D.W. 1988. Effects of acid rain on freshwater ecosystems. *Science* **239**:149–157.

Schindler, D.W. 1990. Experimental perturbations of whole lakes as tests of hypotheses concerning ecosystem structure and function. *Oikos* **57**:25–41.

Schindler, D.W., K.G. Beaty, E.J. Fee, D.R. Cruikshank, E.D. DeBruyn, D.L. Findlay, G.A. Linsey, J.A. Shearer, M.P. Stainton, and M.A. Turner. 1990. Effects of climatic warming on lakes of the central boreal forest. *Science* **250**:967–970.

Schindler, D.W., T.M. Frost, K.H. Mills, P.S.A. Chang, I.J. Davis, F.L. Findlay, D.F. Malley, J.A. Schearer, M.A. Turner, P.J. Garrison, C.J. Watras, K. Webster, J.M. Gunn, P.L. Brezonik, and W.A. Swenson. 1991. Freshwater acidification reversability and recovery: Comparisons of experimental and atmospherically-acidified lakes. In: Acidic Deposition: Its Nature and Impacts, ed. F.T. Last and R. Watling, pp. 193–226. Proc. of the Royal Society of Edinburgh, 97B.

Schindler, D.W., S.E. Kasian, and R.H. Hesslein. 1989. Losses of biota from American aquatic communities due to acid rain. *Environ. Monit. Assess.* **12**:269–285.

Schindler, D.W., K.H. Mills, D.F. Malley, D.L. Findlay, J.A. Schearer, I.J. Davies, M.A. Turner, G.A. Linsey, and D.R. Cruikshank. 1985. Long-term ecosystem stress: The effects of years of experimental acidification on a small lake. *Science* **228**:1359–1401.

Schofield, C.L. 1976. Acid precipitation: Effects on fish. *Ambio* **5**:228–230.

Schofield, K., C.R. Townsend, and A.G. Hildrew. 1988. Predation and the prey community of a headwater stream. *Freshwater Biol.* **20**:85–96.

Segner, H., H. Gebhardt, M. Linnenbach, R. Marthaler, and A. Ness. 1991. Influence of low pH on brown trout, *Salmo trutta*. *Verh. Inter. Verein Limnol.* **24**:2470–2473.

Segner, H., R. Marthaler, and M. Linnebach. 1988. Growth, aluminium uptake and mucous cell morphometrics of early life stages of brown trout, *Salmo trutta*, in low pH water. *Environ. Biol. Fishes* **21**:153–159.

Sevaldrud, I.H., I.P. Muniz, and S. Kalvenes. 1980. Loss of fish populations in southern Norway: Dynamics and magnitude of the problem. In: Ecological Impact of Acid Preciditation, ed. D. Drabløs and A. Tollan, pp. 350–351. Oslo: SNSF Project.

Sevaldrud, I.H., and O.K. Skogheim. 1986. Changes in fish populations in southernmost Norway during the last decade. *Water, Air, Soil Pollut.* **30**:381–386.

Shearer, J.A., E.J. Fee, E.R. DeBruyun, and D.R. DeClercq. 1987. Phytoplankton primary production and light attenuation responses to the experimental acidification of a small Canadian Shield lake. *Can. J. Fish. Aquat. Sci.* **44**:83–90.

Siegfried, C.A., J.A. Bloomfield, and J.W. Sutherland. 1989. Acidity status and phytoplankton species richness, standing crop and community composition in Adirondack, New York, U.S.A. lakes. *Hydrobiologia* **175**:13–32.

Skogheim, O.K., B.O. Rosseland, and I.H. Sevaldrud. 1984. Deaths of spawners of Atlantic salmon in River Ogna, SW Norway, caused by acidified aluminum-rich water. *Rep. Inst. Freshw. Res. Drottningholm* **61**:159–202.

Smith, M.A. 1990. The ecophysiology and epilithic diatom communities of acid lakes in Galloway, southwest Scotland. *Phil. Trans. R. Soc. Lond. B* **327**:251–256.

Smol, J.P. 1981. Problems associated with the use of "species diversity" in paleolimnological studies. *Quaternary Res.* **15**:209–212.

Steinberg, C., H. Hartmann, K. Arzet, and D. Drause-Dellin. 1988. Paleoindication of acidification in Kleiner Arbersee (F.R. Germany, Bavarian Forest) by chydorids, chrysophytes, and diatoms. *J. Paleolimnol.* **1**:149–157.

Stenson, J.A.E., and M.O.G. Eriksson. 1989. Ecological mechanisms important for the biotic changes in acidified lakes in Scandinavia. *Arch. Environ. Contam. Toxicol.* **18**:201–206.

Stoner, J.H., A.S. Gee, and K.R. Wade. 1984. The effects of acidification on the ecology of streams in the Upper Tywi catchment in West Wales. *Environ. Pollut. A* **35**:125–157.

Suzuki, I., U. Dular, and S.C. Kwok. 1974. Ammonia or ammonium ion as substrate for oxidation by *Nitrosomonas europaea* cells and extracts. *J. Bacteriol.* **120**:556–558.

Tam, W.H., J.J. Fryer, I. Ali, M.R. Dallaire, and B. Valentine. 1988. Growth inhibition, gluconeogenesis, and morphometric studies of the pituitary and interrenal cells of acid-exposed brook trout (*Salvelinus fontinalis*). *Can. J. Fish. Aquat. Sci.* **45**:1197–1211.

Turner, M.A., E.T. Howell, M. Summerby, R.H. Hesslein, D.L. Findlay, and M.B. Jackson. 1991. Changes in epilithon and epiphyton associated with experimental acidification of a lake to pH 5. *Limnol. Oceanogr.* **36**:1390–1405.

Tyler, S.J., and S.J. Ormerod. 1992. A review of the likely causal pathways relating the reduced density of breeding Dippers *Cinclus cinclus* to the acidification of upland streams. *Environ. Pollut.* **78**:49–56.

Von Ende, C.N., and D.O. Dempsey. 1981. Apparent exclusion of the Cladoceran *Bosmina longirostris* by invertebrate predator *Chaoborus americanus*. *Am. Midl. Nat.* **105**:240–248.

Weatherly, N.S., S.J. Ormerod, S.P. Thomas, and R.W. Edwards. 1988. The response of macroinvertebrates to experimental episodes of low pH with different forms of aluminum, during a natural spate. *Hydrobiol.* **169**:225–232.

Webster, K.E., T.M. Frost, C.J. Watras, W.A. Swenson, M. Gonzalez, and P.J. Garrison. 1992. Complex biological responses to the experimental acidification of Little Rock Lake, Wisconsin, U.S.A.. *Environ. Pollut.* **78**:73–78.

Webster, K.E., A.D. Newell, L.A. Baker, and P.L. Brezonik. 1990. Climatically induced rapid acidification of a softwater seepage lake. *Nature* **347**:374–376.

Witters, H., S. Van Puymbroeck, and O.L.J. Vanderborght. 1991. Adrenergic response to physiological disturbances in rainbow trout, *Oncorhynchus mykiss*, exposed to aluminum at acid pH. *Can. J. Fish. Aquat. Sci.* **48**:414–420.

Wright, R.F., E. Lotse, and A. Semb. 1993. RAIN Project: Results after 8 years of experimentally reduced acid deposition to a whole catchment. *Can. J. Fish. Aquat. Sci.* **50:**258–268.

Wright, R.F., T. Dale, A. Henriksen, G.R. Hendrey, E.T. Gjessing, M. Johannessen, C. Lysholm, and E. Stren. 1977. Regional Survey of Small Norwegian Lakes. IR33/77. Oslo: SNSF Project.

Xun, L., N.E.R. Campbell, and J.W.M. Rudd. 1987. Measurements of specific rates of net methyl mercury production in the water column and surface sediments of acidified and circumneutral lakes. *Can. J. Fish. Aquat. Sci.* **44**:750–757.

Yan, N.D., and P. Stokes. 1978. Phytoplankton of an acidic lake, and its responses to experimental alterations of pH. *Environ. Cons.* **5**:93–10.

# 20

# Does Reduction in $SO_2$ and $NO_x$ Emissions Lead to Recovery?

M. HAUHS
Bayreuth Institute for Terrestrial Ecosystem Research (BITÖK),
Dr.-Hans-Frisch Str. 1-3, Postfach 10 12 51, 95448 Bayreuth, F.R. Germany

## ABSTRACT

Clear empirical links exist between atmospheric deposition and its indirect soil-mediated effects on aquatic ecosystems. Such links form the basis of conceptual models that are used for emission control policies. One prominent example is the critical load concept. A review of field observations shows that our ability to predict the biological consequences of decreased deposition is still uncertain. This is quantitatively due to the relative short time over which such observations have been made and qualitatively due to the lack of biological recovery in ecosystem-scale experiments with artifically decreased deposition.

The transfer of models derived and tested under various acidification scenarios to the deacidification case should be based on a clear distinction between observations and interpretations. In this chapter, I propose definitions for *ecosystem*, *reversibility*, and *recovery*. These definitions are then used to inspect two alternate interpretations of the available field evidence. Two hypothetical cases are discussed that assess potential causal links between deposition fluxes and the biological response in receiving ecosystems. It is not possible to discard one of these two principally different views. Therefore, the lack of experimental evidence on recovery in manipulative studies with whole ecosystems may be the result of (a) experiments at appropriate scales having not been performed long enough or (b) experiments having asked the wrong questions.

## INTRODUCTION

Over the last decades, aquatic and terrestrial ecosystems in remote, sensitive regions of Europe have shown unexpected biological effects, such as the loss of fish populations or forest decline. Research initiated by these effects has demonstrated clear links between the biological status of whole ecosystems and the spatial and temporal trends in atmospheric deposition (Drabløs and Tollan 1980; Mason 1990; Last and Watling 1991).

*Acidification of Freshwater Ecosystems: Implications for the Future*

The overall picture of acid deposition effects at regional and long-term time scales is relatively simple and empirically predictable (Cosby et al., this volume). Accordingly, acid deposition research succeeded in establishing empirically reliable predictions as a basis for ecosystem management. In contrast to the preacidification situation, these predictions have to include deposition fluxes from the atmosphere. Such predictive models combine estimates of the degree of external disturbance with estimates about the sensitivity of the affected system. They combine two input fluxes into ecosystems: (a) the deposition from the atmosphere as the external load and (b) the input of base cations from silicate weathering as a measure of site-specific sensitivity. A regional assessment of these factors has been initiated across Europe to assess the acidification risk (Hettelingh et al. 1991).

In response to the clear relation between surface water acidification in sensitive waters of central and northern Europe and atmospheric deposition, emission control was installed and gradually led to a decrease in deposition for some key elements. The deposition of acidifying substances from the atmosphere is, to a large extent, a result of $SO_2$ and $NO_x$ emissions. Presumably a reduction of $SO_2$ and $NO_x$ emissions to preindustrial levels will be sufficient to reestablish preindustrial regimes of atmospheric deposition.

Two questions arise: Do the empirical links to biological responses also hold under reversed deposition trends? Do the conceptual models that are able to match the recent acidification history of catchments also hold under this reversed scenario?

During the past decade of acidification research, the measurement of input and output fluxes to whole landscape units, such as catchments, became one of the most important methods of analyzing the impact of atmospheric deposition (Hornung et al. 1990). In this chapter, I use the term *ecosystem* to describe such functional units. The location of the boundaries of a functional ecosystem unit is characterized by the irreversible input and output fluxes (Hauhs 1992). Most fluxes of matter across ecosystem boundaries occur far from thermodynamic equilibrium. Typically, input fluxes (e.g., rainfall and weathering) do not depend on the state of corresponding intensity variables within the ecosystem whereas output fluxes (e.g., runoff and precipitation of secondary minerals) do not depend on the state of the environment.

Measurements in catchment ecosystems have resulted in two key observations, which, in turn, serve as the starting point for discussion of potential future deacidification trends:

1. Acidification of fresh waters (measured as a loss of alkalinity) results from a change in deposition fluxes from the atmosphere.
2. In catchments dominated by soils and bedrock resistant to chemical weathering, this additional, anthropogenically induced acidification of soils led to a negative alkalinity in surface waters with a concurrent decline of fish populations and other changes in the aquatic biota.

The term *acid deposition* summarizes a number of interrelated elements and aspects by which input fluxes into ecosystems have changed over the last century, due to increased air pollution. Simple predictive models use the total proton equivalent or the total sulfate flux as a measure of the external loading.

In acidification research, attempts have often failed to establish the causality underlying the simple ecosystem-scale picture at the level of internal processes (Hauhs 1990a; Christophersen et al. 1993). The concept of critical loads typifies conceptual models that summarize important aspects of the empirical evidence at hand for management purposes. As a result of the incomplete understanding of ecosystems, the critical load concept must combine empirical evidence with implicit assumptions about the underlying causality. Such conceptual simplifications are justified as long as models are not used outside the experience upon which they are based. This may be the case for a reversed trend in deposition or for catchment-scale experiments.

The task of predicting ecosystem development within a deacidification scenario thus requires a reassessment of the results from the acidification phase.

## THE STARTING POINT FOR A REVERSAL OF ACIDIFICATION TRENDS

The response of freshwater ecosystems to acid deposition can be conceptually split into a physical/chemical aspect (water and soil acidification) and its subsequent biological component (decline of fish). Biological effects of acid deposition in terrestrial ecosystems, however, cannot be conceptualized in this same, simple way. Several pathways of positive and negative feedback between abiotic and biotic elements of these ecosystems (e.g., in form of the N-cycle) render such a separation difficult, if not impossible. This is why the relative importance of internal abiotic factors in forest decline (e.g., Al toxicity) is still unclear. As in many other ecosystem-scale problems (e.g., Peters 1991), attempts to reveal the complete internal causality behind simple empirical models have failed. In the case of forest decline, the emerging picture of empirical relations between external loadings, site sensitivity, and forest decline resulted in a trend similar to that of freshwater acidification: predicitive models became gradually independent of a successful causal analysis of the underlying mechanisms.

In the case of historical acidification, the initial (preindustrial) state of affected ecosystems was largely unknown and could usually not be inferred from present observations. The political goal to reverse acidification poses the opposite problem for science. Now, the initial state can be observed, but experience-based models will be few and preliminary. Ecosystem-scale experiments and a theoretical assessment of the acidification problem become critically important.

The potential sources of information about reversed trends in ecosystem acidification are:

1. results from ecosystem monitoring (new, relevant experience);
2. model predictions based on the experience gained during the preceding acidification phases, e.g., the MAGIC model (extrapolated experience);
3. results of ecosystem-scale experiments with manipulated deposition fluxes (new, artifical experience);
4. theoretical assessment of the biological effects from expected or concurrent changes in deposition fluxes (inferred from theoretical understanding).

Several authors have examined the reversibility of acidification (Hutchinson and Havas 1986; Hauhs and Wright 1988; Gunn 1990; Wright and Hauhs 1991; Schindler et al. 1991). Emphasis was given to the first three sources (see above) and to the reversibility of the physical/chemical aspects of acid deposition effects. Here, I wish to update recent empirical data and supplement a discussion of some theoretical problems involved in the fourth item. This theoretical emphasis allows an inclusion of biological recovery, an issue where, even today, empirical data are sparse or missing. The results are summarized in Table 20.1.

An important, yet still unanswered question in acidification research can be phrased as follows: Given the clear evidence by which boundary flux changes (e.g., from the atmosphere) relate to problems in predicting and managing ecosystems (first item, Table 20.1), what is the appropriate level of scientific resolution for a theoretical explanation of the observed effects (fourth item, Table 20.1)? The discussion below includes the extreme view that no such level might exist beyond the ecosystem itself.

**Table 20.1** Summary of the difference between the state of knowledge on acidification and deacidification cases.

| | Increased Deposition | Reversal of Deposition |
|---|---|---|
| Strong empirical evidence of biological effects | yes | yes |
| Conceptual models available | yes | yes |
| Biological response reproduced by ecosystem-scale experiments | yes | no |
| Causality of dose-response relationships revealed in terms of internal processes | no | no |

## THE BASIS FOR PREDICTIONS OF REVERSIBILITY AND RECOVERY

### Evidence of Decreasing Deposition Loads and Their Effect on Surface Water Chemistry

In addition to the examples given by earlier reviews (Hauhs and Wright 1988; Wright and Hauhs 1991), evidence is now also available from areas of high deposition in central Europe. At the Solling and Lange Bramke sites, throughfall fluxes of sulfate underneath Norway spruce canopies decreased by about 20-30% over the last five years (Matzner, pers. comm.; Hauhs, unpubl.). Records of $SO_2$ gas concentration from the nearby Brocken, the highest elevation in the Harz mountains, show a marked decline after 1987. This can be attributed to a series of mild winters and decreasing emissions from former East Germany. For sulfur, the recent emission trends in Germany are reflected in throughfall records from long-term monitoring sites (Gravenhorst, unpubl.).

No data are available that indicate declining N-deposition in any of the regions suffering from acid deposition effects.

At the Lange Bramke catchment, the decrease in deposition fluxes of sulfate has not yet been accompanied by a response in streamwater chemistry. Sulfate concentration levels do not show any trend between the first sampling in 1969 and recent samples. This is also true for the short-term time signals in sulfate export.

The Lange Bramke catchment also responded slowly during the acidification phase. In 1986, prior to the onset of a decreasing deposition trend in sulfate, the corresponding export rate in stream water had only reached 50% of the atmospheric loading. Sulfate export was almost constant with a small superimposed seasonal fluctuation parallel with nitrate (Hauhs 1990b).

The export rates at Lange Bramke are within the lower 15% range for the Harz mountains. A regional survey of 100 springs in the Harz mountains taken in 1991 indicated a small reduction of sulfate relative to a 1974 survey (Ostertag, unpubl.). Given the minor and so far insignificant changes in streamwater chemistry, it is unlikely that biological recovery has yet occurred.

### Monitoring of Recovery

Experiences at Sudbury, Canada, provide us with an example of direct observations of reversing trends in acidification and biological recovery (Keller et al. 1992). Here, the close empirical relation between deposition changes and biota also holds in the reversed direction. In the cases where fish populations had already become extinct or were close to extinction, reversal of the chemical lake status did not result in recovery. In such cases, recovery may largely depend upon the dispersal strategies of the species involved.

Another potential source for recovery observations are those ecosystem-scale experiments where deposition is changed by means of a roof. Such experiments are underway in Risdalsheia, Norway, and at Gårdsjön, Sweden (Wright et al. 1990; Hultberg, unpubl.). These experiments require a catchment size too small to support a perennial headwater stream with a typical range of aquatic biota. As the spatial variability and range of stream chemistry within these highly sensitive areas is much less than the time variability, such headwater experiments can still be used for bioassays (Kroglund and Rosseland 1992). The results from the RAIN Project give no indication of a biological recovery potential, although the manipulated deposition clearly reduced the proportion of strong mineral acids in the stream water.

At Gårdsjön the roof treatment results in an immediate decline of streamwater sulfate (Hultberg, pers. comm.). Here a bioassay has not yet been performed. Both the experiments at Risdalsheia and at Gårdsjön are based on the assumption that the total amount of acid deposition is the most relevant factor and therefore only this factor must be reduced to historical levels. The manipulations, however, have almost the opposite effect with respect to the predictability (complexity) of the input signals. The range of flux rates from the atmosphere is reduced to one application rate with a constant composition. That is why such experiments provide an opportunity to test the hypothesis about alternate links of causality. If the complexity of input fluxes is important in its own right, the manipulation will impose increasing biological stress rather than initiate recovery.

## Model Predictions

The final way to assess the future relations between declining deposition and recovery of affected systems is through modeling. Here empirical relations are interpreted by means of conceptual models of long-term soil acidification (e.g., the MAGIC model; Cosby et al. 1985) and are used to extrapolate the empirical links to the condition of reversed deposition trends.

All existing models of freshwater acidification focus on the input-output budgets and/or equilibrium concepts applied to ecosystems. The widely used MAGIC model has been successfully applied to several manipulated catchments, both for increased and decreased deposition rates (Wright et al. 1990; Norton et al. 1992). However, the interpretation adopted by this model (e.g., annually weighted averages of stream chemistry) has not been shown to be causal for the biological response, although it may be a good empirical indicator for many field situations.

In light of the sparse empirical evidence from the short history of declining deposition rates, it seems necessary to achieve independent support for the far-reaching assumptions that must be built into such models. Future environmental risks, such as climate change, may require mitigating strategies to be effective before *experience-based* models are available at the relevant scale. For this reason, theoretical aspects of ecosystem disturbance may be of greater importance in the development of models.

## TERMS USED TO CHARACTERIZE DEACIDIFICATION TRENDS

The reestablishment of a preacidification status for physical factors[1] will be discussed as *reversibility* of acidification, whereas for biological aspects, the term *recovery* will be used. These two terms, *reversibility* and *recovery*, correspond to the conceptual split between the physical and biological elements of ecosystems.

### Recovery

Recovery is a coined term for organism-like aspects of ecosystems and is defined ambiguously. It is typical of biological structures where, at any time scale considered, the time order seems unique and cannot be reversed. This constitutes a necessary condition for biological memory and adaption. Processes such as growth, succession, or evolution are all irreversible in the strictest sense (Whitrow 1980). No change in their spatial boundary conditions allows the reestablishment of a past "ecosystem state." Repetition in biological systems is only possible by direct interference with the composition of species, e.g., by direct control of the initial state. Moreover state functions for ecosystems have not been identified and use of the term *state* implies a number of yet untested assumptions about the nature of ecosystems.

The concept of ecosystem stability is thus difficult to define in terms of internal processes. It is first through the introduction of, and relative to, a human time scale of interest that many of these irreversible processes are slow enough to be ignored. Unlike other biological integration levels, the relation between structure and function at the ecosystem scale is unknown.

Human utilization of ecosystems typically consists of a periodical resetting of those initial "states" for which empirical prediction models for their future development exist. Ecosystem stability is often confused with the long-term predictability of ecosystem function. The latter concept, however, allows us to define recovery in a pragmatic way, i.e., recovery of affected ecosystems should proceed to a state where ecosystem management and prediction can again ignore atmospheric deposition. For this concept, a sufficient condition for complete recovery is fulfilled if the predictability of aquatic biota within an affected region is no longer impacted by atmospheric deposition fluxes, although the restocked or migrated set of populations may not be the original (locally adapted).

### Reversibility

The terms *state* and *state functions* can, in principle, be used for abiotic systems that enclose an ecosystem. Based on the assumption that the interaction between ecosystems (functional aspects) and their abiotic environments can be completely described by irreversible fluxes of matter and energy, a definition of reversibility can be given:

[1] The term *physical* will be used in a broad sense to include all abiotic, physical, and chemical factors.

the function of an ecosystem, with respect to its abiotic environment, has fully reversed when the flux density (function of space and time at the boundary) of the considered elements have returned to their preacidification state (e.g., when compared to pristine sites).

Many ecosystem models, especially those for ion budgets, introduce additional internal state variables, such as the base saturation of the soil. A definition of reversibility with respect to these internal storages cannot yet be made independent of the underlying model or the operationally defined methods of the respective measurements. The definition of a necessary condition relies on input-output fluxes only. In a simple way it relates to the definition of recovery without using further assumptions about internal variables. It may also prove to be a sufficient condition by future observation. If memory effects, however, can also be demonstrated for physical aspects of ecosystems, it must be supplemented by additional internal measures.

The proposed definition of reversibility relates to flux densities and not to their integrals over space and time. Input fluxes across ecosystem boundaries are usually much more complicated, in terms of patterns in space or time, than the corresponding output fluxes. This is an essential feature that makes ecosystem function predictable (and capable of being modeled). However, most budgeting approaches simplify this difference by integrating input and output fluxes over space and time.

According to this evaluation of flux data, we implicitly assume that the total mass within this flux provides the causal link to biological effects. Although this is only one of the two hypotheses, it is usually used exclusively over several scales (Hauhs 1990a); if it fails at a course resolution, it will be reformulated automatically for the next higher resolution. A typical example is the history of hypotheses regarding the biological significance of inorganic Al in soil solution. The toxic fraction was gradually hypothesized for smaller time and spatial scales, where it is unobservable in the field by current methods.

Despite the above-mentioned difficulties, an interpretation of the internal mechanism by which the deposition effect becomes manifest is indispensible in order to define a useful measure for the reversibility of acidification. For example, the critical load concept assigns a clear meaning to the reversal of acid deposition; its total amount has to approach historical levels.

I would like to mention briefly an alternative interpretation based on the same observations. A more complete discussion is beyond the scope of this chapter. The primary aspect of ecosystem function is seen in the quantitative accumulation of information about the physical environment through the process of mutation and selection of its biological members. In this view, the whole ecosystem can be treated as an information processing device that responds to the information contained in the input signal of boundary matter fluxes within the sensitivity range.[2] The ecosystem

[2] This hypothesis is a computer paradigm to ecosystem functioning. Irreversible erasure and thus compression of information is characteristic of any computer. Although this task requires energy dissipation, its overall function can be completely abstracted from its internal and external energy fluxes. It is not feasible to resolve software malfunctions by exclusively analyzing hardware structures or energy relations.

(or its biological elements) might be adapted to a certain type of signal or a certain predictability of input fluxes. Individual biological elements have no possibility to sense boundary fluxes directly.

The biological effect of a changed input flux may thus be linked to its complexity in space and time (predictability from the viewpoint of the receiving system), while the flux integrals (the total transferred masses) are irrelevant. Only within the ecosystem environment, e.g., the atmosphere, might the total mass transfer and complexity of the flux density be coupled: more simply, predictable input fluxes must be accompanied by higher totals.

The output fluxes from acidified terrestrial ecosystems show that these aspects do not vary independently: acidified waters are not only characterized by gross chemical changes, such as a lowered alkalinity, pH, or an increased sulfate concentration, the autocorrelation among these elements or to the discharge rate is also decreased. In general, the chemistry of acidified waters is less predictable (given the runoff) than it is in a corresponding unaffected system.

In summary, the concept of reversibility requires a quantitative measure of changes in deposition fluxes. For any flux signal, this measure can be chosen from two complementary abstractions: the total mass or the total complexity of the recorded fluxes. The empirical evidence from the past history of acidification does not seem to discriminate between the two possible abstractions. However, any decision on a reversibility experiment or any assessment of critical loads must implicitly select only one of the complementary perspectives. In the past, the mass aspect was exclusively selected for the prediction tasks.

## CONCLUSIONS

After more than a century of acid deposition, its effect on ecosystems has become empirically predictable. A similar ability to predict the biological consequences of a reversed trend in disturbance is still unreliable, with data accumulating at a slow pace.

Extrapolation of existing knowledge into future deposition scenarios depends critically upon basic but untested assumptions as to the causality in disturbed ecosystems. Our lack of understanding is usually regarded as unavoidable but irrelevant in the light of working prediction models for the present deposition. Theoretical understanding may, however, become limiting for extrapolations into the future. For example, the ongoing ecosystem-scale experiments may focus on the wrong aspect of acid deposition inasmuch as they exclusively test changes in the total mass of the manipulated flux densities.

An alternate approach is suggested that regards information processing as the primary function of ecosystems with respect to their physical environment. It can be used to explain the acidification history but predicts further deterioration for the biota under the ongoing roof experiments. Reduction in emissions is still the single most likely factor to lead to ecosystem recovery; however, past predictions might have been correct for the wrong reasons.

## ACKNOWLEDGEMENTS

This review is based on work funded by the EC (EG-EV4V-0032-D) and the German Ministry for Research and Technology (BMFT), Grant No. 0339476A.

## REFERENCES

Christophersen, N., C. Neal, and R.P. Hooper. 1993. Modelling the hydrochemistry of catchments: A challenge for the scientific method. *J. Hydrol.*, in press.

Cosby, J.B., R.F. Wright., G.M. Hornberger, and J. Galloway. 1985. Modelling the effects of acidic deposition: Assessment of a lumped-parameter model of soil and streamwater chemistry. *Water Resour. Res.* **21**:51–63.

Drabløs, D., and A. Tollan, eds. 1980. Ecological Impact of Acid Precipitation. 1432 Ås, Norway: SNSF-Project, 383 pp.

Gunn, J.M. 1990. Biological recovery of an acid lake after reductions in industrial emissions of sulphur. *Nature* **345**:431–433.

Hauhs, M. 1990a. Ecosystem modelling: Science or technology? *J. Hydrol.* **116**:25–33.

Hauhs, M. 1990b. Lange Bramke: An ecosystem study of a forested watershed. In: Acidic Precipitation, vol. 1: Case Studies, ed. D.C. Adriano and M. Havas, pp. 275–305. New York: Springer.

Hauhs, M. 1992. A definition of ecosystems based on information theory. In: Mathematical Modelling of Forest Ecosystems, ed. J. Franke and A. Roeder. Frankfurt/Main: J.D. Sauerländer's Verlag.

Hauhs, M., and R.F. Wright. 1988. Reversibility of soil and water acidification, a review. Air Pollution Report No. 11. Brussels: Comm. European Community.

Hettelingh, J.-P., R.J. Downing, and P.A.M. de Smet, eds. 1991. Mapping critical loads for Europe. RIVM Report No. 90–6960–011–0. Bilthoven: RIVM (Box 1,3270 BA Bilthoven, The Netherlands).

Hornung, M., F. Roda, and S.J. Langan, eds. 1990. A review of small catchment studies in Western Europe producing hydrochemical budgets. Air Pollution Res. Report 28. Brussels: Comm. European Community.

Hutchinson, T.C., and M. Havas. 1986. Recovery of previously acidified lakes near Coniston, Canada following reductions in atmospheric sulphur and metal emissions. *Water, Air, Soil Pollut.* **28**:319–333.

Keller, W., J.M. Gunn, and N.D. Yan. 1992. Evidence of biological recovery in acid-stressed lakes near Sudbury, Canada. *Environ. Pollut.* **78**:79–85.

Kroglund, F., and B.O. Rosseland. 1992. Reversibility of acidification: Fish responses in experiments at Risdfalsheia, Norway. Acid Rain Res. Report 27/1992. Oslo: NIVA (Box 69 Korsvoll, Oslo, Norway).

Last, F.T., and R. Watling, eds. 1991. Acidic deposition, its nature and impacts. *Proc. Roy. Soc. Edinburgh* **97B**:343.

Mason, J., ed. 1990. The Surface Waters Acidification Programme. Cambridge: Cambridge Univ. Press.

Norton, S.A., R.F. Wright, J.S.Kahl, and J.P. Scofield. 1992. The MAGIC simulation of surface water acidification at, and first year results from, the Bear brook watershed manipulation, Maine, USA. *Environ. Pollut.* **77**:279–286.

Peters, R.F. 1991. A Critique for Ecology. Cambridge: Cambridge Univ. Press.

Schindler, D.W., T.M. Frost, K.H. Mills, P.S.S. Chang, I.J. Davies, L. Findlay, D.F. Malley, J.A. Shearer, M.A. Turner, P.J. Garrison, C.J. Watras, K. Webster, J.M. Gunn, P.L. Brezonik, and W.A. Swenson. 1991. Comparisons between experimentally- and atmospherically-acidified lakes during stress and recovery. *Proc. Roy. Soc. Edinburgh* **97B**:193–226.

Whithrow, G.J. 1980. The Natural Philosophy of Time. Oxford: Oxford Univ. Press.

Wright, R.F., and M. Hauhs. 1991. Reversibility of acidification: Soils and surface waters. *Proc. Roy. Soc. Edinburgh* **97B**:169–191.

Wright, R.F., J. Cosby, M.B. Flaten, and J. Reuss. 1990. Evaluation of an acidifcation model with data from manipulated catchments in Norway. *Nature* **343**:53–55.

# 21

# The Influence of Catchment Management Processes in Forests on the Recovery in Fresh Waters

K. KREUTZER
Department of Soil Science, University of Munich,
Hohenbachernstr. 22, 85354 Freising, F.R. Germany

## ABSTRACT

Management processes in forested catchments, which act to increase alkalinity and pH of drainage water, include liming, changing tree species composition, harvesting, and other silvicultural management procedures. Application of 2–4 t $ha^{-1}$ ground limestone on the forest floor may provide an efficient shield against acid precipitation until the lime is dissolved, after about 5–10 years. The change of alkalinity and pH in drainage water, however, decreases with increasing soil depth and is highly dependent on the base-neutralizing capacity of the soil and on the hydrological soil conditions. A large part of the buffer capacity of the lime is consumed by neutralizing the potential acidity of variable charges in the surface humus layer, in part restoring the capacity of the exchange buffer. This type of buffer, however, is less efficient than the carbonate buffer of fine-grained limestone. Liming of forests includes risks for the water quality due to nitrification and mobilization of heavy metal.

Tree species composition affects alkalinity of drainage water. The differences between species are attributed to different filtering efficiencies of air pollutants due to various canopy characteristics, to different capacities for storing organic N and S in the mineral soil (primarily as a result of species-typic fine root distribution pattern), and to different incorporation rates of N in the biomass of the species. Soil hydrological effects such as transient water logging, which favors denitrification, may also affect pH and alkalinity of the drainage water.

Harvesting of forest products influences the water quality via base cation export. Silvicultural management procedures may contribute to this by extending the rotation time, by changing the regeneration system, and by regulating stand density.

Liming of forest soils over large areas cannot be recommended without reserve due to the risks. Deciduous tree species (with the exception of alder, at particular sites, and *Robinia*) that have spatially extended fine-root formation and which are grown in long rotations are advantageous.

Clear-cutting should be avoided. Restoration of freshwater quality by forest management procedures is a slow and long-lasting process. Only under specific conditions, such as soils with predominantly subsurface flow and low buffering capacity, can a rapid effect of forest manipulation on the surface water be expected.

*Acidification of Freshwater Ecosystems: Implications for the Future*
Edited by C.E.W. Steinberg and R.F. Wright 

# INTRODUCTION

Recovery of biota in acidified fresh waters can occur if the preacidification alkalinity and pH is achieved. Alkalinity, an expression for the acid-neutralizing capacity (ANC), can be defined as

$$\text{alk (aq.)} = 2\,[Ca^{2+}] + 2\,[Mg^{2+}] + [K^{+}] + [Na^{+}] + [NH_4^{+}] - 2\,[SO_4^{2-}] - [NO_3^{-}] - [Cl^{-}], \tag{21.1}$$

where the units are mol $L^{-1}$. This definition is termed *charge balance alkalinity* (Hemond 1990). It holds for oxic conditions and a range of pH of about 3.5–8, over which it is possible to identify "strong" or "weak" acid or base species. If appreciable levels of organic acids or aluminum are present, further parameters for ANC must also be considered.

In this chapter, I focus on four catchment management processes, each of which may increase the alkalinity and pH of drainage water in forest ecosystems on acidified sites. These processes are:

1. liming,
2. regulation of the tree species,
3. harvesting, and
4. silvicultural management.

It often takes a long time for the effects of such manipulations to appear in the groundwater or runoff water of the catchment. This may be due to feedbacks and buffering capacities within the forest ecosystem as well as in the deeper seepage zone. In this deeper zone, reactions depend, to a large part, on the geohydrological conditions and may disguise the effects of the management processes for some time. Therefore, it is important to study these effects within the ecosystem, especially at the lower boundary of the rooting zone, and they should also be studied over a longer time period.

A relatively rapid response in surface waters to specific management procedures in catchments may be expected if subsurface flow on steep slopes prevails.

# LIMING

## General

Liming has been a silvicultural practice in central Europe for about 80 years. It has been used to improve growth conditions and, more recently, to compensate specifically for acid deposition, thereby avoiding primarily aluminum toxicity to the tree roots. Techniques developed thus far can be adapted to liming, with the intent of restoring freshwater quality. Well-proven is the application of carbonate lime with a high dolomitic content, usually in the form of ground dolomitic limestone. According to

Gussone (1987), such carbonate lime distributed by helicopter should have a granulation ranging primarily from 0.2 and 2 mm, with 70% below 1 mm in diameter. The main advantages of this method are the easy adaptation of dosing rates to the site conditions and its long-lasting effect.

Siliceous lime and lime resulting from ore-melting processes are also recommended. The advantage of their usage is the lower solubility relative to carbonates. Thus they provide a longer-lasting effect. Most liming involves application to the soil surface; however, there have also been some attempts to put the lime into deeper soil layers.

## Acid Buffering by Lime

### *The Buffer Capacity*

Acid neutralization occurs by dissolution of the lime in two steps:

1. With regard to calcite:

$$CaCO_3 + H^+ \rightarrow HCO_3^- + Ca^{2+}$$
$$HCO_3^- + H^+ \rightarrow CO_2 + H_2O$$

---

$$CaCO_3 + 2H^+ \rightarrow Ca^{2+} + CO_2 + H_2O$$

2. With regard to dolomite:

$$CaMg(CO_3)_2 + 2H^+ \rightarrow 2HCO_3^- + Ca^{2+} + Mg^{2+}$$
$$2HCO_3^- + 2H^+ \rightarrow 2CO_2 + 2H_2O$$

---

$$CaMg(CO_3)_2 + 4H^+ \rightarrow Ca^{2+} + 2Mg^{2+} + 2CO_2 + 2H_2O$$

Liming dose is usually 2–4 t $ha^{-1}$ carbonate content 95–98%, and thus has an ANC of about 40 to 80 kmol $ha^{-1}$.

Since 1850, the total acidic emission in the area covered by preunified West Germany amounts to about 400 kmol $ha^{-1}$ (Ulrich 1989; UBA 1989). Assuming that this acidity has accumulated in the soil, a lime dose of 20 t $ha^{-1}$ would be necessary to neutralize it. In practice, this would mean liming 6–7 times each with 3 t $ha^{-1}$.

### *The Lime Dissolution Rate in Forest Soil*

On one hand, the lime dissolution rate depends on material properties, e.g., the proportion of calcite and dolomite as well as the grain-size distribution. On the other, the rate is influenced by such forest soil properties as (a) the pH of the soil solution, (b) the potential of protonized functional groups, (c) the removal of the dissolution products by drainage, (d) the $CO_2$ content of the soil air, and (e) the soil temperature. Little is known regarding the dissolution behavior of carbonate lime in acidic forest soils.

On the basis of theoretical considerations, Prenzel (1985) calculated a maximum annual dissolution rate of 1000 kg $ha^{-1}$ of carbonate lime. He assumed a surface distribution of the lime, 10 kmole $ha^{-1}$ $yr^{-1}$ proton input, 1000 mm precipitation annually, $CO_2$ partial pressure and a temperature typical for forest soils. He did not, however, make allowance for the base-neutralizing capacity of the protonized functional groups of the organic matter and the removal of dissolution products by way of drainage.

Kreutzer et al. (1991) studied the lime dissolution in the Höglwald field experiment (Figure 21.1) and produced a lime dissolution curve experimentally. Noteworthy conditions underlying this experiment were:

- surface liming with 4000 kg $ha^{-1}$ of dolomitic lime (carbonate portion 3865 kg; Ca:Mg ratio close to 1),
- a mean bulk precipitation of 891 mm,
- a mean throughfall volume 524 mm,
- a mean annual temperature of 7.3,
- a mean proton input of 20 mmol $m^{-2}$ $yr^{-1}$,
- a mor humus layer of 6 cm,
- a pH ($CaCl_2$) of 3.20 in LOf, of 2.91 in Oh, and of 3.20 in Ah.

Figure 21.1 shows that the temporal curve of the remaining amount of dolomite in variation with time follows an exponential function, such that

$$m\,(t) = 3865 \times \ exp. - 0.6 \times \ t, \tag{21.2}$$

where $m$ is the amount of remaining dolomite in kg, and $t$ is the number of years.

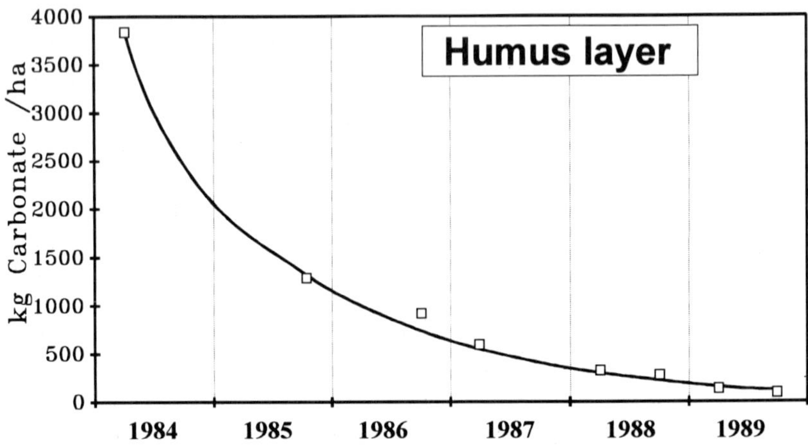


**Figure 21.1** Carbonate dissolution at the Höglwald site. 4 t $ha^{-1}$ lime (3865 kg $ha^{-1}$ carbonate) were applied in April, 1984 on forest floor.

The half-life of the dissolution was 1.15 years; about 2 t $ha^{-1}$ dissolved in the first year. All dolomite was dissolved after 6–7 years, which agrees well with observations made in the forest (Bichlmaier 1985; Aldinger 1987; Reinwald 1987).

### *The Change of the Buffer Substance*

The dissolution of the lime does not necessarily lead to a corresponding reduction of the ANC in the soil. This applies to substrates with variable charge, such as humic layers. Their functional groups, predominantly carboxylic groups, deprotonize with increasing pH and can exchange Ca and Mg ions for protons:

$$R-(COO)H + M^{+} \leftrightarrow R-(COO)M + H^{+}, \tag{21.3}$$

where $M^{+}$ denotes metal ions. This is a reversible $H^{+}$ buffer reaction. Here, in contrast to the carbonate, the buffer substance is not destroyed. The adsorption of Ca and Mg corresponds to the restoration of the buffer. In this way, the applied carbonate can be partially transformed into exchange buffer with variable charge.

The Höglwald experiment also provides insight into the time and the depth distribution of this buffer transformation (see Figure 21.2). The adsorption of Ca is higher than Mg. After complete dissolution of the lime, about 70% of the Ca and 30% of the Mg are still present in the humus layer. About 50% of the ANC of the applied carbonate is retained as an exchange buffer. Down to 40 cm depth, the retention is about 90%.

### *Effects on pH and Alkalinity*

For the different buffer substances, Schwertmann et al. (1987) ascertained principal pH-ranges of buffering (see Table 21.1). Whereas lime particles and solid organic matter are more or less immobile buffer substances, $HCO_3^-$ as well as dissolved organic compounds (complexed with Ca and Mg) serve as mobile buffers. They are translocated by the drainage water, thereby increasing pH at depth. A homogenous deacidification front is not formed; however, in macropores the deacidification proceeds faster than in the soil as a whole.

In the Höglwald experiment, the soil solution shows a pH rise up to the $HCO_3^-$ buffer range at the bottom of the humus layer; however, no change was observed in soil solution at a depth of 20 cm (Figure 21.3). The pH (KCl) of the soil as a whole changed only in the surface humus layer and showed a steep gradient with depth (Figure 21.4). This limited pH increase in the uppermost soil layers is due primarily to deprotonation reactions of humic substances.

The gradient is translocated downward by about 1 cm $yr^{-1}$ (Figure 21.4). This is probably due to buffering as well as to recent litter deposition. There is also a noticeable reacidification from the soil surface due to the fresh litter fall. The original low pH is not reached, however, because earthworms are promoted by liming and mix

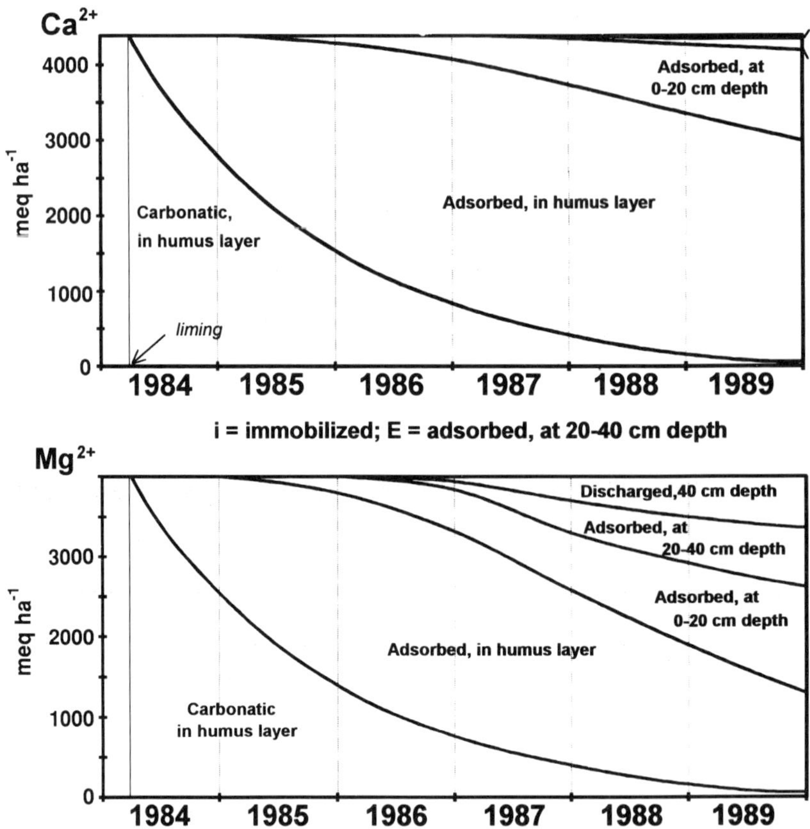


**Figure 21.2** Distribution of $Ca^{2+}$ and $Mg^{2+}$, derived from lime.

**Table 21.1** Principal pH ranges of buffering (Schwertmann et al. 1987).

| | | pH Range |
|---|---|---|
| Carbonates | | |
| 1. Step | $CaCO_3 + H^+$ | 8–6.5 |
| 2. Step | $HCO_3 + H^+$ | 7–4.5 |
| Organic Materials | | |
| Carboxyl groups | R–(COO)M + $H^+$ | 6–<3 |
| Amino groups | R–$NH_2$ + $H^+$ | >7–4 |

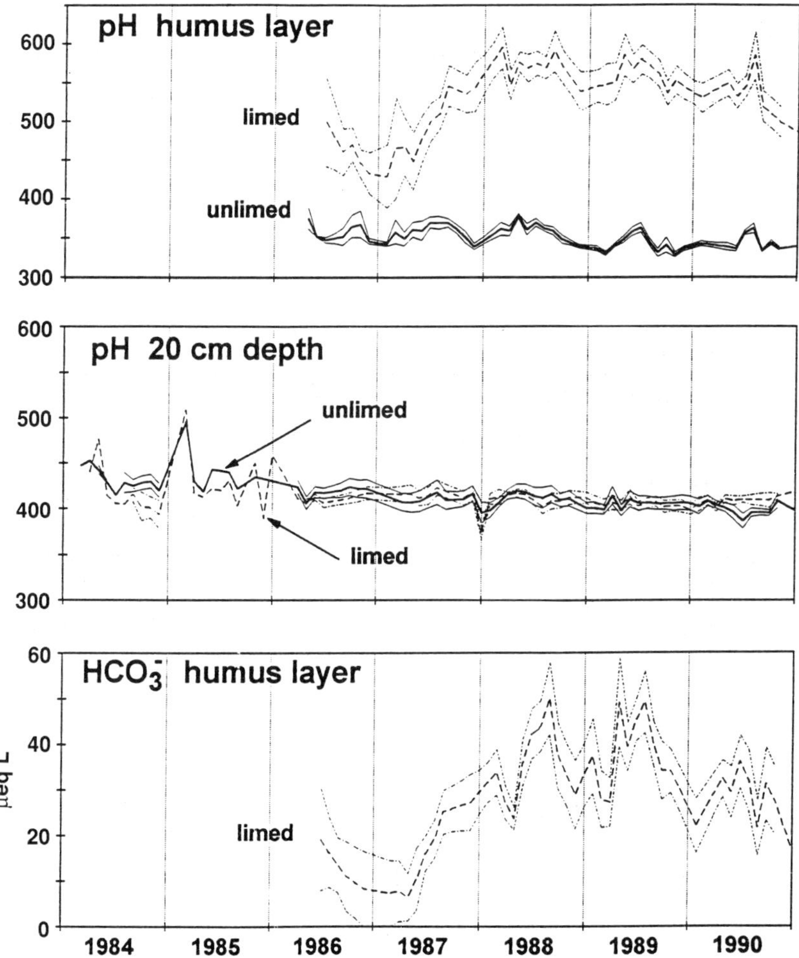


**Figure 21.3** pH and $HCO_3^-$ concentration of soil solution on the control (unlimed) and the limed plot at the Höglwald site (mean and standard deviation; humus lysimeters installed in 1986).

material from the deacidified layer with the fresh litter (Makeschin 1991). The downward mixing is rather small since earthworms avoid the very low pH values in the Oh layer, thus maintaining the steep gradient on the largest part of the area. In macropores, earthworms move primarily downward, where deacidification has occurred.

The effect of liming on the alkalinity of the soil solution is clearly visible at 20 cm (Figure 21.5), due mainly to mobilized Mg (increasing the Mg/Al ratio) (Figure 21.6).

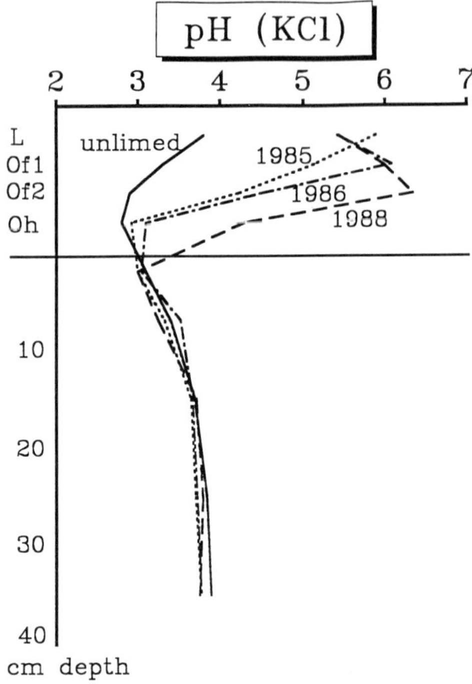


**Figure 21.4** Soil pH on the control (unlimed) and the limed plot.

Since 1984 the additional acid irrigation with about 400 meq $m^{-2}$ $yr^{-1}$ protons (as $H_2SO_4$, pH 2.6–2.7) on the limed sites has not resulted in a decrease of the alkalinity of the soil solution at the bottom of the humus layer (Figure 21.7). No effect is seen on the $H^+$ concentration until 1988 (Figure 21.8). Since 1989, $H^+$ concentration has increased, but it is quantitatively too small to be reflected in an alkalinity decrease. The increase in $H^+$ concentration coincides with the consumption of the lime.

Surface liming creates an efficient protection against high acid deposition, at least as long as carbonate is present. Furthermore, the restored exchange buffer is somewhat less efficient than fine-grained carbonate. This lower efficiency is assumed to be connected with the wide $H^+$ buffer range in the humus (pH 7 until 3), which is caused by different binding strengths of the $NH_4$–exchangeable base cations due to various forms of binding including complexation. The accessability of the exchange sites may also play a role.

*Risks*

Liming of acid soils constitutes a drastic impact on the existing microbiological conditions and elemental turnover. This may include risks for the water quality due to nitrification and heavy metal mobilization.

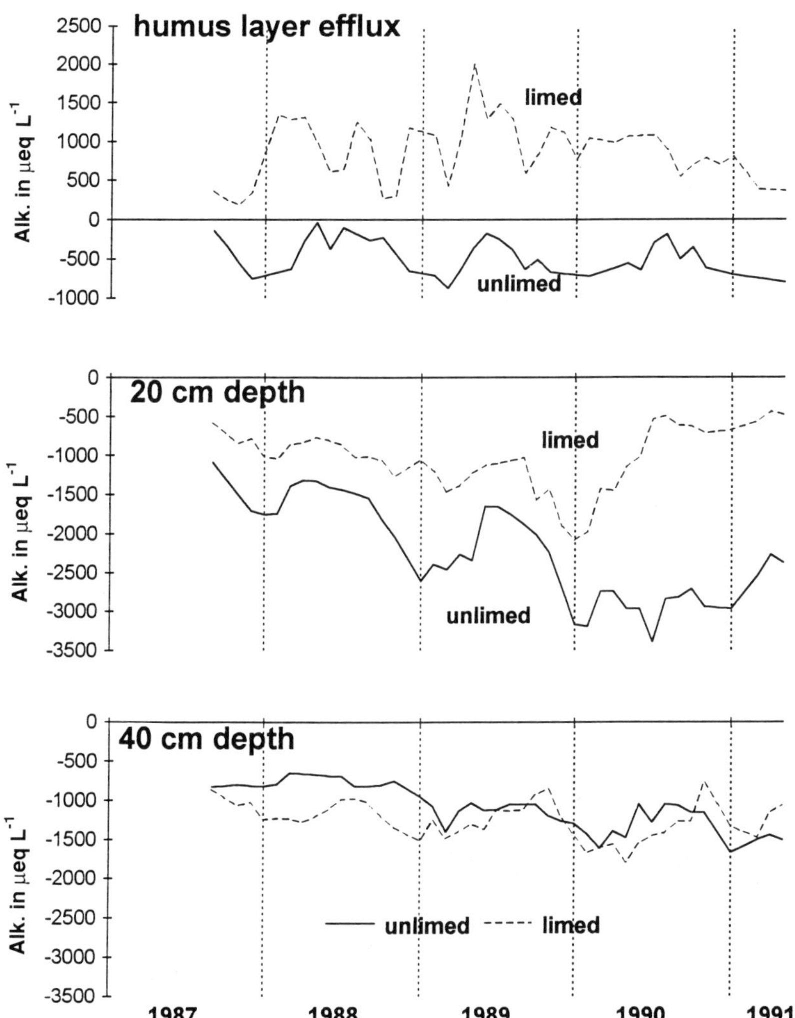


**Figure 21.5** Alkalinity (Alk.) of the soil solution on the control (unlimed) and the limed plots at the Höglwald site.

1. *Nitrification*: Liming stimulates the microbial acitivity in the soil and leads to increased mineralization of the organic matter. Thus, more ammonium becomes available and nitrification increases. On nitrogen-deficient sites, this additional nitrogen supply is largely consumed by plants and microorganisms, so that nitrate leaching is negligible. However, as a result of the increasing input of plant-available nitrogen forms from the air, many areas are no longer deficient in N. This means that liming may lead to an excess in nitrate due to

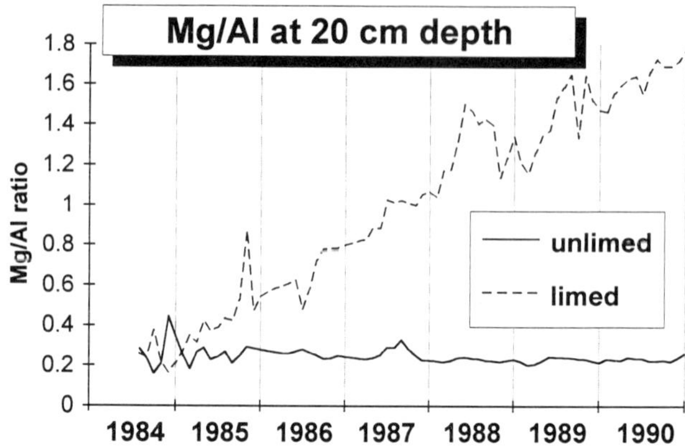


**Figure 21.6** Molar ratio of Mg/Al on the control (unlimed) and the limed plot.

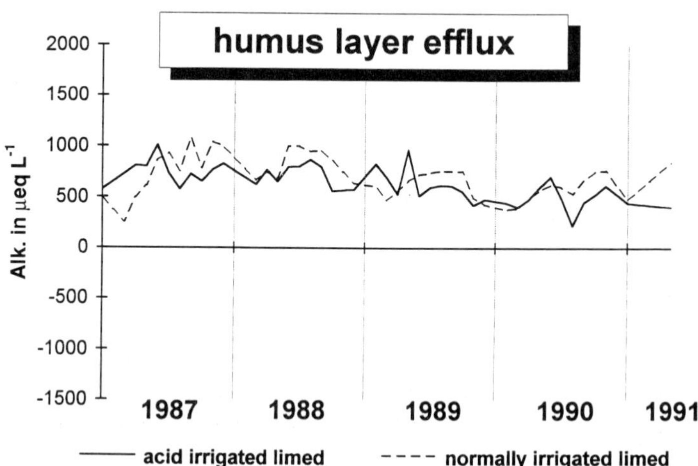


**Figure 21.7** Alkalinity of the soil solution on limed plots with and without acid irrigation (acid irrigation compared to normal irrigation).

nitrification. At nitrogen-saturated systems, e.g., in the Höglwald, liming may raise the nitrate content to more than 1500 μeq $L^{-1}$ with peaks exceeding 3000 μeq $L^{-1}$ (Figure 21.9). The surface humus layer of forest soils contains 3500–11,000 mmol N $m^{-2}$ organically bound nitrogen, some of which may be mobilized by successive liming. This may cause deterioration of groundwater quality (Kreutzer 1989; Sauter and Meiwes 1990).

Nitrification also reduces the deacidifiction effect of the lime. The portion of the lime-induced reduction amounted to about 200 meq $m^{-2}$ $yr^{-1}$ in the Höglwald.

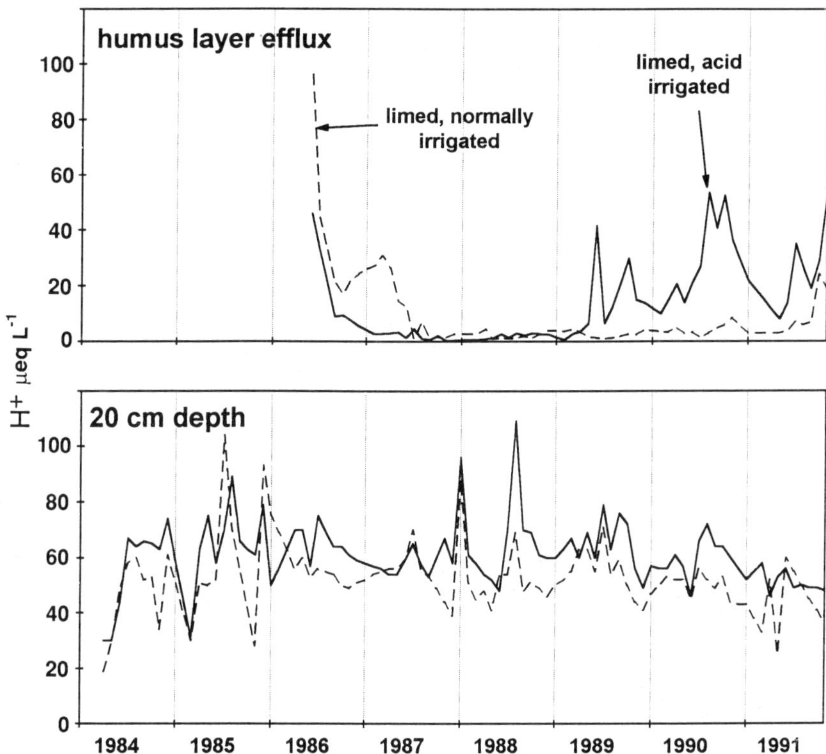


**Figure 21.8** $H^+$ concentration of the soil solution on limed plots with and without acid irrigation (acid irrigation compared to normal irrigation).

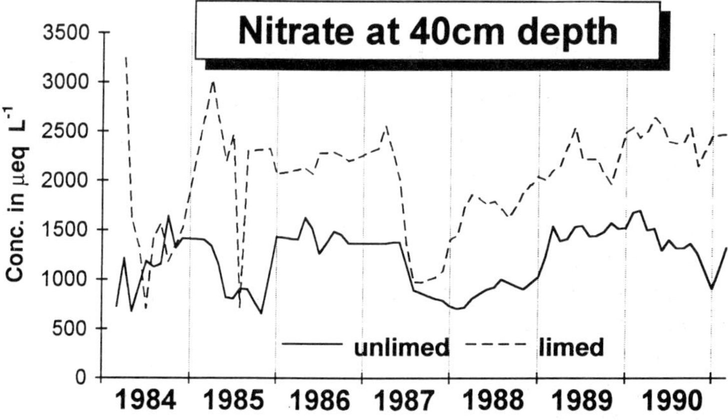


**Figure 21.9** Nitrate concentration of the soil solution on the control (unlimed) and the limed plot.

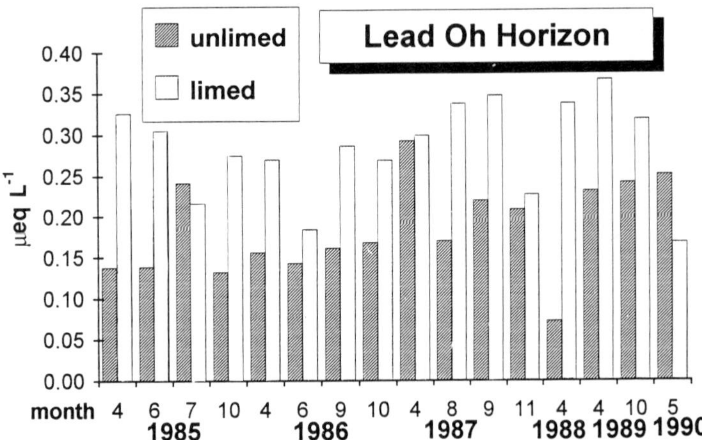


**Figure 21.10** Lead concentration in aqueous soil extracts from the control (unlimed) and the limed plot.

2. *Heavy metal mobilization*: Liming causes a drastic increase of dissolved organic carbon. This organic fraction comprises a number of complexing agents that may form stable metal-organic complexes. In the Höglwald, concentrations of organically complexed Cu, Pb, and Fe increased significantly in the soil solution after liming (Schierl and Kreutzer 1991; see Figure 21.10). The concentration of complexed aluminium also increased.

   Such liming effects may be harmful when the surface humus is contaminated due to copper and/or lead deposition. At sites where the water regime is dominated by subsurface (or surface) flow, e.g., on slopes in regions with high precipitation, the mobilized heavy metals may be leached into streams. There the contamination effect increases, especially when the metal ions are released by decomplexation or decomposition of the organic ligands.

## Soil Restoration

The deacidification front caused by surface liming moves downward very slowly at sites with a high base-neutralizing capacity. To accelerate the deacidification, a so-called soil restoration is recommended for soils that are strongly acidified down to the deeper subsoil (Eder and Pohlmeyer 1989; Benecke et al. 1992; Benecke 1992). This procedure calls for:

1. Mixing relatively large amounts of lime into the deeper soil layers by deep ploughing or by machines that loosen the ground and break up the soil, so that lime may be applied at depth.

2. Mixing organic matter (such as slash, stumps, and other materials) into the soil, also by deep ploughing. This organic matter should have a wide C:N ratio. The aim is to increase the nitrogen uptake capacity by microbes in the deeper soil layers and thereby to provide conditions for formation of durable N-rich humus substances. This accumulation process should diminish nitrate leaching.
3. Favoring a type of ground vegetation characterized by a high nitrogen uptake capacity and easily decomposable litter. This should contribute to the creation of "better" humus forms, such as mull or mull-like mor.

The application of this procedure seems to be limited to clearcut areas, although there is evidence that some techniques do not seriously damage the roots in mature stands. More information is needed, however, before this procedure can be generally recommended.

## CHANGING TREE SPECIES COMPOSITION

Trees not only influence the internal cycling of elements in the ecosystem, they also affect the external turnover. There are great variations between the different species at the same site due to canopy structure, root distribution, growth, etc. The specific effect of the tree species is greatest in mature stands.

### Filtering of Air Pollutants

There are great differences in the filtering of air pollutants between tree species (Kreutzer et al. 1986; Bredemeier 1987; Matzner 1988). These differences are primarily caused by specific foliar surfaces and by the length of foliation. One can generally state that evergreen conifers, e.g., instance Norway spruce (*Picea abies*), silver fir (*Abies alba*), Douglas fir (*Pseudotsuga Menziesii*) and, to a lesser degree, Scots pine (*Pinus sylvestris*), are more active filters than deciduous tree species.

Filtering comprises various processes such as scavenging of dust, fog, and cloud droplets from the air as well as adsorption and absorption of gases (Hultberg and Grenfelt 1986; Lindberg et al. 1986). The filtering effect of the species also depends on the quality of the accepting surface (e.g., epidermal waxes, roughness, hairs). Further differences between species are caused by specific efficiencies of the leaves/needles for metabolizing a part of the deposited nutrients in the canopy as well as for buffering of deposited acidity. Such physiological influences may change the correlation between ions resulting from physicochemical processes during deposition.

### Humus Form and Accumulation of Organic N and S

The accumulation of organic N and S in the ecosystem may occur by the formation of surface humus due to low decomposition rates of tree litter. This kind of accumulation, however, is unfavorable with respect to increase in the drainage water alkalinity, especially when inactive raw humus is formed, for the following reasons:

1. Organic acids accumulate in the top soil.
2. Release of base cations stored in the humus is blocked.
3. Acidification pulses occur due to excess nitrification, especially during the regeneration phase.

This type of humus is also prone to leach dissolved organic acids directly into surface waters, especially at sites with high precipitation if stormflow or rapid snow melting occurs and if runoff is dominated by lateral subsurface flow from slopes.

Another kind of organic N and S accumulation via trees, which does not show these disadvantages, occurs in the mineral soil. There, N and S are primarily bound in stable organic compounds having a low C:N ratio (Johnson 1992). These humus compounds are the result of advanced humification of root debris and leaf litter. The base cations contained in the litter are released by this humification process as plant available base cations.

The distribution pattern of N and S in the soil depends on the tree species due to (a) the decomposability of their litter by microorganisms and to (b) their fine root distribution in the soil.

Species that tend to have a rather homogenous fine-root distribution down to the deeper mineral soil layers, not only below the stumps but also in the space between the trees, show a higher N- and S-storing capacity in the mineral soil as compared with tree species that are shallow rooting on the interspace. Such N- and S-storing species include beeches, oaks, lime trees, ash trees, and maples, in contrast to Norway spruce. In addition, broadleaf species, above all maples, elms, lime and ash trees, show a better microbiological litter decomposition than do conifers on the same site. Consequently, the earthworm population increases, mixing surface litter into the mineral soil.

The difference in the storage capacity of N in the mineral soil between broadleaves and conifers may amount to tens of thousands meq $m^{-2}$ (Kreutzer 1988). However, little is known about the capacity limits and the accumulation rates at various sites.

## Growth

Fast-growing species may acidify the soil more than slow-growing ones (Nilsson and Miller 1982; Holstener-Jørgensen 1988). Fast-growing species store more base cations per time in the biomass. These cations are largely taken up by a corresponding release of protons from the roots into the rhizosphere, thus decreasing the ANC.

A similar effect occurs, within the same species, during the period of maximum growth, mostly at an age of 20 to 40 years. Differences also exist between species due to the form and sharpness of the maximum. Species showing relatively slow growth and a less-pronounced maximum are *Quercus petraea*, *Q. robur*, and *Carpinus betulus*.

### Nitrogen Fixation by Symbionts

Nitrogen fixation per se is not combined with a net $H^+$ transfer. However, it may contribute to nitrogen saturation and subsequent excess nitrification. The alders and the legume *Robinia pseudacacia* belong to these species. The sites where the black alder, an important species in central Europe silviculture, is native and where its cultivation is emphasized are characterized by a high but fluctuating groundwater table. Thus a significant part of the formed nitrate is normally reduced again (Pröbstle 1987). The alkalinity of the soil solution, therefore, often shows a large seasonal oscillation. Only the uppermost part of groundwater enriched with nitrate is usually involved with nitrate translocation into surface waters. On well-drained soils, however, alders can lead to exessive acidification, as shown by Van Migroet and Cole (1984).

### Hydrological Effects

These effects are due to differences in the interceptional evaporation of species. From the canopy of evergreen conifers much more intercepted precipitation can evaporate than from deciduous tree species. Under comparable conditions, this means that the soil under broadleaves receives more water than under conifers. This affects (a) the ion concentration in the soil solution, and by that the pH and the alkalinity, and (b) the denitrification conditions. The latter effect, however, is confined to soils which show transient wet conditions. In such soils the redox potentials tend to fall temporarily below 300 mV. During wet phases, the nitrate content may be reduced by microorganisms to gaseous nitrogen compounds leaving the ecosystem by volatilization. As the denitrification consumes protons, the alkalinity increases. These hydrological effects may contribute to the fact that the pH and alkalinity of the soil solution is at a higher level under deciduous species than under conifers at the same site.

## HARVESTING

The removal of organic matter by harvesting may contribute significantly to internal soil acidification via base cation exports. Major exports occur by brushwood and bark removal in addition to stemwood harvesting, since these compartments contain relatively high concentrations of base cations (Kreutzer 1979). The former removal of litter also contributed significantly to a decrease in cation supply. Critical levels of low base cation supply occurred on primarily poor sites, impoverished by intensive utilization over centuries (Feger 1988). On such sites, the acidification was furthermore increased by the cultivation of Scots pine and the invasion of heather into the forest stands.

For freshwater restoration purposes, harvesting should be restricted to the removal of stemwood. The wood refuse, felling debris, slash, etc. should be left in a more or less even distribution.

## SILVICULTURAL MANAGEMENT

Silvicultural management procedures may contribute to increase the alkalinity via:

1. *Extension of the rotation time.* If the rotation time exceeds the economic one, the mean increment decreases. The fast-growing age classes, which are the most acidifying, then cover a relatively smaller part of the catchment area. This means that the mean alkalinity of the soil solution in the catchment increases.
2. *Choice of the regeneration system.* Regeneration of forest stands by clear-cutting and subsequent planting (clear-cutting system) should be avoided because on clearcut areas the humus mineralization and the conversion of $NH_4^+$ to $NO_3^-$ are usually stronger stimulated than on sheltered regeneration areas, due to increased warming of the forest floor. Since, at the same time, the plant uptake capacity for nutrients is extremely reduced on clearcuts, an excess nitrate production may occur, connected with an equivalent acidification effect in the soil. This increase in nitrate concentration after felling is widely reported (reviewed by Vitousek and Melillo 1979; Vitousek 1981; Stevens and Hornung 1988, 1990). The magnitude, however, varies to a large extent, depending on the site conditions and the development of the vegetation. Transient water logging, which increases on flat cleared areas with fine-grained soils, diminishes the nitrate concentration due to denitrification. The new weeds on clearcut areas also tend to decrease the nitrate concentration, since they take up a considerable part of that nitrate. Nonetheless, they induce further problems, such as competition with forest plantations (light!) and subsequent weed control. On very acidic forest soils, however, excess nitrification may occur nearly as well as on neutral ones (Kreutzer 1983), dominated on acidic soils by heterotrophic nitrifiers in contrast to chemolithotrophic ones on neutral soils (Adams 1986; Killham 1987; Papen et al. 1991).
3. *Regulation of stand density.* Thinning procedures that regulate stand density may significantly influence the filtering effect of the canopy. Spacious stands have a reduced needle/leaf surface. This may lead to a smaller interception deposition. In windy climates, however, the inverse may be true, because the winds have a better access to spacious stands and the filtering effect as a product from pollutant concentration in the air and wind speed may be higher.

## SUMMARY AND CONCLUSIONS

In this chapter, I have focused on central European experiences in rotation forestry. The question of afforestation of agricultural land has been excluded.

1. Surface liming produces an efficient protection against airborne acid deposition, as long as carbonate is available in the soils.

2. The deacidification front induced by liming moves downward at speeds that depend upon the capacity and efficiency of the base-neutralizing potential of the soil. In the surface humus, a large amount of acidity is stored in the protonated functional groups. For deacidification of the whole soil profile, repeated liming would be necessary in most cases. It is estimated that, on average, 40,000 meq $m^{-2}$ of acidity has accumulated in forest soils of West Germany since 1850. This corresponds to an ANC of about 20,000 kg lime $ha^{-1}$.
3. The dissolution of the lime partly restores the exchange buffer especially in the humus layer. On average, this buffer seems to have a lower efficiency than the fine-grained carbonate, so moderate reacidification may occur immediately after lime dissolution, in spite of relatively high base saturation.
4. Liming stimulates nitrification, which may result in increased nitrate concentrations in seepage water. This risk increases due to high deposition rates of nitrogen. The mobilizable nitrogen stored in the surface humus layer is in a range of 3,500–11,000 meq $m^{-2}$. A further risk of liming is mobilization of Cu and Pb as organic complexes. This may be harmful at contaminated sites with surface or subsurface runoff. Because of the risks, surface liming cannot be recommended without reserve. It is necessary to weigh the priorities.
5. "Soil restoration" by deep ploughing or soil loosening in combination with mixing of lime and of organic matter with wide C:N ratio into the deeper soil layer has not yet been fully tested.
6. Recommendations for silviculture are:
   a. Preference should be given to broadleaf species that are deep-rooting with spatially extended fine-root formation, and with easily decomposable litter.
   b. Slow-growing species are preferred. The rotation time should be extended as far as possible. Thinnings should provide moderate to spacious stand densities (but this may be questionable in windy climates). An important principle is to establish stable stands. Vegetation with high N uptake is desirable.
   c. Species with nitrogen-fixing symbionts as well as clearcutting and extensive thinning should be avoided.
   d. Harvesting should be limited to stemwood without bark, and the cutting refuse should be left evenly distributed on the cut area.
7. All management processes should be strongly related to site conditions (including to the deposition load).

These statements and recommendations are general and preliminary. Investigations are needed to examine the species effect on the nitrogen cycle under different site conditions, in order to elucidate more precisely the decisive factors. More information is necessary because of the increasing nitrogen deposition. The effect of mixed stands should also be investigated.

We also need to translate the knowledge into practice by working out site/species-related criteria for practical solutions; site mapping should be included.

A further point to be clarified involves the question: Under what conditions does liming pose significant risks? What are the factors or observable site properties to be checked before decisions are made for liming campaigns?

## ACKNOWLEDGEMENTS

Financial support was provided by the Bavarian Forest Administration, Munich, and the Federal Department of Research and Technology, Bonn. I would like to thank M. Fuchs, J. Kitzbichler, and H. Koch for their help and valuable comments.

## REFERENCES

Adams, J.A. 1986. Identification of heterotrophic nitrification in strongly acid larch humus. *Soil Biol. Biochem.* **18**:339–341.

Aldinger, E. 1987. Elementgehalte im Boden und in Nadeln verschieden stark geschädigter Fichten-Tannen-Bestände auf Praxiskalkungsflächen im Buntsandstein-Schwarzwald. Dissertation. Freiburg i. Br.: Albert-Ludwigs-Universität.

Benecke, P. 1992. Gedanken zur Waldbodenrestaurierung mit Bodenbearbeitung. *AFZ* **10**:542–545.

Benecke, P., C. Eberl, and M. Marbach. 1992. Bestandesbegründung mit Bodenbearbeitung, Kalkung und Hilfspflanzenanbau. *AFZ* **10**:546–550.

Bichlmaier, K. 1985. Auswirkungen von Kalkungen auf Nährstoffgehalte von Böden und Nadeln. Diplomarbeit Forstw. Fak. München: Ludwigs-Maximilians-Universität.

Bredemeier, M. 1987. Forest canopy transformation of atmospheric deposition. *Water, Air, Soil Pollut.* **40**:121–138.

Eder, W., and H. Pohlmeyer. 1989. Waldbauliche Konsequenzen aus fortschreitender Bodenversauerung und Bestandesschäden. *AFZ* **35–36**:970–973.

Feger, K.H. 1988. Historical changes in catchment use. In: Effects of Land Use in Catchments on the Acidity and Ecology of Natural Surface Waters, ed. H. Barth, pp. 65–74. Air Pollution Report 13. Commission of the European Communities. Brussels: E. Guyot S.A.

Gussone, H.A. 1987. Kompensationskalkungen und die Anwendung von Düngemitteln im Walde. *Der Forst- und Holzwirt* **6**:158–163.

Hemond, F.H. 1990. Acid-neutralizing Capacity, Alkalinity, and Acid Base Status of Natural Waters Containing Organic Acids. *Environ. Sci. Technol.* **24**:1486–1489.

Holstener-Jørgensen, H., M. Krag, and H.C. Olsen. 1988. The influence of 12 tree species on the acidification of the upper soil horizons. *Det forstlige forsgsv sen i Danmark* (*The Forest Research in Denmark*) **42**:15–25.

Hultberg, H., and P. Grenfelt. 1986. Gardsjon Projekt: Lake acidification, chemistry in catchment runoff, lake liming and microcatchment manipulations. *Water, Air, Soil Pollut.* **30**:31–46.

Johnson, D.W. 1992. Nitrogen Retention in Forest Soils. *J. Environ. Qual.* **21**:1–12.

Killham, K. 1987. A new perfusion system for the measurement and characterisation of potential rates of soil nitrification. *Plant and Soil* **97**:267–272.

Kreutzer, K. 1979. Ökologische Fragen zur Vollbaumernte. *Forstw. Cbl.* **98**:298–308.

Kreutzer, K. 1983. Stickstoffaustrag in Abhängigkeit von Kulturart und Nutzungsintensität in der Forstwirtschaft. In: Nitrat: Ein Problem für die Trinkwasserversorgung, ed. Deutsche Landwirtschaftsgesellschaft, vol. 177, pp. 69–83. Frankfurt/Main: DLG–Verlag.

Kreutzer, K. 1988. The impact of forest management practices on the soil acidification in established forests. In: Effects of Land Use in Catchments on the Acidity and Ecology of Natural Surface Waters, ed. H. Barth, pp. 75–90. Air Pollution Report 13. Commission of the European Communities. Brussels: E. Guyot S.A.

Kreutzer, K. 1989. Änderungen im Stickstoffhaushalt der Wälder und die dadurch verursachten Auswirkungen auf die Qualität des Sickerwassers. *DVWK-Mitt.* **17**:121–132.

Kreutzer, K., E. Deschu, and G. Hösl. 1986. Vergleichende Untersuchungen über den Einfluß von Fichte (*Picea abies* [L.] Karst.) und Buche (*Fagus sylvatica* L.) auf die Sickerwasserqualität. *Forstw. Cbl.* **105**:346–371.

Kreutzer, K., A. Göttlein, and P. Pröbstle. 1991. Dynamik und chemische Auswirkungen der Auflösung von Dolomitkalk unter Fichte (*Picea abies* [L.] Karst.). *Forstw. Forschungen* **39**:186–209.

Lindberg, S.E., G.M. Lovett, D.R. Richter, and D.W. Johnson. 1986. Atmospheric deposition and canopy interaction of major ions in a forest. *Science* **231**:141–145.

Makeschin, F. 1991. Auswirkungen von saurer Beregnung und Kalkung auf die Regenwurmfauna (Lumbricidae, Oligochatae) im Fichtenaltbestand Höglwald. *Forstw. Forschungen* **39**:117–127.

Matzner, E. 1988. Der Stoffumsatz zweier Waldökosysteme im Solling. Berichte des Forschungszentrums Waldökosysteme/Waldsterben, ed. B. Ulrich, Series A 40. Göttingen: Selbstverlag des Forschungszentrums Waldökosysteme/Waldsterben der Universität Göttingen, 217 pp.

Nilsson, S.I., And H.G. Miller. 1982. Forest growth as a possible cause of soil and water acidification. An examination of concepts. *Oikos* **39**:40–49.

Papen, H., R. v. Berg, B. Hellmann, and H. Rennenberg. 1991. Einfluß von saurer Beregnung und Kalkung auf chemolithotrophe und heterotrophe Nitrifikation in Böden des Höglwaldes. *Forstw. Forschungen* **39**:111–116.

Prenzel, J. 1985. Die maximale Löslichkeit von oberflächlich ausgebrachtem Kalk. *AFZ* **43**:1142.

Pröbstle, P. 1987. Vergleichende Untersuchungen über den Einfluß von Schwarzerlen- und Fichtenbeständen auf die Ionenkonzentration im Bodenwasser. Diplomarbeit Forstw. Fak. München: Ludwigs-Maximilians-Universität.

Reinwald, H. 1987. Einfluß der Kalkung auf bodenchemische Parameter und Saugkerzenwässer. Diplomarbeit Forstw. Fak. München: Ludwigs-Maximilians-Universität.

Sauter, U., and K.J. Meiwes. 1990. Auswirkungen der Kalkung auf den Stoffaustrag aus Waldökosystemen mit dem Sickerwasser. *Forst und Holz* **20**:605–610.

Schierl, R., and K. Kreutzer. 1991. Einfluß von saurer Beregnung und Kalkung auf die Schwermetalldynamik im Höglwaldexperiment. *Forstw. Forschungen* **39**:204–211.

Schwertmann, U., P. Süsser, and L. Nätscher. 1987. Protonenpuffersubstanzen in Böden. *Z. Pflanzenernähr. Bodenk.* **150**:174–178.

Stevens, P.A., and M. Hornung. 1988. Nitrate leaching from a felled Sitka spruce plantation in Beddgelert Forest, North Wales. *Soil Use Manag.* **4**:3–9.

Stevens, P.A., and M. Hornung. 1990. Effect of harvest intensity and ground flora establishment on inorganic-N leaching from a sitka spruce plantation in north Wales, U.K. *Biogeochem.* **10**:53–65.

Ulrich, B. 1989. Immissionsbelastung des Waldes und seiner Böden-Gefahr für die Gewässer *DVWK-Mitt.* **17**:7–24.

UBA (Umweltbundesamt). 1989. Daten zur Umwelt 1988/89. Hrsg. Umweltbundesamt. Berlin: Erich Schmidt Verlag.

Van Migroet, H., and D.W. Cole. 1984. The impact of nitrification on soil acidification and cation leaching in a Red Alder Ecosystem. *J. Environ. Qual.* **13**:586–590.

Vitousek, P.M. 1981. Clear cutting and the nitrogen cycle. In: Terrestrial Nitrogen Cycles. Processes, Ecosystem Strategies and Management Impacts, ed. F.E. Clark and T. Rosswall. *Ecol Bull.* **33**:631–642.

Vitousek, P.M., and J.M. Melillo. 1979. Nitrate losses from disturbed forests: Patterns and mechanisms. *For. Sci.* **25**:605–619.

# 22

# Invertebrate Community Changes Caused by Reduced Acidification

G.G. RADDUM and A. FJELLHEIM
Department of Animal Ecology, Museum of Zoology,
University of Bergen, 5007 Bergen, Norway

## ABSTRACT

Decreased acidity of streams and lakes due to reduced acid deposition or liming results in recovery of sensitive invertebrate species. Examples of both high and low degrees of recovery have been reported. Rapid response in the fauna is generally observed among the most mobile insect larvae in running waters. In lakes situated within large acidified regions, natural recovery usually takes a longer time. The presence of refuges, holding undamaged fauna within acidified areas, seems to be the most important factor for the recovery process. Another important factor appears to be related to the structure of the acid community, which has developed due to direct toxic effects on sensitive species. A change in this structure to a less acidic community structure, after improved water quality, will depend on biological interactions. Reappearance of key top predators, important for the formation of a nonacidified community, will therefore determine the success and time needed for the recovery of invertebrates.

## INTRODUCTION

Investigations on acidification and its effect on freshwater ecosystems have demonstrated that all trophic levels are influenced by acidity (Drabløs and Tollan 1980; Martin 1986; Schindler 1988; Last and Walting 1991). There is a multitude of mechanisms leading to the responses seen among the affected species. The two main groups of effects are of a direct (primary) or indirect (secondary) nature. The direct effects are connected to toxicity of increased concentrations of $H^+$, toxic metals (especially different compounds of aluminum), and other substances in acidic water that can exceed the tolerance limits of sensitive species. The result is mortality or sublethal stress on the species, which is detected as reproduction failure, reduced growth or feeding rate, lowered survival and respiration rate, increased catastrophic drift in running water, etc. (Økland and Økland 1986; Locke 1991; Merrett et al. 1991;

*Acidification of Freshwater Ecosystems: Implications for the Future*
Edited by C.E.W. Steinberg and R.F. Wright 

Wren and Stephenson 1991). The indirect effects are mostly connected to biological interactions in the predator-prey system, in the grazing system, or other relations within the food web. Other secondary effects modifying the balance in freshwater communities are increased parasitism and reduced health of organisms (Nyman et al. 1985; Stenson 1985; Økland and Økland 1986; Schindler et al. 1991; Muniz 1991).

In areas where pH has increased in acidified water, due to decreased acid deposition or liming, recovery of aquatic fauna has been documented to varying degrees (Eriksson et al. 1983; Raddum and Fjellheim 1984; Raddum et al. 1986; MacIsaac et al. 1986; Gunn and Keller 1990; Keller and Yen 1991; Fjellheim and Raddum 1992). Investigations on experimental acidification and recovery have been carried out by Schindler et al. (1991). These studies demonstrate that the degree of recovery is variable, especially the reestablishment of the expected original balance between species in the ecosystem. Important factors are the length of the acidification period, level of acidity (both of the locality itself and of the surroundings), original fauna complexity, structure of the community, and eutrophic status of the site during the previous nonacidic state.

## CASE STUDIES FROM NORWAY

During the period of 1980 to 1990, sulfur emission has decreased by about 50% in western Europe (Iversen et al. 1991). In southern Norway, the concentration of anthropogenic sulfate in precipitation declined by about 25–30% (SFT 1991). Wet acid deposition, however, changed little, due to increased precipitation over the same time period. In fact, elevated acidity was recorded on the west coast of southern Norway during the very wet years of 1989–1990, due to high deposition of sea salts. The sea salts accounted for much of the increased acidity during these years (SFT 1992). Since 1980, the monitoring of water chemistry in this part of Norway showed a small increase in pH from 1983 to 1987, followed by a decrease during 1988–1990 and an increase in 1991 (Figure 22.1). The acid-neutralizing capacity (ANC) followed the same pattern. The only period that showed negative values was in 1990. Recovery of ANC was quick in the dryer year, 1991, and showed a "normal" concentration during this period.

Since 1981, biological monitoring has been conducted in southern Norway, both in undisturbed and in limed watersheds. The monitoring of unlimed watersheds includes strongly-to-less-acidified localities with varying acid deposition. The limed watersheds are all situated in areas with a high load of acids.

### Recovery of Fauna with Regard to Increase/Decrease in pH or ANC

The monitored watersheds, Vikedal (southwestern coast) and Nausta (northwestern coast), were exposed to the general changes in pH mentioned above. All watersheds have pHs in the range of 5 to 6 and have an ANC of 0 to 20 μeq $L^{-1}$, which is lowest in Vikedal and highest in Nausta. The southwestern watersheds near Farsund have also

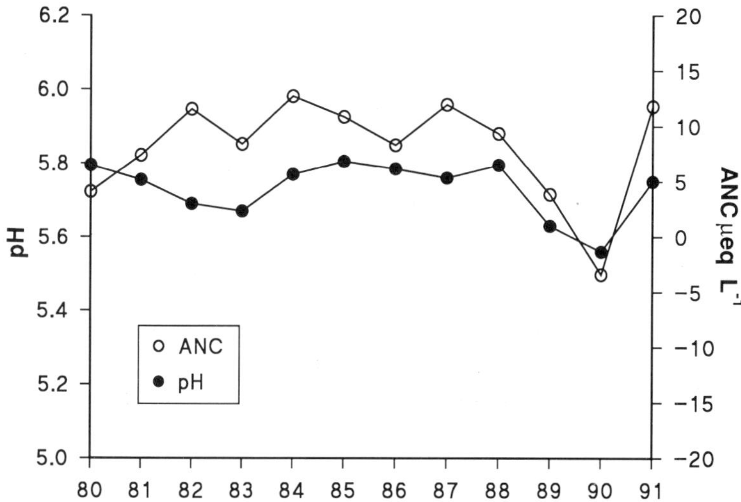


**Figure 22.1** Mean pH and ANC alkalinity in watersheds in the western part of southern Norway during the period 1980–1991 (data from SFT 1992).

been exposed to lowered sulfur deposition due to their geographical location; however, these areas generally receive much higher acid deposition than Vikedal and Nausta. All watersheds are oligotrophic and have simple ecosystems compared to more complex systems having higher alkalinity and nutrition level. Most of the investigated sites are running water localities.

The acidity and general water quality of Vikedal and Nausta is in a range where many sensitive invertebrates have their tolerance limits (Fjellheim and Raddum 1990; Lien et al. 1992). Due to this, small changes in pH can strongly affect the presence or absence of sensitive species. Furthermore, the occurrence of acidic episodes is relatively high in these low buffered systems. Rapid changes in the acidification score (based on tolerance of species) were recorded (Raddum et al. 1988; Fjellheim and Raddum 1990; see Figure 22.2).

The more chronically acidified watersheds (Farsund) show a low score or a high acidification level. The acidity of the watersheds has so far exceeded the tolerance limits of most of the sensitive species. Possible improvements in water quality, therefore, have not reached the levels necessary for recovery of sensitive fauna.

The species *Baetis rhodani*, *Diura nanseni*, *Lepidostoma hirtum*, *Isoperla* spp., *Hydropsyche* spp., and *Apatania* spp. are most frequently observed to recover during periods of improving water quality, which indicates a change in toxicity from lethal to sub- or nonlethal levels. These species usually have the highest influence on the acidification score. For example, the increase in acidity during 1988–1990 (Figure 22.1) gave an immediate decline in the score, due to extinction of these species from many of the sites with the lowest buffer capacity.

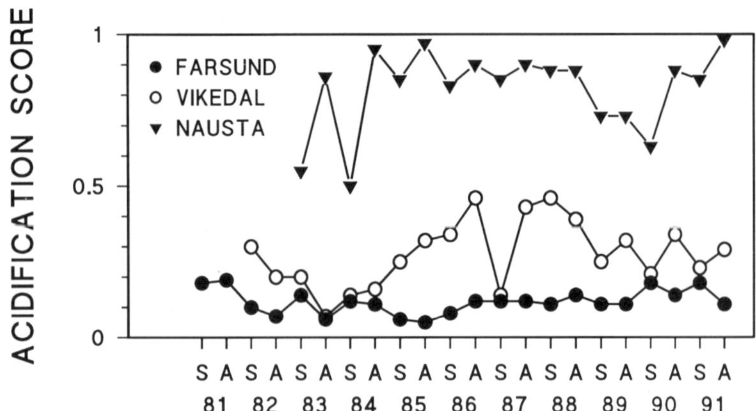


**Figure 22.2** Mean acidification score for watersheds in the western part of South Norway (S: spring, A: autumn).

This example focuses on tolerance, i.e., responses due to changes in the direct factor system. At this basic level, recovery and deterioration of sensitive species is closely related to acidity and the degree of acidification, while biotic effects or interactions are of lesser importance. It is possible, however, to detect sublethal effects on a sensitive species before extinction by comparing densities of sensitive and tolerant species occupying the same habitat. In Nausta, a significant reduction in the ratio between the density of the sensitive mayfly *B. rhodani* and tolerant stoneflies was recorded in the samples from 1989 and spring 1990 (Figure 22.3). The ratio is well correlated with the ANC level during the period and indicates sublethal stress on the mayfly community in sites where extinction has not occurred. In this case, recovery from sublethal toxic stress is very rapid and occurs in the next generation or season. This is due to the high reproduction and spreading potential of *B. rhodani*. Biological changes connected to sublethal stress obviously exist, but they seem to be of minor importance in the recovery of populations. Our experience indicates that recovery of fauna from moderately acidic streams and rivers is very good.

## Liming Lakes

Liming in Norway includes strongly as well as moderately acidified watersheds. The aim of liming is to increase pH and reduce Al to nontoxic levels. Through this treatment, the content of Ca will increase above the original content and the water chemistry at the locality will be different from that of preacidification. Limed waters, however, seldom contain concentrations of Ca above 250 μeq $L^{-1}$.

A strong increase in the density of benthic animals, mostly chironomids, was observed 2–3 years after liming of Lake Hovvatn (Raddum et al. 1986). Lake Hovvatn is a second-order headwater lake situated within the area with the stongest acidification

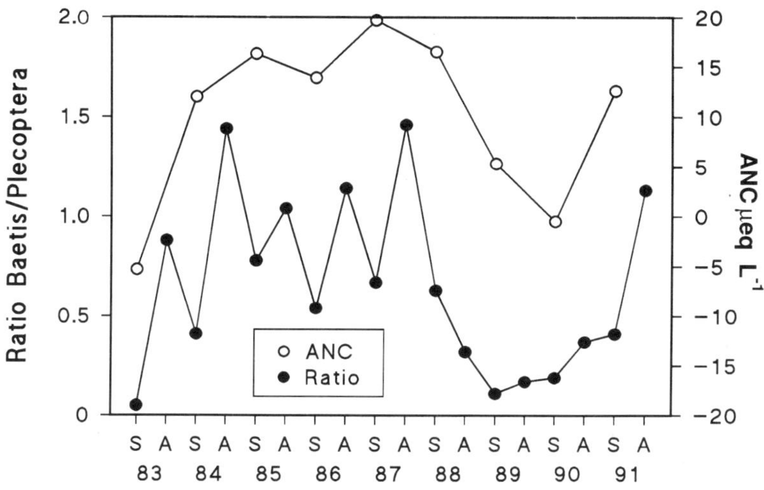


**Figure 22.3** Mean ANC alkalinity and the ratio Baetis/Plecoptera for River Nausta during the period 1983–1991 (ANC data from SFT [1992]; S: spring, A: autumn).

in Norway. No pools of better water quality exist in the catchment. The highest response was seen in the density of *Tanytarsus pallidicornis*, *Heterotanytarsus apicalis*, and *Pagastiella orophila*. These species almost disappeared again after reacidification in 1986. A detailed study of the chironomids from 1980 to 1991 shows that the presence of species was very similar, both before and after liming (Schnell, in prep.). No new species occurred; however, a few species disappeared from the samples after liming, probably due to the establishment of new relationships in the competition between species.

In the littoral zone, acid-tolerant mayflies, *Leptophlebia vespertina* and *L. marginata*, increased significantly in abundance after the first liming, despite predation from restocked fish. This indicates that the *Leptophlebia* spp. were highly favored in the limed water and that fish predation did not harm the population.

Reliming of the lake in 1987 and 1989 kept the pH at a relatively high level until now. Only one example of recovery of a species can be found from the records, namely that of the moderately sensitive mayfly, *Siphlonurus lacustris*: Two larvae of the species were found for the first time in 1989. In 1991, the species were common at several stations in the lake.

In the zooplankton community the proportion of the most common species changed; however, no sensitive species, e.g., *Daphnia* spp., occurred after liming.

So far the liming experiment at Lake Hovvatn has demonstrated that recovery of fauna to an expected natural composition of species has not occurred. The most conspicuous effect observed, except for colonization of *S. lacustris*, involves quantitative changes within the existing zooplankton and benthic communities. The tolerance of chironomid species is poorly known. The ongoing work shows that the species can be divided into pH preference groups (Schnell, in prep.). By doing this, the chironomids can be treated statistically in the same way as has been done for diatoms. In conclusion, a true recovery of the fauna in Lake Hovvatn seems to take a long time, as the first recording of a moderate acid-sensitive species occurred 10 years after the initial liming. The reasons for this are obviously coupled to the situation of the lake, which is in the middle of a major acidification area.

## Limed Rivers

At present, monitoring of invertebrates is carried out in the limed parts of the watersheds Audna, Ogna, Vikedal, and Yndesdal. In these watersheds, recovery of sensitive species took place within a period of 0.5 to 2.5 years after treatment, with the fastest recovery in rivers with lowest acidification. The longest monitoring record is from Audna, where the liming started in the spring of 1985. Invertebrate data exist from 1982 and regularly from 1985. Significant recovery of sensitive species was first observed in the autumn of 1987 (Figure 22.4; Fjellheim and Raddum 1992). A number of new species were recorded at limed sites during the period 1987–1991. Of these *B. rhodani*, *S. lacustris*, *Heptagenia sulphurea*, *Caenis horaria*, *Isoperla obscura*, *Hydropsyche pellucidula*, *H. siltalai*, and *Lepidostoma hirtum* were most common. Moderately sensitive species that expanded in limed localities, and that were also able to live in a few unlimed sites, were *Diura nanseni*, *I. grammatica*, and *Apatania sp.* Similar recovery is going on in Ogna, Vikedal, and Yndesdal. The stage of recovery in these watersheds has reached different levels dependent on the limed period.

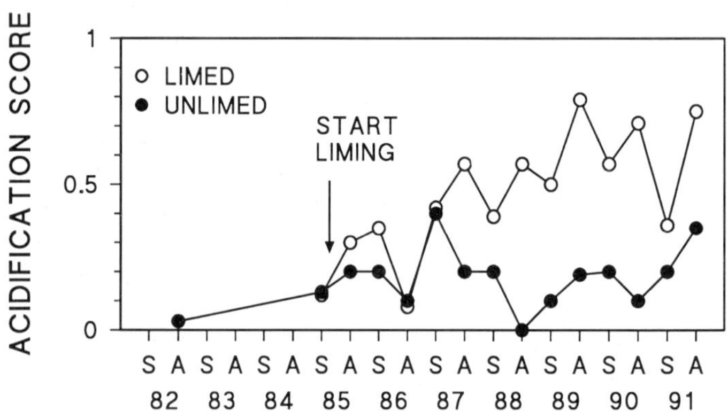


**Figure 22.4** Mean acidification score for limed and unlimed sites in River Audna.

In conclusion, the faunal recovery in western Norway, due to liming or reduced acid deposition in moderately acidified watersheds, is rapid and very similar in different geographical areas. The cause or effect seems mostly to be connected with reduced toxicity of the water. Secondary biological interactions have, most probably, a low influence on this process. In strongly acidified localities recovery, in the sense of immigration of new species, seems to take a much longer time. The reasons for this seem to be (a) the long distance to source populations of sensitive species and (b) the profound changes in the balance and structure between species caused by acidification over time.

## DISCUSSION

The cause-effect pathways of acidification in freshwater communities are many. During the recovery process the question is: to what extent are these pathways reversible? In this regard, it is important to discriminate between the effect of the reduction of direct toxic effects of water quality and the effects connected to biological interactions. Furthermore, changes in the biological interactions cannot occur without a previous direct effect.

From the literature it is astonishing how similar the responses of acidic water are on invertebrate communities, with reduced species richness, presence of similar tolerant and sensitive groups, toxicity in humic and clearwater sites, importance of high and low ionic content, etc. (Raddum and Fjellheim 1984; Otto and Svenson 1983; Økland and Økland 1986; Hämäläinen and Huttunen 1990; Schindler et al. 1991; Muniz 1991). Exact predictions of acidic effects are, however, difficult and will depend on the fauna of the geographical area, its richness, specific characteristics of each locality, and how energy moves through the ecosystem.

The best recovery towards an expected original condition can be seen in localities with a short acidic history, moderate acidification, and simple relationships in the cause-effect pathways, and in dynamic systems as oligotrophic running water. The reasons for this might be that both the fauna composition and structure of the ecosystem in moderately acidic localities still have parts of the original characteristics present somewhere. This means that the chance of finding refuges exists and that a nonacidified system can be found in restricted areas inside the watershed, "a biological memory." Through this, the ability to restore fauna is kept intact. In addition, in dynamic systems, such as running water, the fauna is adapted to higher variations in abiotic parameters than in lakes. Recovery from acidic water episodes can probably be considered parallel to recovery from some other lethal abiotic factor for which the system is adapted. Furthermore, the possibility for presence of natural source communities or species in small streams and tributaries to rivers is high due to natural variation in catchments. The chance of finding source communities in lakes is smaller. The drift of invertebrates from refuges may be the most obviuos reason for the rapid recovery observed in running water.

In strongly acidified watersheds, only tolerant communities can exist. The chance to find sensitive species or an original community in this type of water is very small. In addition, over time the acid communities will create energy pathways adapted for these ecosystems. Examples of this are the shift in top predators from different fish species to invertebrates (Erickson et al. 1980; Stenson and Oscarson 1985). Such changes are of a conservative nature and will make it difficult to reestablish the original community (see Nyberg 1984 and Yen et al. 1991). The recovery after decreased acidic toxicity will, therefore, take a long time since the structure and balance of species in the systems are the main determining factors and not the toxic water quality. As they immigrate, acid-sensitive species will probably meet stronger competition in this ecosystem than in the original nonacidified system. In Lake Hovvatn this hypothesis, together with the fact that there are few refuges in the catchment, explain the near absence of real recovery.

The productivity both of "acid invertebrate communities" and of restocked fish in deacidified lakes can, however, be very high (Raddum et al. 1986; Barlaup et al. 1989). Therefore, neutralizing acidic water is valuable.

As for the relationship between water chemistry and biology in general, there is also a good correlation between ANC, pH, and Al and the acidification score used in the biological monitoring (Lien et al. 1992). The score gives information about the presence or absence of species. In areas where the general knowledge of the fauna is high, the score will give a good understanding of the fauna composition and can be very predictable with respect to increased or decreased acidity, as seen in running water in western Norway. Most likely, the score will also be very simple to incorporate in modeling work and can, to a high extent, predict the major species composition of a site. The ability of aquatic invertebrates to predict pH in Norway, by using weighed averaging regression and calibration with error estimation by bootstrapping (Birks et al., in prep.) shows a prediction of 0.3 pH units. This information coupled with the score is suggested to improve models of acidification. The important task will be to discriminate between the complexity of different ecosystems in lentic and lotic water, degree and duration of the acidification, and to what extent direct or indirect effects will determine the recovery process.

## REFERENCES

Barlaup, B., Å. Åtland, and G.G. Raddum. 1989. Improved growth in stunted brown trout (*Salmo trutta* L.) after reliming of Lake Hovvatn, southern Norway. *Water, Air, Soil Pollut.* **47**:139–151.

Drabløs, D., and A. Tollan, eds. 1980. Ecological Impact of Acid Precipitation. Proc. Int. Conf. Sandefjord. Oslo-Ås, Norway, March 11–14, 1980. Oslo: SNSF Project, Box 61, 1432 Ås-NLH, Norway.

Eriksson, F., E. Hørnstrøm, P. Mossberg, and P. Nyberg. 1983. Ecological effects of lime treatment of acidified lakes and rivers in Sweden. *Hydrobiol.* **101**:145–164.

Eriksson, M.O.G., L. Henrikson, B.I. Nilsson, G. Nyman, H.G. Oscarson, A.E. Stenson, and K. Larsson. 1980. Predator-prey relations important for the biothic changes in acidified lakes. *Ambio* **9**:248–249.

Fjellheim, A., and G.G. Raddum. 1990. Acid precipitation: Biological monitoring of streams and lakes. *Sci. Total Environ.* **96**:57–66.

Fjellheim, A., and G.G. Raddum. 1992. Recovery of acid-sensitive species of *Ephemeroptera*, *Plecoptera* and *Trichoptera* in River Audna after liming. *Environ. Pollut.* **78**:173–178.

Gunn, J.M., and W. Keller. 1990. Biological recovery of an acid lake after reductions in industrial emissions of sulfur. *Nature* **345**:431–433.

Hämäläinen, H., and P. Huttunen. 1990. Estimation of acidity in streams by means of benthic invertebrates: Evaluation of two methods. In: Acidification in Finland, ed. P. Kauppi, P. Anttila, and K. Kenttämies, pp. 1050–1070. Berlin: Springer.

Iversen, T., N.E. Halvorsen, S. Mylona, and H. Sandnes. 1991. Calculated budgets for airborne sulfur and nitrogen in Europe, 1985, 1987, 1988, 1989 and 1990. EMER/ESC–W Rep. 1/91. Oslo: Norwegian Meterological Institute.

Keller, W., and N.D. Yen. 1991. Recovery of crustacean zooplankton species richness in Sudbury area lakes following water quality improvments. *Can. J. Fish. Aquat. Sci.* **48**:1635–1644.

Last, T.F., and R. Walting, eds. 1991. Acidic Deposition: Its Nature and Impacts. *Proc. Roy. Soc. Edinburgh*, vol. 97B.

Lien, L., G.G. Raddum, and A. Fjellheim. 1992. Critical loads of acidity to freshwater fish and invertebrates. Res. Rep. 23. Oslo: Norwegian Institute for Water Research.

Locke, A. 1991. Zooplankton responses to acidification: A review of laboratory bioassays. *Water, Air, Soil Pollut.* **60**:135–148.

Maclsaac, H.J., W. Keller, T.C. Hutchinson, and N.D. Yen. 1986. Natural changes in the planktonic Rotifera of a small acid lake near Sudbury, Ontario, following water quality improvments. *Water, Air, Soil Pollut.* **31**:791–797.

Martin, H.C., ed. 1986. Acidic Precipitation. Proc. Int. Symp. on Acid Rain. *Water, Air, Soil Pollut.*, vols. 30 and 31.

Merrett, W.J., G.P. Rutt, N.S. Weatherly, S.P. Thomas, and S.J. Ormerod. 1991. The response of macroinvertebrates to low pH and increased aluminum concentrations in Welsh streams: Multiple episodes and chronic exposure. *Arch. Hydrobiol.* **121**:115–125.

Muniz, I.P. 1991. Freshwater acidification: Its effects on species and communities of freshwater microbes, plants and animals. *Proc. Roy. Soc. Edinburgh* **97B**:227–254.

Nyberg, P. 1984. Impact of *Chaoborus* predation on planktonic crustacean communities in some acidified and limed forest lakes in Sweden. *Rep. Inst. Freshw. Res. Drottningholm* **61**:154–166

Nyman, H.G., H.G. Oscarson, and J.A.E. Stenson. 1985. Impact of invertebrate predators on the zooplankton composition in acid forest lakes. *Ecol. Bull.* **37**:239–243.

Økland, J., and K.A. Økland. 1986. The effects of acid deposition on benthic animals in lakes and streams. *Experientia* **42**:471–486.

Otto, C., and B.S. Svenson. 1983. Properties of brown water streams in South Sweden. *Arch. Hydrobiol.* **99**:15–36.

Raddum, G.G., P. Brettum, D. Matzov, J.P. Nilssen, A. Skov, T. Sveälv, and R.F. Wright. 1986. Liming the acid Lake Hovvatn, Norway: A whole-ecosystem study. *Water, Air, Soil Pollut.* **31**:721–763.

Raddum, G.G., and A. Fjellheim. 1984. Acidification and early warning organisms in freshwater in westren Norway. *Verh. Inter. Verein Limnol.* **22**:1973–1980.

Raddum, G.G., A. Fjellheim, and T. Hesthagen. 1988. Monitoring of acidification through the use of aquatic organisms. *Verh. Inter. Verein Limnol.* **23**:2291–2297.

Raddum, G.G., G. Hagenlund, and G.A. Halvorsen. 1984. Effects of lime treatment on the benthos of Lake Søndre Boksjø. *Rep. Inst. Freshw. Res. Drottningholm* **61**:167–176.

Schindler, D.W. 1988. Effects of acid rain on freshwater ecosystems. *Science* **239**:149–157.

Schindler, D.W., T.M. Frost, K.H. Mills, P.S.S. Chang, I.J. Davies, L. Findlay, D.F. Malley, J.A. Shearer, M.A. Turner, P.J. Garrison, C.J. Watras, K. Webster, J.M. Gunn, P.L. Brezonik, and W.A. Swenson. 1991. Comparisons between experimentally and atmospherically acidified lakes during stress and recovery. *Proc. Roy. Soc. Edinburgh* **97B**:193–226.

SFT (The Norwegian State Pollution Control Authority). 1991. Overvåking av forurenset luft og nedbør. Report 466/91. Oslo: Statens forurensningstilsyn (State Pollution Control).

SFT (The Norwegian State Pollution Control Authority). 1992. Overvåking av forurenset luft og nedbør. Oslo: Statens forurensningstilsyn (State Pollution Control), in press.

Stenson, J.A.E. 1985. Biothic structures and relations in the acidified Lake Gårdsjøn system: A synthesis. *Ecol. Bull.* **37**:319–326.

Stenson, J.A.E., and H.G. Oscarson. 1985. Crustacean zooplankton in the acidified Lake Gårdsjøn system. *Ecol. Bull. Stockholm* **37**:224–231.

Wren, C.D., and G.L. Stephenson. 1991. The effect of acidification on the accumulation and toxicity of metals to freshwater invertebrates. *Environ. Pollut.* **71(2–4)**:205–242.

Yen, N.D., W. Keller, H.J. MacIsaac, and L.J. McEachern. 1991. Regulation of zooplankton community structure in an acidic lake by *Chaoborus*. *Ecol. Appl.* **1**: 52–65.

# 23

# Predicting Recovery of Freshwater Ecosystems: Trout in Norwegian Lakes

B.J. COSBY[1], A.J. BULGER[1], and R.F. WRIGHT[2]
[1]Dept. of Environmental Sciences, Clark Hall, University of Virginia, Charlottesville, VA 22901, U.S.A.
[2]Norwegian Institute for Water Research, Box 69 Korsvoll, 0808 Oslo, Norway

## ABSTRACT

Prediction of the recovery of freshwater ecosystems is largely based upon the simulation of future hydrogeochemical conditions that are compatible with healthy, reproducing biological populations. Direct simulation of the status of biological populations requires the coupling of often dissimilar modeling approaches and thus has not been often attempted. In this chapter, we examine the nature and severity of problems associated with developing coupled models of ecosystem response to acidic deposition by combining simple and complex statistical models of brown trout status with simple and complex process-based models of catchment geochemical response. Comparisons among the models (after calibrations to fisheries and water quality data from Norway) are based on estimated critical loads for recovery or protection of the trout populations at 50 years in the future. The comparisons demonstrate that combinations of models give approximately similar results.

## INTRODUCTION

The ultimate goal of modeling the effects of acidic deposition is to predict the status of the important biological components of ecosystems under different deposition regimes. Most acidic deposition modeling activities, however, have focused on predicting components of soil and surface water hydrology and chemistry rather than components of biological populations. Implicit in the focus on hydrogeochemical responses has been the assumption that biota is ultimately controlled by its hydrogeochemical setting. For instance, in aquatic ecosystems there is little inherent economic or esthetic value in a pH value of 5.5 or an alkalinity value of 0.0. The reason that

*Acidification of Freshwater Ecosystems: Implications for the Future*
Edited by C.E.W. Steinberg and R.F. Wright 

these particular values are the focus of attention in establishing deposition limits is that they have been determined to be indications of threshholds of water quality below which fish and invertebrate populations are damaged.

Prediction of the recovery of freshwater ecosystems, therefore, is based largely upon the simulation of future hydrogeochemical conditions that are compatible with healthy, reproducing biological populations. Simulation of the dynamics of hydrogeochemical variables is usually accomplished using either empirical or mechanistic (process-based) mathematical models. Establishing the compatibility of biological populations with hydrogeochemical conditions is usually done empirically using statistical models. It is frequently the case that the most efficient statistical models of biological relationships to water quality rely on variables that are not accurately or precisely simulated by (or even included in) the process-based hydrogeochemical models (and vice versa).

Any approach to the overall problem of predicting recovery of freshwater ecosystems must, therefore, begin with a consideration of the problems inherent in linking hydrogeochemical models with biological status/response models. Here we examine the nature and severity of these problems based on a study of acidic deposition in southern Norway and its effects on the chemistry and brown trout populations of the lakes in that region. A number of statistical models of trout status and process-based models of lake geochemical response will be examined for compatibility and efficiency of linking. The models represent a range of complexity in structure, simulated variables, and analysis techniques. Comparisons among the approaches are made using estimated critical loads for recovery or protection of the trout populations at 50 years in the future.

## HYDROGEOCHEMICAL MODELS

### Overview

At present there are a number of methods for making projections of future water chemistry for areas receiving acidic deposition that cover a spectrum of complexity. The simplest methods are based on an empirical approach, whereby extrapolations from present conditions are made using assumed relationships among water quality variables (e.g., the F-factor approach of Henriksen [1984]). At the opposite extreme are methods that utilize large, complex, process-oriented models ("pseudophysical models") of hydrology, geochemistry, and/or nutrient cycling to make the quantitative linkage between atmospheric deposition and surface water quality (e.g., the ILWAS model; Goldstein et al. 1984). There also exists an intermediate approach, complimentary to the other two and drawing on information from each. This third approach seeks to define predictive algorithms that retain, in large part, the simplicity of the empirical models but that have mechanistic, process-oriented explanations incorporated in their structure. The MAGIC (Model of Acidification of Groundwater In Catchments) model is an example of this type of intermediate-complexity, process-oriented model (e.g., Cosby et al. 1985a, b, c).

Hydrogeochemical response models thus cover a full range of complexity. Each approach has certain advantages, yet the shortcomings of each must also be considered when selecting a model. The simplified methods can be criticized on the basis that they are strictly empirical and a theoretical basis for establishing confidence in forecasts is therefore lacking. On the other hand, the complex, multi-parameter, whole-system models suffer from a lack of robustness of identifiability. In the words of Young (1980):

> It is well known that a large and complex simulation model, of the kind that abound in current ecological and environmental systems analysis, has enormous explanatory potential and can usually be fitted easily to the meagre time-series data often used as the basis for such analysis. Yet even deterministic sensitivity analysis will reveal the limitations of the resulting model; many of the "estimated" parameters are found to be ill-defined and only a comparatively small subset are seen to be important in explaining the *observed* system behavior.

## Selection and Calibration of Hydrogeochemical Models

The limitations of each modeling approach can result in unexpected (or undiscerned) effects on the prediction of ecosystem recovery. For this reason, we chose two models: the F-factor model as the "simple" model and MAGIC as the "complex" model. The use of two models incorporates a range of complexity in simulating ecosystem recovery, and comparison of the results should yield information concerning similarities and differences that are attributable to model complexity.

### *F-factor Model*

The F-factor model of Henriksen (1984; see also Henriksen and Brakke 1988; Henriksen et al. 1988) is a steady-state water chemistry model for application to lakes and streams. The model was initially applied to Norwegian lakes but it has also been used extensively elsewhere. The model requires calibration in that values of the F-factor must be specified for each site or a distribution of values must be specified for a region. The F-factor is defined as the fraction of change in sulfate concentration that is balanced by change in base cation concentration. We used the algorithm that bases F on observed values of base cations (Brakke et al. 1990). The F-factor has been used to calculate critical loads for surface waters throughout Fenno-scandinavia (Henriksen et al. 1992). Here we follow the procedures summarized by Henriksen et al. (1992) to use the F-factor model to calculate critical loads of sulfate.

*MAGIC Model*

The MAGIC model combines a number of key soil processes lumped at the catchment scale to simulate soil and surface water chemistry (Cosby et al. 1985a, b, c). MAGIC is a dynamic model and can be applied for site-specific studies or for regional surveys. As implemented here, MAGIC was calibrated to the 310 lakes of the 1986 Norwegian Thousand Lakes Survey (Henriksen et al. 1988) that are located in the area of southern Norway bounded by the 150 × 150 km "Birkenes" square of the EMEP deposition model. The calibration procedures are essentially the same as those described by Hornberger et al. (1989) and Wright et al. (1991) for previous MAGIC regional applications to southern Norway.

The regional application of MAGIC relies on Monte-Carlo techniques to simulate statistical distributions of soil and surface water chemical variables. Inputs and parameters for the model are drawn from specified initial joint distributions. Multiple simulations are performed using the chosen inputs and parameters to produce an ensemble of simulated soil and surface water chemistry outputs. The regional model is "calibrated" by adjusting the joint distributions of inputs until the joint distributions of outputs match the observed survey data. The ensemble of calibrated parameter sets may then be used for predicting regional responses to changes in deposition or other inputs.

The calibrated regional MAGIC used here reproduced the means and variances of the individual observed variables to better than 10% for all measures (Table 23.1). In addition, the MAGIC simulations produced pairwise correlation coefficients among the variables that were all within ± 0.2 of the equivalent pairwise correlation coefficients derived from the observed data (more than 90% of the coefficients agreed to at least ± 0.1).

## FISH RESPONSE MODELS

### Overview

Trout status models are statistical. Different levels of complexity are introduced by varying (a) the number of categories in trout population status (dependent variables), (b) the number and permutations of water quality parameters (independent variables), and (c) the type of statistical model constructed. The complex model uses four statistically distinct categories of fish population status; the simple model uses two. For each model, seven different groups of measured water quality parameters were used as independent variables to construct both regression and discriminant function models. Model success in each analysis was judged by the number of correct predictions achieved when the calibrated model was re-applied to the training data set. For both, the four category (complex) and two category (simple) fish status models and the combination of analysis technique and independent parameter group that gave the greatest success was selected for coupling to the hydrogeochemical models.

**Table 23.1** Comparison of means and standard deviations of observed and simulated variables for the calibration of MAGIC to the Norwegian Thousand Lakes survey data.

| Variable | Mean | | Standard Deviation | |
|---|---|---|---|---|
| | Observed | Simulated | Observed | Simulated |
| Lake Chemistry | | | | |
| Ca | 30 | 30 | 18 | 15 |
| Mg | 19 | 19 | 12 | 11 |
| Na | 55 | 55 | 39 | 37 |
| K | 4 | 4 | 3 | 2 |
| SBC | 109 | 109 | 64 | 57 |
| $SO_4$ | 64 | 64 | 24 | 25 |
| Cl | 62 | 62 | 47 | 43 |
| $NO_3$ | 9 | 9 | 5 | 5 |
| SAA | 135 | 135 | 67 | 62 |
| pH | 5.0 | 4.9 | 0.4 | 0.4 |
| ANC | −26 | −26 | 21 | 20 |
| H | 14 | 15 | 9 | 7 |
| $Al_i$ | 12 | 12 | 7 | 11 |
| Soil Chemistry | | | | |
| Exch. Ca | 22 | 22 | 15 | 13 |
| Exch. Mg | 6 | 6 | 3 | 3 |
| Exch. Na | 3 | 3 | 1 | 1 |
| Exch. K | 4 | 4 | 2 | 1 |
| Base Saturation | 35 | 35 | 17 | 16 |

Lake chemistry: aqueous concentrations in $\mu$eq $L^{-1}$ (n observed = 310, n simulated = 500).
SBC = sum of base cation concentrations = Ca + Mg + Na + K
SAA = sum of acid anion concentrations = $SO_4$ + Cl + $NO_3$
Soil chemistry: Exchangeable cations as percent of total cation exchange capacity.
(n observed = 84, n simulated = 500)

## Selection and Calibration of Fish Response Models

The information on chemistry and trout status used to develop the trout response models came from the Norwegian Thousand Lake Survey (Henriksen et al. 1988, 1989). Information on chemistry, and the present status and change in status of brown trout populations was available for 584 lakes, all of which had brown trout populations in the 1940s. Lakes with any significant human influence other than fishing were avoided. The lakes are located in bedrock areas expected to yield low-ionic strength surface waters (roughly 45% of the surface area of Norway is underlain by such bedrock); the lakes are distributed throughout Norway, but the majority are located in the southern half of the country.

The fish status information, in the form of interviews, was collected by local fisheries authorities in each of Norway's counties; it was thereafter verified by fisheries experts at the Norwegian Institute for Water Research (NIVA) and elsewhere (Henriksen et al. 1989). In this and other brown trout surveys in Norway (Wright and Snekvik 1978; Sevaldrud and Skogheim 1986), agreement between test-fishing results and interviews exceeded 90%, e.g., none or very few fish are caught in lakes reported to be barren, many fish are caught in lakes reported to be healthy, and marginal lakes produce intermediate numbers of fish.

The original trout status data consisted of two categories with integer scores: trout status and change in trout status.

1. *Trout Status*: The condition of a lake's trout population at the time of the Thousand Lake Survey in 1986 (updated for a few lakes in 1990); 3 levels are scored (Rosseland et al. 1980).
   a. Healthy: successful recruitment common, as evidenced by the presence of many year classes and normal population structure; density of fish high enough to support satisfactory yield.
   b. Marginal: sparse population, either historically so or reduced by acidic deposition; density of fish too low to support practical exploitation. Historically sparse populations typically have many year classes but low density; in populations damaged by acid deposition, recruitment failure is common, as evidenced by the presence of only a few year classes as well as low density.
   c. Extinct: population lost since the 1940s apparently due to acidic deposition.
2. *Change in Trout Status*: The change (if any) in the status of a lake's trout population between the 1940s and 1986 (updated for a few lakes in 1990); 4 levels are scored for analysis.
   a. Improved: yield, recruitment success (number of year classes) and/or density of population has increased since the 1940s.
   b. Unchanged: no significant change in yield, recruitment, and/or density since the 1940s.
   c. Declined: yield, recruitment, and/or density reduced since the 1940s, apparently by acidic deposition.
   d. Extinct: population lost since the 1940s apparently due to acidic deposition.

Values for both trout variables are given for each lake in the data set. The definitions of the categories in the two variables would appear likely to introduce a large correlation between the paired responses. Indeed, of the twelve possible combinations of these variables, only six combinations are actually represented in the data while six combinations are not observed in the data set (Table 23.2). In deciding which trout variable to use for predicting ecosystem recovery, it is important to consider the apparent nonindependence of the measured variables and to define a new variable(s) that is more nearly unique.

**Table 23.2** Cross comparison (contingency table) of the number of cases in each of the trout categories for the Norwegian Thousand Lakes Survey. The four statistically distinct categories (with respect to lake chemistry) are indicated by the solid lines (Bulger et al. 1993).

| | | Status Categories | | | |
|---|---|---|---|---|---|
| | | Healthy | Marginal | Extinct | TOTAL |
| Change in | Improved | 24 | 4 | 0 | 28 |
| Status | Unchanged | 166 | 64 | 0 | 230 |
| Categories | Declined | 0 | 98 | 0 | 98 |
| | Extinct | 0 | 0 | 228 | 228 |
| | TOTAL | 190 | 166 | 228 | 584 |

*Specification of Dependent Variables*

The trout variables *status* and *change in status* were analyzed by Bulger et al. (1993) with respect to the following chemical variables: major ions (Ca, Mg, Na, K, Cl, $SO_4$, $NO_3$); ANC (acid-neutralizing capacity, calculated as ANC = Ca + Mg + Na + K – Cl – $SO_4$ – $NO_3$); pH; TOC (total organic carbon); $Al_a$ total monomeric aluminum); and $Al_o$ (organically bound monomeric aluminum concentration). Aluminum fractions were operationally defined (Henriksen et al. 1988).

Bulger et al. (1993) concluded that there were really only four categories of trout status/response that were statistically distinct with respect to the chemistry of these lakes. These four categories are formed by combining several categories of the original *status* or *change in status* variables (Table 23.2). Lakes in the healthy-improved and healthy-unchanged categories were combined into a single "healthy" category. Lakes in the marginal-improved and marginal-unchanged categories were combined into a single "marginal-unchanged" category. "Marginal-declined" and "extinct" categories remained the same. These four distinct categories are used here to define the "complex" fish response model. Combining the healthy and marginal-unchanged categories into an "unaffected" category and the marginal-declined and extinct categories into an "affected" category produces the two categories used to define the "simple" fish response model.

*Specification of Independent Variables*

A number of observed water quality variables are available in the survey data base to be used as independent variables for predicting trout response: pH and the concentrations of ANC, Ca, Mg, Na, K, Cl, $SO_4$, $NO_3$, and inorganic monomeric aluminum ($Al_i$). The selection of groups of these variables for use in predicting trout response can be guided by two criteria: (a) the intended use of the resulting model (i.e., coupling to a geochemical model) or (b) theoretical and empirical understanding of fish physiological response to water quality. For the latter criterion, the variables pH, $Al_i$, and Ca are commonly thought to be the most important water quality variables affecting physiological response of fishes and should be considered as one grouping of independent variables (see Rosseland and Stuarnes, this volume).

For the first criterion, the two different structures of the geochemical models must be considered. The F-factor model simulates only ANC. Therefore, the single variable ANC must be considered as a second grouping of independent variables. On the other hand, MAGIC simulates most of the variables available in the fish data base, so a third grouping of independent variables that must be considered is composed of the fish physiological variables (pH, $Al_i$, and Ca) plus the F-factor variable (ANC) plus the remaining measured ions.

This suggests that definition of three basic groups of variables, with all permutations of combinations of these three groups used in the statistical model building exercise, should cover the range of possible "good" models of trout response. The groups of independent variables are thus defined as:

Group 1 – ANC (calculated as Ca + Mg + Na + K – $SO_4$ – $NO_3$ – Cl);
Group 2 – Physiological Ions (pH, $Al_i$, Ca);
Group 3 – ANC plus Physiological Ions;
Group 4 – Major Ions (Ca, Mg, Na, K, Cl, $SO_4$, $NO_3$, Cl);
Group 5 – ANC plus Major Ions;
Group 6 – Physiological Ions plus Major Ions;
Group 7 – ANC plus Physiological Ions plus Major Ions.

*Specification of Model Type*

Only a single dependent variable is considered: trout response. For the simple model, this variable has two categories; for the complex model, the variable has four categories. The categories can be assigned numerical values: 1 or 2 for the simple model, 1 through 4 for the complex model (note that assigning values to the complex model implies some linear ordering of the categories).

The problem of predicting category membership as a function of water quality variables can be approached through multiple regression or discriminant analysis techniques. It is not strictly appropriate to use regression analysis with discontinuous y-variables. However, previous studies have characterized the relationships between lake chemistry and fish status in Norway using regression models (e.g., Wright and Snekvik 1978; Muniz and Walloe 1990). We included regressions in this analysis to facilitate comparison to earlier work. Discriminant function analysis is designed to predict group membership from a set of variables. It is conceptually and mathematically related to ANOVA: group membership serves as the independent variable in ANOVA and the dependent variable in discriminant analysis. Variables are often referred to as "predictors" in discriminant analysis. The major question in discriminant analysis is whether combinations of predictors (called discriminant functions) can predict group membership reliably (Tabachnik and Fidell 1989).

*Choice of "Best" Models*

Both multiple regression and discriminant models were constructed for both the complex and simple response models, using each of the seven groupings of independent variables for each model. The robustness or goodness-of-fit of the resulting models can be assessd by "predicting" group membership for each lake in the data base, using values of the independent variables and the discriminant or regression functions, and then computing the number of correct group assignments.

The selection of the "best" structure for the simple and complex response models must be based on some criteria. In this instance, the goodness-of-fit in predicting category membership (per cent correct classification, PCCC) is an obvious choice; however, there are ambiguities. For instance, the PCCC for different response categories can vary widely within a given model structure (Figures 23.1 and 23.2). It will be necessary to select "best" model structures based on the PCCC of only one category within each model.

The model is being developed to predict ecosystem recovery as defined by the critical load required to recover (or protect) the trout population. Therefore, the model with the highest PCCC for the "extinct" or "unaffected" categories would be preferable to any other (preferable even to a model that has a higher "overall" PCCC). This criterion was used for the selection of final trout response models for coupling to the geochemical models. The results for this criterion have been extracted from Figures 23.1 and 23.2 and are presented seperately in Figure 23.3 for clarity.

For the simple response model, the largest PCCC for prediction of the unaffected category is produced by a regression equation using Group 5 variables (major ions + ANC). For the complex response model, the largest PCCC for prediction of the extinct category is produced by a discriminant function using either Group 4 (major ions) or Group 5 (major ions + ANC) variables. Both models predicted 81.6% correctly. For comparability with the simple model, the discriminat function based on Group 5 variables was arbitrarily selected.

These choices of "best" model are made based on the a priori criterion of largest value of PCCC. It is apparent from Figures 23.1 through 23.3 that many of the models give strikingly similar PCCC values. Although confidence levels are not presented for these results, from a pragmatic point of view there is probably no good *objective* way to select among many of the models.

The two response models just selected are the "most" efficient statistical predictors of trout response in Norwegian lakes, and they are based on variables simulated by MAGIC; thus they are also amenable to coupling with that hydrogeochemical model. The F-factor model, however, simulates only ANC, so appropriate response models (based only on ANC, the Group 1 independent variable) must be identified for coupling

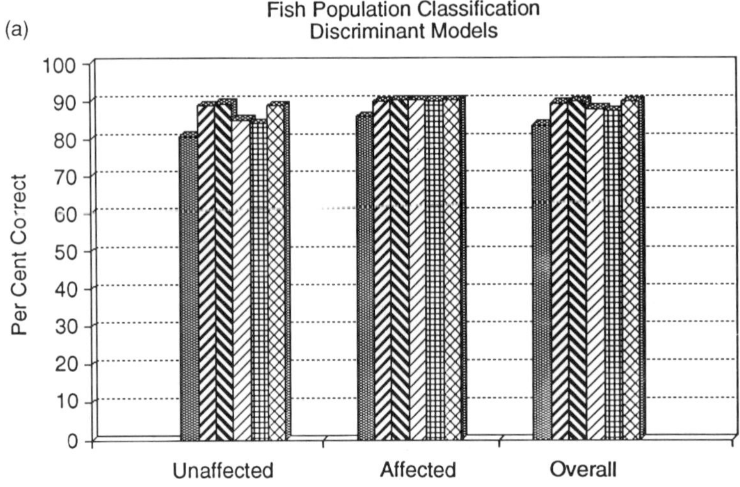


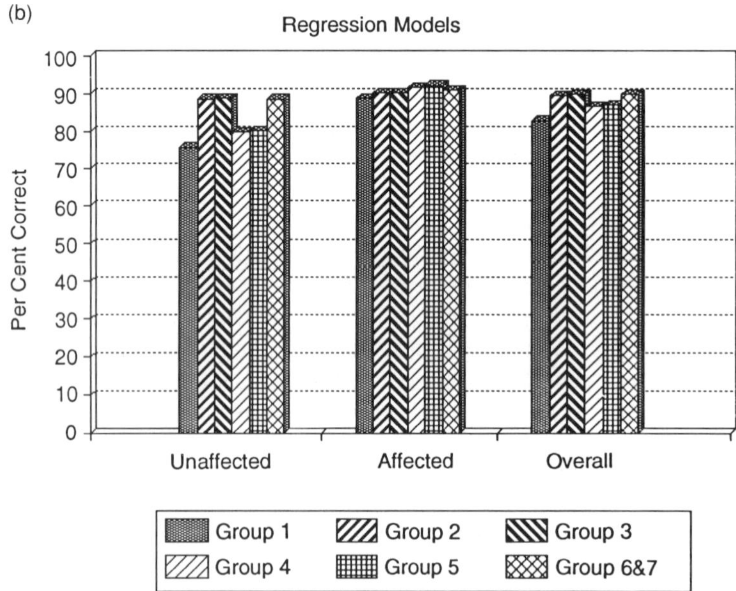


**Figure 23.1** Model goodness-of-fit for discriminant models (upper panel) and regression models (lower panel) for the "simple" fish response variable for which two categories are possible: "affected" and "unaffected." Goodness-of-fit is measured as the per cent correct classification (PCCC, see text). Values of PCCC are plotted for each category as well as for the weighted sum of all correct classifications ("overall"). The bars in each case represent the different groups of independent variables used (see text).

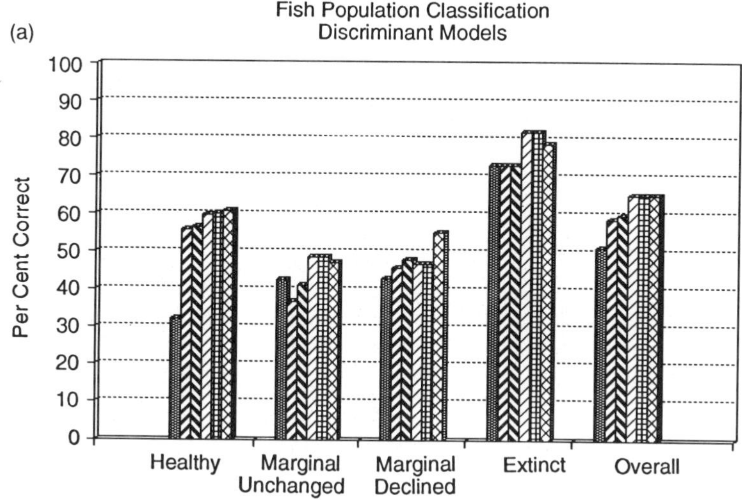


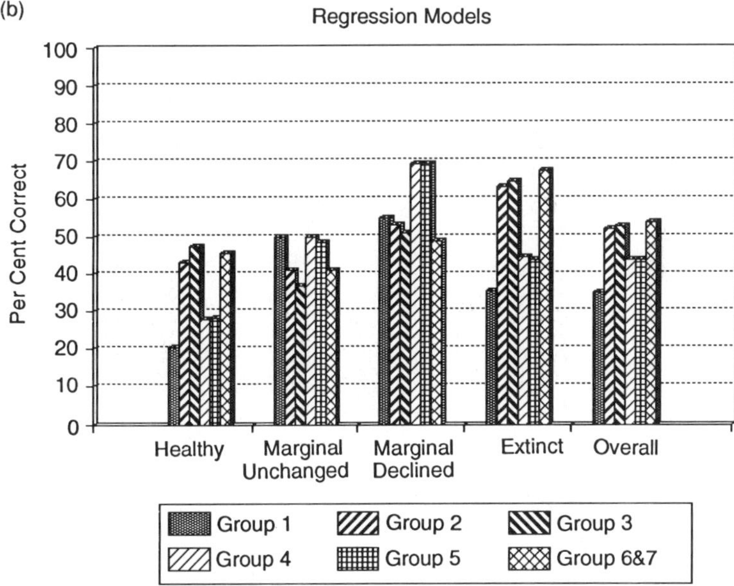


**Figure 23.2** Model goodness-of-fit for discriminant models (upper panel) and regression models (lower panel) for the "complex" fish response dependent variable which has four categories are possible: "healthy," "marginal-unchanged," "marginal-declined," and "extinct." Interpretation is as in Figure 23.1.

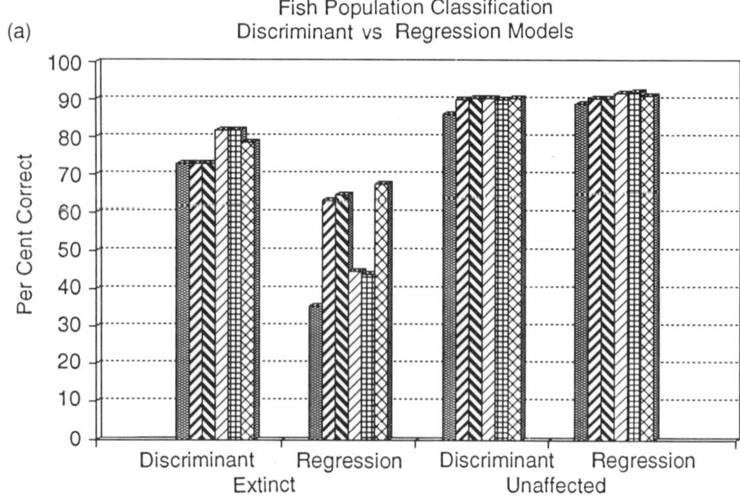


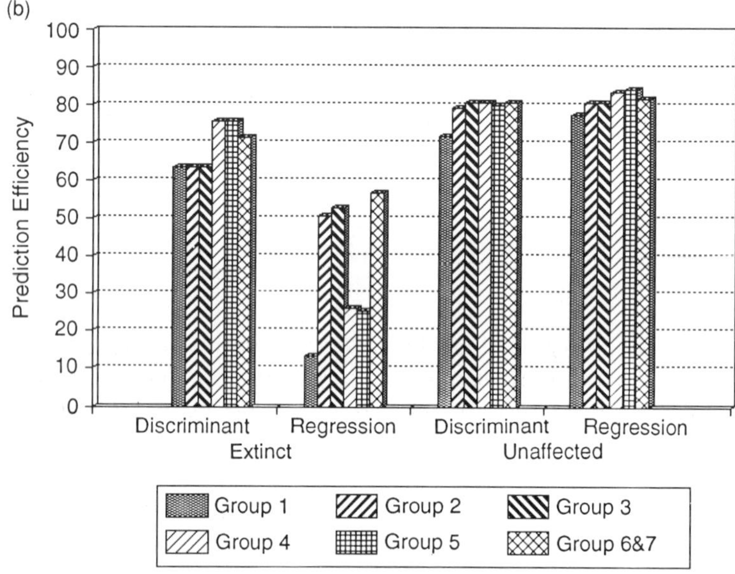


**Figure 23.3** Model goodness-of-fit for "damaged" categories only. The upper panel displays per cent correct classification (PCCC) for the "complex" model (left two groups of bars) and the "simple" model (right two groups of bars), for both regression and discriminant models. The lower panel displays the results as prediction efficiency defined as 1 – (the number of incorrect classifications *in excess of those expected by chance*)/(the number of incorrect classifications expected by chance alone), and presented as a percent. Bar codes are interpreted as in Figures 23.1 and 23.2.

to the simple geochemical model. Considering only ANC, the "best" model for predicting the extinct category in the complex response model is the discriminant model. For prediction of unaffected lakes in the simple response model, a regression structure is "best."

## EVALUATION OF ECOSYSTEM RECOVERY

### Overview

One measure of ecosystem recovery that may be of particular interest to planners and policymakers is the theoretical deposition load that will bring the biological population to the edge of a particular response category without actually entering the category. This concept is, of course, hypothetical because real-world populations move along a continuum of responses rather than between discrete states. Nevertheless, the coupling of fish response models with geochemical models can produce estimates of critical loads for biological population responses. Comparing critical loads among different systems can provide a measure of the sensitivity of the resource in the region of interest.

For instance, an ecosystem with a higher critical load than some reference system might be considered "more protected" than the reference if current deposition is lower than the critical load, or it might be considered "more likely to recover" than the reference if current deposition is in exceedance of the critical load. When applied to many lakes, a distribution function of numbers of resources protected or recovered as a function of deposition critical loads can be constructed. Evaluation of these distributions over a large geographical region will provide the kind of information necessary to evaluate the effects on ecosystem recovery of planned or hypothesized deposition changes.

### Application of Coupled Models

For this study, four coupled models will be used to predict critical loads for recovery (or protection) of the brown trout resource in southern Norway. The four coupled models represent combinations of simple and complex fish response and geochemical models:

1. the "simple" (F-factor) geochemical model coupled with a "simple" fish response model (two categories with "unaffected" predicted by regression from ANC);
2. the "simple" (F-factor) geochemical model coupled with a "complex" fish response model (four categories with "extinct" predicted by a discriminant function model based on ANC);
3. the "complex" (MAGIC) geochemical model coupled with a "simple" fish response model (two categories with "unaffected" predicted by regression from ANC and major ions); and

4. the "complex" (MAGIC) geochemical model coupled with a "complex" fish response model (four categories with "extinct" predicted by a discriminant function model based on ANC and major ions).

For each of the four coupled models, the same protocol was followed. Five hundred values of critical loads were derived using the distributions of water quality variables simulated by MAGIC (that matched the distributions observed in southern Norway in 1986; see Table 23.1 and previous sections) and used as starting points for critical loads calculations. For a given lake chosen from the 500 available, the calculation of critical loads was an iterative procedure. Using the initial water quality variables, a trout response status was calculated. If the trout response status was already in the damaged category (either "affected" or "extinct"), the deposition of $SO_4$ was decreased, the response of the water quality variables calculated, and a new trout response status was derived. This procedure was repeated until the predicted trout response emerged from the "damaged" category. The value of $SO_4$ deposition at that point was designated the critical load. For lakes where the initial fish response was outside the "damaged" category, $SO_4$ deposition was increased iteratively until the status just entered the "damaged" category. The value of $SO_4$ deposition at that point was designated the critical load.

The F-factor model is a steady-state model; therefore, the pattern and timing of increases or decreases in $SO_4$ deposition do not need to be specified. MAGIC, however, is a dynamic model and the patterns do need to be specified. For calculations using MAGIC, changes in deposition were implemented using a 50-year "square wave": increases or decreases to new deposition levels were assumed to occur "instantly," with the resulting deposition remaining constant at the new level for 50 years. Simulated values of water quality variables at the end of the 50-year period were used to evaluate the trout response functions.

The critical loads were expressed as absolute deposition (meq $m^{-2}$ $yr^{-1}$ of $SO_4$) or as a percentage of current deposition at each site (values in excess of 100% indicate exceedance of critical loads).

## Comparison of Predicted Recovery

The distribution of critical loads of $SO_4$ deposition for southern Norway as a function of the percent of lakes protected or recovered is very similar for three of the four coupled modeling approaches (Figure 23.4). There is particularly good agreement among three of the approaches in the region of the curve representing levels needed for recovery or protection of 50–90% of the lakes. This region is most likely the most important for policy and management purposes. The curves all begin to diverge (poor agreement among approaches) in the region representing recovery or protection for less than 15% of the lakes; however, decisions are not likely to be taken for protection of such a small number of resources.

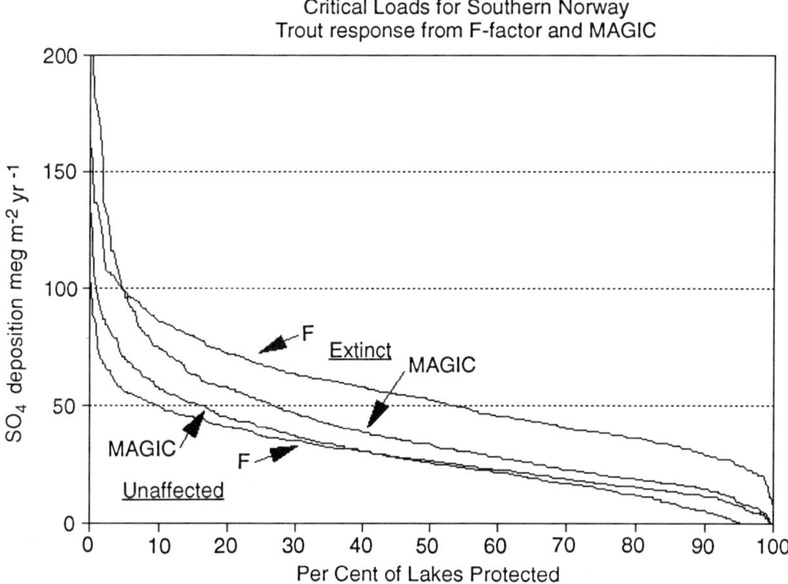


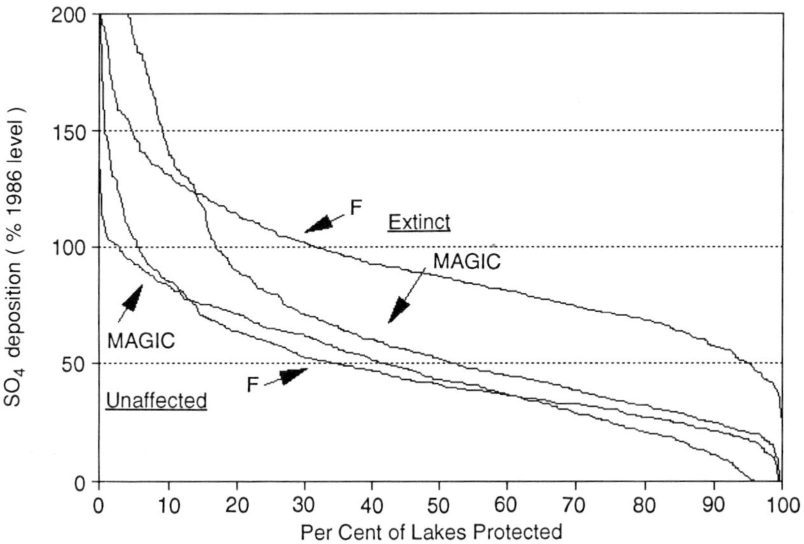


**Figure 23.4** Critical loads as a function of the percentage of lakes to be recovered or protected for the four modeling approaches described in the text.

The one approach that differs most is the "simple" geochemical model coupled with the "complex" fish response model. It generally gives higher critical loads for the protection or recovery of the same number of lakes. This difference appears to arise from the fact that MAGIC is a dynamic model while the F-factor model is not. Predictions of future water quality values from MAGIC for short periods into the future will be more strongly conditioned by the past history of the systems than will predictions for longer periods. That is, MAGIC "knows" that the lakes in southern Norway have been receiving acidic deposition for a number of years. The critical load calculated to just avoid "extinction" of trout 50 years in the future will, therefore, be lower than that calculated for longer periods in the future, because soils will not have fully recovered after only 50 years. The F-factor model has no such temporal dependence of recovery because it is a steady-state model. Thus, the critical loads produced by the F-factor model are lower than the 50-year estimates produced by MAGIC.

Other comparisons can be made which suggest that the coupled model approach is robust. Both MAGIC and the F-factor models predict higher critical loads for "extinct" vs. "unaffected" trout responses. In other words, to avoid any effect on trout populations, the critical load must be lower than that which would produce extinction. The difference between the two deposition levels is greater for the F-factor model than for MAGIC, probably for the reasons discussed above.

## DISCUSSION

In general, agreement among the methods is rather good. This implies that, at least from a regional perspective, simple models may be as adequate as more complex models for certain decision-making processes. This last statement, however, cannot be substantiated without guidance from decision-makers as to what level of uncertainty is acceptable or what degree of concordance is desired in predicting ecosystem recovery.

For example, the goal might be to protect or recover 70% of the lakes with respect to trout populations. If it is desired that the trout populations be totally unaffected, then MAGIC predicts a critical load of 17 meq $m^{-2}$ $yr^{-1}$ while the F-factor model predicts a critical load of 20 meq $m^{-2}$ $yr^{-1}$ (Figure 23.4). Present-day deposition is 100 to 200 meq $m^{-2}$ $yr^{-1}$. Is the difference between these two estimates too great to allow a decision to be made? Are the risks involved in taking a wrong decision based on this level of difference too large? Is the difference smaller than the uncertainty in specifying the level of deposition that might result from proposed regulatory actions?

If, on the other hand, the desire is only to prevent extinction in 70% of the lakes, the predicted critical loads from MAGIC and the F-factor approach are 23 and 43 meq $m^{-2}$ $yr^{-1}$, respectively (Figure 23.4). The risks involved in taking a wrong decision are potentially much greater. The two methods disagree by almost 100%. The extra costs involved in emissions controls to reduce deposition from 43 to 23 meq $m^{-2}$ $yr^{-1}$ may be prohibitive.

The question of agreement or disagreement among these models relates to the estimation of uncertainty in model results. In this case, we have attempted to demonstrate empirically (by employing several different approaches simaltaneously) what the level of uncertainty might be in attempting to predict ecosystem recovery. Most modelers (one hopes) are similarly diligent in their work, providing (when requested) analyses of uncertainty produced by their models. These uncertainty estimates are often viewed with horror (and the modelers subsequently castigated) by managers and decision-makers who want to use the model results. In many cases, the models are rejected by the decision-makers as being worse than no information at all.

We suggest that this frequent impasse is not entirely the fault (or problem) of the modelers alone. For a model to be truly useful, the ultimate application and mode of implementation of the model must be known when the model is being initially designed. This, in turn, means that the end users (managers and decision-makers) *must* be involved in the process from the beginning.

As demonstrated in this chapter, there are usually a number of different modeling approaches that can be used to address a problem. The approaches differ in complexity of model structure, data requirements, and assumptions. It often happens that several of the approaches converge and the choice among several alternatives must be made arbitrarily. It is also often the case that a priori expectations concerning the appropriateness of certain approaches are mistaken and those approaches must be abandoned (e.g., the "best" model for prediction of trout response in this chapter was not the one based on the theoretically most "correct" structure, which uses the physiological response variables pH, $Al_i$, and Ca).

While it may be true that, from a scientific point of view, there are times when arbitrary selections can be made or certain approaches abandoned without apparent loss of rigor, these decisions could affect ultimate acceptance of the model results outside of the scientific community. It is because of this possibility that we encourage the inclusion of secondary criteria of model "goodness" during the model-building process. These criteria should be derived from the needs and concerns of the end users of the models.

## CONCLUSIONS

We conclude that currently applied models of geochemical response can be succesfully coupled with empirical models of fish response to produce useful (at least in a comparative sense) forecasts of freshwater ecosystem recovery. We further conclude that, while differences do exist in the recovery predicted by the different approaches, the differences are generally small. We acknowledge that our description of the differences as being "small" is subjective and suggest that an important component in evaluating models used to predict ecosystem recovery is missing. That component is a definition of the intended use of the recovery models, particularly some indication from the ultimate users of the models as to what levels of uncertainty are acceptable.

## REFERENCES

Brakke, D.F., A. Henriksen, and S. Norton. 1990. A variable F-factor to explain changes in base cation concentrations as a function of strong acid deposition. *Verh. Inter. Verein Limnol.* **24**:146–169.

Bulger, A.J., L. Lien, B.J. Cosby, and A. Henriksen. 1993. Brown trout status and chemistry from the Norwegian Thousand Lake Survey: Statistical analysis. *Can. J. Fish. Aquat. Sci.,* **50(3)**:575–585.

Cosby, B.J., R.F. Wright, G.M. Hornberger, and J.N. Galloway. 1985a. Modeling the effects of acidic deposition: Assessment of a lumped-parameter model of soil and streamwater chemistry. *Water Resour. Res.* **21**:51–63.

Cosby, B.J., R.F. Wright, G.M. Hornberger, and J.N. Galloway. 1985b. Modeling the effects of acidic deposition: Estimation of long-term water quality responses in a small forested catchment. *Water Resour. Res.* **21**:1591–1601.

Cosby, B.J., G.M. Hornberger, J.N. Galloway, and R.F. Wright. 1985c. Time scales of catchment acidification. *Environ. Sci. Tech.* **19**:1144–1149.

Goldstein, R.A., S.A. Gherini, C.W. Chen, L. Mak, and R.J.M. Hudson. 1984. Integrated acidification study (ILWAS): A mechanistic ecosystem analysis. *Phil. Trans. R. Soc. Lond. B* **305**:409–425.

Henriksen, A. 1984. Changes in base cation concentrations due to freshwater acidification. *Verh. Inter. Verein Limnol.* **22**:692–698.

Henriksen, A., and D.F. Brakke. 1988. Sulfate deposition to surface waters. *Environ. Sci. Tech.* **22**:8–14.

Henriksen, A., J. Kamari, M. Posch, and A. Wilander. 1992. Critical loads of acidity: Nordic surface waters. *Ambio* **21**:356–363.

Henriksen, A., L. Lien, B.O. Rosseland, T.S. Traaen, and I.S. Sevaldrud. 1989. Lake acidification in Norway: Present and predicted fish status. *Ambio* **18**:314–321.

Henriksen, A., L. Lien, T.S. Traaen, I.S. Sevaldrud, and D.F. Brakke. 1988. Lake acidification in Norway: Present and predicted chemical status. *Ambio* **17**:259–266.

Hornberger, G.M., B.J. Cosby, and R.F. Wright. 1989. Historical reconstructions and future forecasts of regional surface water acidification in southernmost Norway. *Water Resour. Res.* **25**:2009–2018.

Muniz, I.P., and L. Walloe. 1990. The influence of water quality and catchment characteristics on the survival of fish populations. In: The Surface Waters Acidification Programme, ed. B.J. Mason. Cambridge: Cambridge Univ. Press.

Rosseland, B.O., I.H. Sevaldrud, D. Svalastog, and I.P. Muniz. 1980. Studies on freshwater fish populations: Effects of acidification on reproduction, population structure, growth and food selection. In: Ecological Impacts of Acid Precipitation, Proc. of an International Conference, ed. D. Drabløs and A. Tollan, pp. 336–337. Sandefjord, Oslo-Ås, Norway, March 11–14, 1980. OSLO: SNSF Project, Box 61, 1423 Ås-NLH, Norway.

Sverdrup, H., W. de Vries, and A. Henriksen. 1990. Mapping Critical Loads. Tech. Rep. of Nordic Council of Ministers, Box 19506, 104 32 Stockholm.

Tabachnick, B.G., and L.S. Fidell. 1989. Using Multivariate Statistics. New York: Harper and Row, 746 pp.

Wright, R.F., B.J. Cosby, and G.M. Hornberger. 1991. A regional model of lake acidification in southernmost Norway. *Ambio* **20**:222–225.

Wright, R.F., and E. Snekvik. 1978. Acid precipitation: Chemistry and fish populations in 700 lakes southernmost Norway. *Verh. Inter. Verein Limnol.* **20**:765–755.

Young, P.C. 1980. The validity and credibility of models for badly defined systems. In: Uncertainty and Forecasting of Water Quality, ed. Beck and van Straten, pp. 69–98. Berlin: Springer.

Standing left to right:
Wolfgang Ahlf, Jack Cosby, Gunnar Raddum, Jan Fott, Karl Kreutzer

Seated, left to right:
Nancy Dise, Michael Hauhs, Dick Wright, Gerhard Brahmer, Ingrid Jüttner

# 24

# Group Report: Are Chemical and Biological Changes Reversible?

N. DISE, Rapporteur

W. AHLF, G. BRAHMER, B.J. COSBY, J. FOTT, M. HAUHS, I. JÜTTNER, K. KREUTZER, G.G. RADDUM, R.F. WRIGHT

## INTRODUCTION

International efforts to reduce emissions of the acid precursors $SO_2$ and $NO_x$ are based, in part, on the premise that reductions in acid deposition will lead to recovery of acidified freshwater ecosystems and protect threatened ecosystems. The recovery process can be delayed or even become irreversible at several links along the causal chain from emissions to biological response in the aquatic ecosystems (Figure 24.1).

Our group discussed definitions of reversibility and recovery, the usefulness of the measures we use to describe them, and the factors determining their rates. We then assessed important future influences on recovery, including the effectiveness and drawbacks of liming and the potential of other environmental stresses, such as climate change, to hinder or reverse recovery from acidification. We reserved the term *reversible* for abiotic processes (usually in the terrestrial ecosystem) that influence recovery of the aquatic ecosystem (Figure 24.1). *Recovery* was used within the context of the biotic response.

## STATE OF KNOWLEDGE

### What Is Recovery?

The definition we choose for recovery depends upon on whether our goal is fish production for humans or more intangible "ecological/conservation" purposes. Among the different definitions of recovery possible, our group discussed the following:

*Acidification of Freshwater Ecosystems: Implications for the Future*
Edited by C.E.W. Steinberg and R.F. Wright © 1994 John Wiley & Sons Ltd.

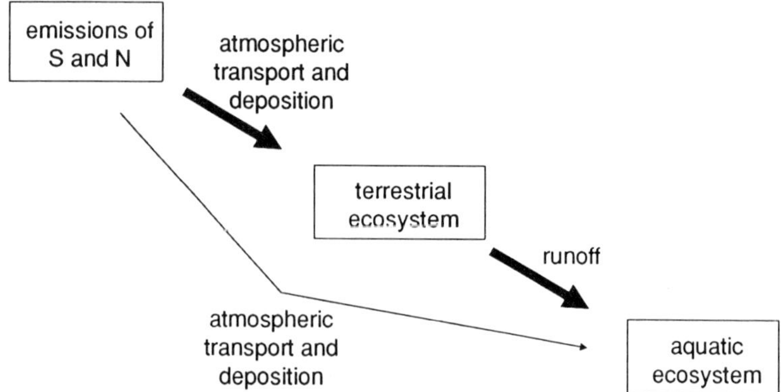


**Figure 24.1** The causal chain linking emissions of $SO_2$ and $NO_x$ to effects in aquatic ecosystems.

1. Return to the "preacidification" community structure as defined by species and their abundance.
2. Return of the "preacidification" species.
3. Return to a community showing normal functions.
4. Return of key "indicator species."
5. Presence of a healthy fish population.
6. A population of indicator species whose biological status may be predicted from empirical models without needing reference to the rate or intensity of acid deposition.

These measures are consistent based on present evidence from aquatic ecosystems currently undergoing recovery.

The first five definitions range from the most "ecological/conservation"-oriented to the most "human interest"-oriented. The first, a return to the exact preacidification community structure and function, would probably rarely occur due to stochastic/founder effects and inevitable changes to the ecosystem due to acid deposition. In addition, it is difficult or impossible to determine the preacidification status of most fresh waters. Inference about this may be derived from information on nonacidified lakes or from paleolimnological records. The latter are only available for some groups, e.g., diatoms, chrysophytes, Cladocera, chironomids, and chaoborids. Some relatively pristine areas, such as northern Scandinavia, may contain intact ecological communities that can be used as references for other high-altitude freshwater ecosystems; however, this is an exception.

A return to the preacidification species, without necessarily a return to the original community structure and function, may be possible in places that have not been severely affected and have ready sources of recruitment. It may, however, be difficult to determine the preacidification species, given natural variability in neighboring lakes and streams. It may also be impossible to achieve preacidification status if there are impacts from other environmental disturbances.

Defining a recovered community simply on the basis of a healthy fish population was too restrictive for us, especially since our goal, as scientists, is to view the ecosystem as a whole from the standpoint of sustainability and conservation. While the problem of fish decline is certainly the popular reason for the intense international interest in surface water acidification, we believe that a simple restocking of a commercially important fish species to a stream or lake creates an ecological community that—if potentially much changed from the original—is ultimately unstable. Despite this problem, uncontrolled stocking of lakes by humans is widespread. The sixth definition of recovery could be tested by reference to existing empirical relations between diversity, species richness, and lake morphometry in Canada (Gunn, pers. comm.). Because of the scarcity of such regional correlations, however, it cannot be currently applied on a widespread basis.

Our group proposed the "conservation-oriented" goal of recovery from acidification to be the return of a self-sustaining ecological community showing normal functions. Therefore, we concluded that more scientific study is desirable on the functional abilities of the ecosystems. This includes stable size-class distribution within the plankton community, predator-prey relationships, interaction between planktonic and benthic communities, and, in "fish lakes," a self-sustaining, reproducing fish population with stable age structures. While this is the most desirable definition, relevant functions may be difficult to define and/or measure (productivity, grazing, etc.). Useful parameters may be size, structure, and diversity indices. In practice, however, identification of stable communities may be difficult without some reference or "key species," which can be considered indicators of a stable ecological community. We define "key species" differently depending upon whether our goals are human use or ecological conservation. If we are simply interested in human use, a key species or population is that species of greatest economical/recreational value, usually fish. If ecological conservation is the goal, then key species are the most sensitive fish and invertebrate species that would be expected in that ecosystem.

Key organisms, such as Mysis, gammarids, snails, mayflies, stoneflies, caddisflies, and cyprinids, are usually associated with "healthy" diverse communities (Fjellheim and Raddum 1990; Schindler et al. 1991). Many species in these groups are acid-intolerant and would only be found in circumneutral or basic water. Because they are often at the top of the food web in the lake or stream, some fish species can be indicators of both a stable ecosystem and suitable water chemistry. One of the main problems of this approach is that the use of only a relatively few select species may leave other important aspects uninvestigated.

Based on the above, our group decided on the following broad working definitions for recovery:

1. Recovery in streams and lakes has taken place if a healthy key species of fish and key species chosen from other components of the community (e.g., among sensitive phytoplankton, zooplankton, or benthic species) have returned.
2. In streams and lakes that have not historically had fish populations, recovery has taken place if identified key species (see examples above) have returned.

These definitions have been shown to be useful in practice to describe recoveries from acidification in both Sudbury, Ontario (Gunn and Keller 1990) and Norway (Fjellheim and Raddum 1990).

*Dispersal and Recovery*

In many cases, it is difficult to decide whether a sensitive species has, in fact, disappeared from a lake, stream, or watershed. Microbes (bacteria, cyanobacteria, algae, fungi, protozoa) and invertebrates may be present either at resting stages or in very low densities, but fail to appear in samples. Some taxa also have effective means of dispersal (wind, water birds). We may assume that microbes (including algae and protozoans) are always potentially present as inocula and that their dispersal need not be taken into consideration.

In the case of fast reversals of very degraded lakes, natural recolonization of invertebrates could be speeded up by intentional introduction. Introduction of molluscs to limed lakes is being carried out in Sweden, although not in any systematic manner.

In most cases, regional reductions in acid deposition and resultant reversal of surface water acidification will probably proceed so slowly that dispersal of invertebrates to sites from which they disappeared would keep pace with the reversal. For many fish species, on the other hand, natural migration to lakes will probably be quite slow, especially for lake-dwelling and -spawning species; these most likely will need to be reintroduced.

## What Is Acidification and Its Reversibility?

Acidification of surface waters is commonly defined as a reduction in the acid-neutralizing capacity (ANC). ANC can be defined as the difference in the sum of concentrations (equivalents per liter) of base cations ($Ca^{2+}$, $Mg^{2+}$, $K^+$, $Na^+$) less the sum of concentrations of strong acid anions ($SO_4^{2-}$, $NO_3^-$, $Cl^-$) (Reuss and Johnson 1986). From the ionic balance it can be shown that through this definition ANC is also equal to the sum of concentrations of weak acid anions ($HCO_3^-$, $OH^-$, and organic anions) less the sum of concentrations of acid cations (H and inorganic Al species). Acidification, the decrease in ANC, might thus be manifest by an increase in sulfate or nitrate concentrations and also by increases in hydrogen ion and aluminum concentrations.

Reversibility is the increase in ANC following a reduction in acid deposition. There is no rate or degree implied in this definition; reversibility could be partial or complete and could occur over days or decades. A system may be said to be permanently impaired ("memory") if some effects of acidification are still evident, even long after the acidic stress has been removed.

*Measures to Assess Reversibility*

It was the opinion of our working group that ANC is the single best abiotic factor providing a link to biological responses. A number of studies have shown high correlations among ANC and measures of both vertebrate and invertebrate community status or viability. In particular, it was felt that ANC is useful in establishing thresholds between classes of biological response (healthy, transitional, extinct populations, etc.). While no cause-and-effect relationships can be inferred from this linkage, the usefulness of such a schema has been repeatedly demonstrated.

The most common indicators of reversibility (increases in ANC) are decreases in concentrations of sulfate, nitrate, $H^+$, and inorganic Al species. The group discussed whether these are the best indicators of reversibility of acidification and if they are of the most biological importance. Indicators appropriate for one ecosystem (e.g., uplands) may not necessarily be appropriate for others (e.g., wetlands). The importance of using values of flux versus concentration was also debated. While the latter is biologically more important because it reflects the environment—including "pulses" of acidity—to which biological species react, it is not clear which concentrations are the most appropriate measurements (e.g., base flow versus highest maximum concentration). For streams, fluxes were chosen for our purposes since they are integrative and since single concentration values can be recovered from calculations of flux. For lakes, concentrations are more appropriate.

*Reversibility of ANC*

ANC is currently used as the primary measure of sensitivity and reversibility of acidic effects. The advantages of using ANC are that it is a pragmatic parameter, readily measured, highly correlated to the biologically meaningful parameters pH and inorganic Al, and reliably predicted in process-oriented models.

One of the disadvantages of ANC is that it may be inappropriate in cases where species are damaged by acidic pulses. The group suggested that criteria for "significant" acidic pulses could be constructed for different ecosystems (Ormerod and Jenkins, this volume). As a rough estimate, an ANC drop below $-20\ \mu eq\ l^{-1}$ for several days was proposed as critical. Empirical evidence exists from mountain streams in Virginia, U.S.A., indicating a positive relationship between baseflow ANC and the minimum pH or ANC observed during a storm event (Marshall 1993). Such criteria and relationships merit further study. ANC may also be inappropriate in highly polluted areas, where toxic effects from metals become important. Here, concentrations of trace metals in addition to ANC may be better indicators of reversibility of stress.

Empirical data from several regions suggest that deceases in ANC due to acid deposition are reversible if acid deposition is reduced at an adequate rate to a sufficient degree (Wright and Hauhs 1991). However, the complete return to preacidification ANC levels is limited by supply rates of base cations from weathering and atmospheric deposition. Our group was in some disagreement as to whether the "hysteresis" of acidification and reversal is greater for shallow, base-poor soils, with only a low rate of cation resupply, or deep soils that had been acidified to some depth. Comparative studies were suggested.

### *Reversibility of Base Cations*

The most important sources of base cations for fresh waters are deposition from the atmosphere and chemical weathering in the soils of the terrestrial portion of the watershed. Weathering and ion-exchange processes in lake sediments can also supply base cations; this source is important in lakes with long water-retention times. Restoration of water quality parameters related to base cations (e.g., ANC) will, therefore, depend upon the reversibility of the effects of acidic deposition on the base cations in catchment soils. During the acidification process, base cations are leached from exchange sites on soils. This cation exchange process buffers the acidification in the drainage waters and thus has a beneficial effect during the onset of acid deposition. Following reductions in deposition, however, the deficit of soil base cations previously removed must be rebuilt before the complete restoration of surface water quality is possible. Replenishment of base cations on soils depends on mineral weathering (and deposition from the atmosphere) and is a slow process, especially in thin soils derived from siliceous parent material. Relative to the reversibility of sulfate in surface water, base cation reversibility is often a much slower process. Thus the initial response (i.e., the first several years) of freshwater systems to moderate decreases in acid deposition is a decrease in sulfate concentration and a decrease in base cation concentration as well, resulting in little early reversibility of ANC.

### *Reversibility of Sulfur*

Changes in the concentrations of sulfate in surface waters in response to changes in deposition are controlled through reactions of sulfur in the terrestrial portion of the watershed and at the sediment-water interface in lakes. Aquatic systems, which include extensive wetlands or bogs, may behave differently from systems without wetlands. Sulfur retention in soils is mediated biologically through incorporation into organic matter, or chemically, through adsorption or precipitation reactions. If the retained sulfur is permanently bound within soils, reversibility of surface water sulfate concentrations will be rapid following changes in deposition. If the terrestrial retention is nonpermanent (i.e., soil ad-/desorption), then reversibility of surface water concentrations may be significantly delayed following decreases in deposition. The presence of wetlands in a catchment introduces delays; the sulfur accumulated over long time periods may be released slowly by oxidation. Current knowledge and observations suggest that sulfate concentrations in surface waters are completely reversible, given adequate time.

In-lake processes such as sulfate reduction are relatively more important in lakes with long water-retention times (greater than about 3 years). Sulfate reduction removes sulfate from the water and causes an equivalent increase in ANC. Acidification can influence the rate of sulfate removal through in-lake processes by, for example, changing the development of metaphytic algae (Kelly, this volume). More research is needed on the influence of in-lake processes during the reversibility of acidification and on the factors controlling these processes.

There is evidence of a "memory" effect resulting from nonpermanent sulfate adsorption from Germany, where no changes in pH, Al, or sulfate concentrations in stream water at Lange Bramke, Harz mountains, have been observed for several years following 30% emission reductions (Hauhs, this volume). In contrast, results from the RAIN project in Norway show that sulfate concentrations decrease following reductions in sulfur deposition (Wright et al. 1993); similarly, lakes in the Sudbury region promptly exhibited decreases in sulfate concentrations following decreases in sulfur deposition (Gunn and Keller 1990). The general decrease is punctuated by pulses of sulfate following dry periods and during the first phases of snowmelt, presumably due to release of sulfur stored in the soil. This pattern is typical of many catchments in areas with thin soils developed from parent material from Pleistocene glaciation. Immediate responses to reductions in emissions were also reported elsewhere (for a review, see Wright and Hauhs 1991).

### *Reversibility of Nitrogen*

Nitrate is generally released from the terrestrial to the aquatic ecosystem in pulses; concentrations fluctuate more widely over time relative to sulfate concentrations. In acidic systems, these pulses of nitrate are often accompanied by episodic decreases in ANC and pH, and increases in Al. For reversibility, this implies that for a similar decline in weighted annual average concentration, the episodicity will be greater for nitrate than for sulfate.

In contrast to the situation with sulfur, there are apparently no cases in which nitrogen deposition has been reduced and nitrate-induced acidification reversed. Several large-scale manipulations provide the best information at present. Large roofed plots at four sites in the Netherlands (van Dijk et al. 1992), for example, show that nitrate concentrations in soil solutions decrease within the first year of experimental reduction in nitrogen (and sulfur) deposition to coniferous forest stands.

Similarly, at the roofed catchment of the RAIN project in Norway, both nitrate and ammonium concentrations in runoff decreased within weeks after inputs were excluded (Wright et al. 1993). Nitrate thus appears to respond more quickly than sulfate following reduction in deposition, probably because at this site, sulfur is much more affected by soil processes such as adsorption and oxidation/reduction. Precisely this characteristic, however, implies that nitrate concentrations in runoff show much larger short-term variations when compared to sulfate. Because it is the episodes of acid and toxic aluminum (i.e., ANC minimum) that cause much of the biological damage, recovery may take longer after reduction in nitrate concentrations relative to sulfate.

Nitrogen in lakes is also affected by in-lake processes (Kelly et al. 1990; Kelly, this volume). The two major processes involved are incorporation into biomass (mainly phytoplankton) and microbial nitrate reduction (mainly in the sediments). Both processes remove nitrate and thus act to increase ANC. Algal removal of nitrate is a much more efficient process than microbial nitrate reduction. Thus inputs of small to moderate amounts of nitrate from terrestrial catchments will generally not result in significant increases in nitrate concentrations in receiving lakes until the algal nitrogen requirements are exceeded.

### *Combined Effects of Nitrogen and Sulfur*

How do changes in nitrogen affect sulfur and vice versa? In many regions, ammonium and sulfate in deposition are closely correlated, especially in throughfall. This combined filtering effect is probably due to trapping of ammonia and sulfur dioxide. The enhanced deposition of ammonium may lead to excess nitrification in the terrestrial ecosystem and result in soil and water acidification.

A large body of experimental work suggests that sulfate adsorption increases with decreasing pH (to a lower limit), all other factors being equal. Thus, there is a possibility that acidic pulses in soil from nitrification, especially during rewetting after a drought, could increase the adsorption of sulfate and decrease soil sulfate concentrations (Mitchell 1992). For sites that leak nitrogen, the relation between nitrate and sulfate concentrations in runoff often shows regular patterns, but these differ among sites: some are in-phase, some out-of-phase, some shifted (e.g., Kreutzer et al. 1991; Matzner 1988).

The importance of this phenomenon (in light of increasing nitrate levels in soils) has not been assessed. A buildup of organic matter by increased nitrogen deposition, predicted for many forest types, could increase the organic storage of sulfur, providing a relatively long-term sink for sulfate. However, rapid oxidation of the organic matter after, for example, a drought or fire could release acidic pulses to receiving waters (Bayley et al. 1993). In general, the relative interactions between the two main acid anions have been inadequately studied, especially in light of current nitrogen increases and sulfur decreases in deposition.

### *Recommendations for Future Research*

1. Investigate reasons for different patterns of nitrate and sulfate in runoff.
2. Conduct more manipulative studies on the effects of varying nitrogen on sulfur levels in soil solutions and runoff.

3. Investigate the "special" relationship between nitrate and inorganic aluminum in soil solution and surface waters.
4. Investigate the factors controlling in-lake processes (especially sulfate and nitrate reduction), and the influence of these factors on reversibility.

**Sources of Evidence for Reversibility and Recovery**

The group agreed on the relative importance of the following forms of knowledge: (a) field evidence from ecosystem observations, either as regional surveys or long-term time series at selected monitoring sites; (b) experimentally manipulated ecosystems; and (c) modeling. Paleolimnological studies provide a link between changes in pH and the effects on diatom assemblages (Battarbee and Charles, this volume).

Each entry in this hierarchy has to be tested and validated against the previous (higher) entry. Thus, field observations from ecosystems provide the "hardest" evidence and are used to evaluate the general relevance and applicability of manipulative studies of acidification (and to some extent also for deacidification). Whole ecosystem manipulations make the extrapolation into future deposition or management scenarios possible, where no field evidence is available. For this to be achieved, it must be shown that the biological acidification response of lakes and headwater streams can be realistically simulated through application of simulated acid deposition, indirectly via the terrestrial or directly to the aquatic ecosystem. When neither field evidence or suitable manipulation experiments are available, models must be used. Here future deposition and management scenarios would be modeled. For this step, a conceptual link between changes in chemical variables and the biological status of the system must be selected. The group proposed to use ANC for this linkage.

The validation step for any model is a test of whether the selected linkage variable can be matched for the two preceding levels in the above hierarchy of criteria. One model that has passed many such tests is the MAGIC model (Cosby et al. 1985a, 1985b).

*Field Evidence of Empirical Linkages*

Cases where a (nonmanipulative) reversal of surface water ANC has led to biological recovery are available from the Experimental Lakes Area (ELA) in Canada (Schindler et al. 1991), the Sudbury region in Canada (Gunn and Keller 1990; Hutchinson and Havas 1985), and the Galloway area in the U.K. (Allott et al. 1992). In all of these cases, biological recovery as defined above (key species return) was relatively fast—2–3 years in streams and about 5 years in lakes—unless there were migration problems.

Nonetheless, the reversibility of surface water ANC depends critically on rate-limiting processes in the upstream terrestrial system. In the above-listed cases, the terrestrial system did not cause major lag times between reversing deposition trends and increasing surface water ANC. However, other cases (such as in the southeastern

United States and in central Europe) exist where decreases for sulfate deposition did not result in reversibility of surface waters and where no signs of any recovery occurred (Hauhs, this volume). In the former cases, the deposition decrease was the rate-limiting factor whereas in the latter it was the reversibility of the terrestrial response. The group could not decide on site characteristics that would make it possible to predict the critical rate-limiting step for recovery prior to the collection of empirical evidence (or to what extent the few observed cases could be regarded as representative for larger areas). Capacity factors such as soil thickness, preceding changes in base saturation, and reversible sulfate pools are likely to be important for such predictions.

### *Evidence from Manipulative Studies*

Experimental manipulations of whole ecosystems to obtain information bearing on reversibility and recovery have involved both catchments (e.g., RAIN project; Wright et al. 1993) and whole lakes (e.g., Lake 223 ELA; Schindler et al. 1991). Catchment manipulations focus on terrestrial influences linking the chemical composition of runoff (i.e., ANC) to responses in aquatic biota. Whole-lake experiments focus on biological recovery and in-lake processes. Liming of terrestrial catchments and aquatic systems also provides information related to biological recovery.

The RAIN project experiment at Risdalsheia (Norway) has reduced the load of acid deposition for 8 years to a small, sparsely forested catchment. The chemical concentrations of sulfate, nitrate, base cations, acid, and aluminum all decreased in runoff. ANC increased (Wright et al. 1993). The first year's results from a similar experiment with a roof beneath the canopy of a mature spruce forest catchment at Gårdsjön, Sweden, show a similar decline in sulfate concentrations in runoff (Hultberg, pers. comm.).

For whole-lake experiments the approach has been to acidify the pristine lake ecosystem experimentally, then allow the system to reverse to pretreatment conditions. At the ELA experiments the focus has been on biological response to acidification and recovery (Schindler et al. 1991; Brakke et al., this volume). Empirical data on biological recovery are sparse. For this reason, the whole-lake manipulations cannot yet be validated against the monitoring evidence. Unlike the case of acidification, the reversibility and subsequent recovery of whole catchment-lake ecosystems has not been successfully simulated so far. Single aspects, however, have been experimentally reproduced (e.g., ANC response).

### *Evidence from Modeling*

Most freshwater ecosystems are strongly influenced by their terrestrial catchments. Exceptions are some seepage lakes and lakes with relatively long water-retention times. Recovery of the aquatic biota cannot proceed until the deterioration of the quality of drainage waters is reversed. Prediction of the recovery of freshwater ecosystems, therefore, is largely based upon the simulation of future hydrogeochemical conditions that are compatible with healthy, reproducing biological populations.

Simulation of the dynamics of hydrogeochemical variables is usually accomplished using either empirical or mechanistic (process-based) mathematical models. Linking biological populations with hydrogeochemical conditions is usually done empirically with statistical models. It is frequently the case that the most efficient statistical models of biological relationships to water quality rely on variables that are not accurately or precisely simulated by (or even included in) the process-based hydrogeochemical models (and vice versa) (Cosby et al., this volume).

Any approach to the overall problem of predicting recovery of freshwater ecosystems must, therefore, begin with a consideration of the problems inherent in linking hydrogeochemical models with biological status/response models. This topic is covered in detail by Cosby et al. (this volume).

*Theoretical Aspects*

In addition to the validity resulting from empirical testing, the level of confidence of model predictions can also be based on the degree of understanding that is incorporated into the model. In the past, however, this criterion did not prove especially useful in deciding among conflicting interpretations of ecosystem-scale changes, although it was very effective in accounting for physical and chemical changes. For ecosystems, the relation between structure (i.e., species diversity) and function (i.e., the energy and mass budgets) is not theoretically understood. There is also no agreement concerning the extent to which higher resolutions can overcome this problem. The various aspects and potential reasons for the weakness of a theoretical approach are addressed by Hauhs (this volume).

We were unable to identify an example where an attempt to understand internal processes below the ecosystem level resulted in an accepted criterion for deciding among conflicting interpretations of ecosystem responses. Since no empirical evidence is available and there are no alternate theoretical explanations at the ecosystem scale itself, this issue remains an open problem.

*Recommendation for Future Research*

1. More research is needed on methods and models to scale up in time and space between internal ecosystem processes and whole-ecosystem response, as well as the extrapolation from individual aquatic ecosystems to entire landscapes and regions.

## ASSESSMENT OF FUTURE DEVELOPMENTS

### Mitigative Strategies: Liming

Liming studies provide further information that bears on the degree and rates of recovery of aquatic ecosystems. Liming causes increases in ANC and decreases in toxic $H^+$ and inorganic Al; the water is thus less toxic to aquatic organisms. Extensive information on the biological response to liming is available and this can be used to infer responses after decreased acid deposition.

Many lakes and streams will not recover for years without liming. When acidification is so extensive that it has caused depletion of base cation reserves in the soils, runoff may still have negative ANC and be toxic to aquatic organisms for many years to come, even after major reductions in acid deposition.

Liming is presently practiced over wide areas of Scandinavia (mainly to raise ANC of waters) and Germany (mainly to raise ANC in forest soils), and to a lesser extent in North America.

Liming will likely continue and increase in use in the future. For streams, there is evidence from Norway and Sweden of rapid species return in upstream recruitment. For lakes, recovery is not as fast due to problems with species recruitment and often large invertebrate population fluctuations in the first few years. Fish (restocked) generally do well.

### Mitigation by Nutrient Addition

The addition of phosphorus alone or as a supplement to liming may help biological recovery of acidic lakes. Mild fertilization can increase primary production in lakes to preacidification levels. In lakes where nitrate is a significant anion, addition of P could result in a decease of nitrate levels with a commensurate increase in ANC; here, the mechanism is nitrate uptake by increased algal growth (Kelly et al. 1990). On the other hand, the addition of phosphorus to lakes where $NH_4$ concentrations are high could result in decreased ANC (Schindler et al. 1985). Properly applied nutrient addition could thus be used to control the rate of ANC change more precisely than liming alone and to promote biological recovery. More research, however, is needed.

### What Are the Undesirable Effects of Direct Liming to Lakes and Streams? When Is Liming Inappropriate?

During acidification, trace metals bound or adsorbed on solids can be released into the aqueous phase. If there is enough suspended matter, the released metals can be readsorbed, primarily to organic material. The net result is a metal transfer from inorganic to organic solids. The order of binding stability and affinity of humic acids for heavy metals is Cu > Pb > Cd > Zn. Considering these basic mechanisms, increasing pH could affect trace metal behavior twice. Trace metals released to the water phase are scavenged by freshly precipitated oxyhydrates and become part of the sediment. The second case is characterized by an increase of DOC from decomposition. The DOC is now a carrier of trace metals into the water column.

Other undesirable effects from liming include:

1. changes in species composition and structure;
2. highly toxic Al conditions (especially for fish at sensitive physiological stages) when mixed with acid tributary or littoral-zone water (Rosseland et al. 1992);

3. phytoplankton blooms and pulses of filamentous algae, generally lower stability of biotic community, more peaks and crashes;
4. dramatic species shifts, death of sensitive species such as *Sphagnum*, especially in liming of wetlands.

## What Are the Undesirable Effects of Liming Catchments?

Liming of forested catchments is practiced to counteract airborne acid deposition by raising thc ANC in soil solution. The effect on aquatic ecosystems depends on the acidic potential (i.e., base-neutralizing capacity and rate) of the seepage water in the soil as well as that in the transition zone between the soil and the freshwater system.

Liming is normally executed in forestry as surface liming. This may lead to undesirable side effects, e.g., humus mineralization with subsequent surplus nitrification and mobilization of heavy metals, such as Pb, Cu, and Hg, connected with increased DOC. These heavy metals may form stable organic complexes that can be released into fresh waters by macropore flow, especially on slopes with prevailing lateral subsurface flows. Critical situations may occur on sites with surface contamination, especially storm flows.

Deep soil liming with high amounts of lime is intended to avoid the disadvantages of surface liming, yet it often results in root damage.

Although several studies show nitrogen appearing in lysimeter water after liming (Matzner 1985; Beese and Prenzel 1985; Kreutzer et al. 1991), the group knew of no evidence of elevated nitrate concentrations in runoff following terrestrial liming. This remains an important research point. It was noted that episodic release of nitrogen during storm events may still be important, even in these systems, and there is a risk of elevated nitrate concentrations in groundwater.

## Biological Response to Liming Differs from Recovery Following Reductions in Acid Deposition

While liming represents a shock to the ecosystem, reductions in acid deposition will probably occur slowly over many years. Both terrestrial and aquatic biota will thus have much longer to react to the changes in the chemical environment. The instability signs often seen after liming (e.g., the explosive increases in zoobenthos) will probably not occur following reduction in acid deposition. Indeed, information from the few sites in which recovery has been reported indicates an orderly procession from acidified to nonacidified conditions (Allott et al. 1992; Gunn and Keller 1990).

## Future Research Questions

1. Are sediments generally a sink for precipitated metals in limed lakes, as seen in Sweden?

2. Mobilization of nitrate by liming: does it appear in runoff?
3. Is biological recovery enhanced by nutrient addition with or without liming?
4. Response of Pb, Cu, and Hg to liming: are they linked to DOC?

## Potential Effects of Changes in Other Environmental Processes on Reversibility and Recovery of Acidification

### *Climate Change*

Climate change represents a new driving variable of precipitation and temperature change superimposed on possible acid deposition declines. How might this affect reversibility and recovery and our predictions?

The group could only speculate on these interactions because of a general lack of experimental or correlative data. One of the few sets of relevant data comes from the long-term records at the ELA, Canada, which span a 17-year natural cycle of climate change and provide perhaps an analog of effects that might be expected in the future, albeit in otherwise pristine, unacidifed lakes (Schindler et al. 1990). Changes in mineralization (and mobilization of acids and Al), flow regimes, drainage water biology, spawning success, and erosion in catchments were suggested. In the aquatic environment, changes in primary productivity due to increased $CO_2$ levels accompanied by increased uptake of nitrate and direct effects of increased UV radiation are among other possibilities.

There is evidence from the Experimental Lakes Area, Canada (Schindler et al. 1990) that fire frequency may increase with temperature. This might reduce the DOC input to lakes and increase lake clarity, which could cause additional stress on sensitive species.

### *Land Use*

Land-use change is another important confounding factor that influences reversibility and recovery (see Feger, Harriman et al., and Kreutzer, all this volume). Forestry practice has a considerable effect: trees filter gaseous and particulate pollutants out of the air, generally increase evapotranspiration, hence causing increasing concentrations of ions in runoff, and deplete the soil of base cations needed for growth (Harriman et al., this volume).

Shifts from coniferous to deciduous trees will probably decrease acid throughfall. Deep-rooting tree species versus shallow-rooting spruce stands will result in some beneficial aspects, since nitrate leaching from the topsoil may be retained in the deeper soil and not always transferred to stream water. Uptake of base cations in the deeper mineral soil will increase the base saturation of the topsoil via litterfall and mineralization (base pump) (Kreutzer, this volume).

Afforestation of nonforested areas enhances acidification. Soil acidification from base cation uptake of an even-aged forest stand will result in a peak acidification at the maximum aggrading phase. Forest growth itself will be a dominant factor in future changes in runoff nitrate concentrations, especially under sustained nitrogen emissions. All management processes have to be strongly related to site considerations. These interactions between acid deposition and forest management have been modeled using MAGIC (Cosby et al. 1990).

*Recommendations for Future Research*

In general, there is little information on these linkages. Current warm and dry years could be used as surrogates to investigate climate change. Regions where land use changes while acid deposition remains constant should be investigated to establish "end points." Lake sediment cores should be analyzed for past evidence of correlations between changes in dominant tree species and changes in water chemistry.

## CONCLUSIONS: ARE CHEMICAL AND BIOLOGICAL CHANGES REVERSIBLE?

Some uncertainty exists concerning the definitions of appropriate measures of reversibility and recovery in freshwater ecosystems. Our working group generally agreed, however, that several conclusions are supported by current knowledge and observation:

1. Changes in the abiotic responses of ecosystems to acidic deposition are reversible when acidic deposition is decreased.
2. As abiotic factors improve, the aquatic ecosystems will move towards recovery. In some cases, significant lags are involved or irreversible changes have occurred.
3. In areas with young, thin soils (generally glaciated during the Pleistocene) reversibility and recovery, following deposition decreases, are relatively rapid compared to the time required to plan and implement reductions in deposition. The rate of recovery depends on the rate of change in emissions of $SO_2$ and $NO_x$, lags due to terrestrial processes such as sulfate adsorption and base cation weathering, and finally to processes in the aquatic ecosystem.
4. Mitigation strategies (e.g., liming) that attempt to restore ecosystems without reducing deposition are only partially successful in restoring water quality and recovering biological populations.
5. In the future, assessment of ecosystem recovery following planned and/or implemented deposition reductions may be obscured by other environmental perturbations such as climate change or modified land-use practices.

## REFERENCES

Allott, T.E.H., R. Harriman, and R.W. Battarbee. 1992. Reversibility of lake acidification at the Round Loch of Glenhead, Galloway, Scotland. *Environ. Pollut.* **77**:219–225.

Bayley, S.E., D.W. Schindler, B.R. Parker, M.P. Stainton, and K.G. Beaty. 1993. Effects of forest fire and drought on acidity of a base-poor boreal forest stream: Similarities between climatic warming and acidic precipitation. *Biogeochem.* **17**:191–204.

Beese, F., and J. Prenzel. 1985. Das Verhalten von Ionen in Buchenwald-Ökosystemen auf podsoliger Braunerde mit und ohne Kalkung. *Allg. Forstz.* **40**:1162–1164.

Cosby, B.J., G.M. Hornberger, J.N. Galloway, and R.F. Wright. 1985a. Modelling the effects of acid deposition: Assessment of a lumped-parameter model of soil water and streamwater chemistry. *Water Resour. Res.* **21**:51–63.

Cosby, B.J., A. Jenkins, J.D. Milles, R.C. Ferrier, and T.A.B. Walker. 1990. Modelling stream acidification in forested catchments: Long-term reconstruction at two sites in central Scotland. *J. Hydrol.* **120**:143–162.

Cosby, B.J., R.F. Wright, G.M. Hornberger, and J.N. Galloway. 1985b. Modelling the effects of acid deposition: Estimation of long-term water quality responses in a small forested catchment. *Water Resour. Res.* **21**:1591–1601.

Fjellheim, A., and G.G. Raddum. 1990. Acid precipitation: Biological monitoring of streams and lakes. *Sci. Total Environ.* **96**:57–66.

Gunn, J.M., and W. Keller. 1990. Biological recovery of an acid lake after reductions in industrial emissions of sulphur. *Nature* **345**:431–433.

Hutchinson, T.C., and M. Havas. 1985. Recovery of previously acidified lakes near Coniston, Canada following reductions in atmospheric sulphur and metal emissions. *Water, Air, Soil Pollut.* **28**:319–333.

Kelly, C.A., J.W.M. Rudd, and D.W. Schindler. 1990. Lake acidification by nitric acid: Future considerations. *Water, Air, Soil Pollut.* **50**:49–61.

Kreutzer, K., A. Göttlein, and P. Pröbstle. 1991. Dynamik und chemische Auswirkungen der Auflösung von Dolomitkalk unter Fichte (*Picea abies L. Karst.*). *Forstw. Forschungen* **39**:186–209.

Marshall, L.M. 1993. Mechanisms controlling variation in stream chemical composition during hydrologic episodes in the Shenandoah National Park, Virginia. M.S. thesis. Charlotteville, VA: Univ. of Virginia.

Matzner, E. 1985. Auswirkung von Düngung und Kalkung auf den Elementumsatz und die Elementverteilung in zwei Waldökosystemen im Solling. *Allg. Forstz.* **40**:1143–1147.

Matzner, E. 1988. Der Stoffumsatz zweier Waldökosysteme im Solling. Berichte des Forschungszentrums Waldökosysteme/Waldsterben, ed. B. Ulrich, Series A 40. Göttingen: Selbstverlag des Forschungszentrums Waldökosysteme/Waldsterben der Universität Göttingen.

Mitchell, M.J. 1992. Retention or loss of sulfur at IFS sites and evaluation of relative importance of processes. In: Atmospheric Deposition and Forest Nutrient Cycling, ed. D.W. Johnson and S.E. Lindberg, pp. 129–133. New York: Springer.

Reuss, J.O., and D.W. Johnson. 1986. Acid Deposition and the Acidification of Soils and Waters. New York: Springer.

Rosseland, B.O., I.A. Blakar, A. Bulger, F. Kroglund, A. Kvellstad, E. Lydersen, D.H. Oughton, B. Salbu, M. Staurnes, and R. Vogt. 1992. The mixing zone between limed and acidic river waters: Complex aluminium chemistry and extreme toxicity for salmonids. *Environ. Pollut.* **78**:3–8.

Schindler, D.W., K.G. Beaty, E.J. Fee, D.R. Cruikshank, E.R. DeBruyn, D.L. Findlay, G.A. Lindsay, J.A. Shearer, M.P. Stainton, and M.A. Turner. 1990. Effects of climatic warming on lakes of the central boreal forest. *Science* **250**:967–970.

Schindler, D.W., T.M. Forst, K.H. Mills, P.S.S. Chang, I.J. Davies, L. Findlay, D.F. Malley, J.A.Shearer, M.A. Turner, P.J. Garrison, C.J. Watras, K. Webster, J.M. Gunn, P.L. Brezonik, and W.A. Swenson. 1991. Comparisons between experimentally- and atmospherically-acidified lakes during stress and recovery. *Proc. Roy. Soc. Edinburgh B* **97**:193–266.

Schindler, D.W., M.A. Turner, and R.H. Hesslein. 1985. Acidification and alkalization of lakes by experimental addition of nitrogen compounds. *Biogeochem.* **1**:117–133.

van Dijk, H.F.G., A.W. Boxman, and J.G.M. Roelofs. 1992. Effects of a decrease in atmospheric deposition of nitrogen and sulphur on the mineral balance and vitality of a Scots pine and a Douglas fir stand in the Netherlands. Interim project report: 1988–1991. Nijmegen, The Netherlands: Dept. of Ecology, Univ. of Nijmegen, 43 pp.

Wright, R.F., and M. Hauhs. 1991. Reversibility of acidification: Soils and surface waters. *Proc. Roy. Soc. Edinburgh B* **97**:169–191.

Wright, R.F., E. Lotse, and A. Semb. 1993. RAIN project: Results after 8 years of experimentally reduced acid deposition to a whole catchment. *Can. J. Fish. Aquat. Sci.* **50**:258–268.

# Author Index

# Subject Index

Note: Page numbers in *italic* refer to figures; those in **bold** refer to tables

*Index compiled by Annette Musker*